Engineering Materials and Processes

Series Editor

Brian Derby

For further volumes:
http://www.springer.com/series/4604

John M. Dealy · Jian Wang

Melt Rheology and its Applications in the Plastics Industry

Second Edition

Springer

John M. Dealy
Montreal, QC
Canada

Jian Wang
Freeport, TX
USA

ISSN 1619-0181
ISBN 978-94-007-6394-4 ISBN 978-94-007-6395-1 (eBook)
DOI 10.1007/978-94-007-6395-1
Springer Dordrecht Heidelberg New York London

Library of Congress Control Number: 2013932769

This is a completely revised and updated second edition of the earlier work Rheology and its Role in Plastics Processing, by John M. Dealy and Kurt Wissbrun, ISBN 978-0-412-73910-1, Kluwer Academic Publishers, Dordrecht, The Netherlands, 1999. First published by Van Nostrand Reinhold 1990, reprinted by Chapman and Hall 1995, 1996.

Printed on acid-free paper

Springer is part of Springer Science+Business Media (www.springer.com)

Preface

This is the second edition of Melt Rheology and its Role in Plastics Processing, although the title has changed a bit to indicate its broadened scope. Advances in rheometer technology and polymer science have greatly enhanced the usefulness of rheology in the plastics industry. It is now possible to design polymers having specific molecular structures and to predict the flow properties of melts having those structures. In addition, rheological properties now provide more precise information about molecular structure.

Our primary interest is in practical applications, not basic polymer science, which is the subject of several excellent books. But an understanding of key aspects of polymer physics is essential to be an intelligent user of rheology, and we provide the information necessary to build this understanding. And extensive references are provided for those wishing to pursue certain issues in greater depth.

Thus, while our primary audience is applied polymer scientists and plastics engineers, the book should also be of use to postgraduate students in polymer science and engineering, and the first edition has been used as a text for graduate courses.

After an introductory chapter we start our discussion of melt rheology with a chapter mainly about viscosity, because most readers will be familiar with this concept and some of its applications. Discussions involving tensors that were present in several chapters of the first edition are now concentrated in Chap. 5, and mastery of this material is not required for an understanding of the other chapters.

The rheology of filled polymers is now too big a subject to deal with in a chapter and is the subject of entire books, including the three listed at the bottom of this page.

Kurt Wissbrun, coauthor of the first edition, is retired from his successful career as an industrial scientist, and the applied science expertise essential to the goals of this book is now provided by Jian Wang. The authors are grateful to many people who provided advice and information that added in important ways to the usefulness of the book. These include Tom Butler, Chan Chung, Albert Co, Stephane Costeux, Willem deGroot, César García-Franco, Jeffrey Giacomin, Kathy Jackson, Teresa Karjala, Helmut Münstedt, Martin Sentmanat, and John Vlachopoulos.

Montreal, QC John M. Dealy
Freeport, TX Jian Wang

References

Leblanc JL (2010) Filled polymers: science and industrial applications. CRC Press, Boca Raton
Han CD (2007) Rheology and processing of polymeric materials. Polymer rheology, vol 1 Chapter 12. Oxford Univ Press, Oxford
Carreau PJ, Kee DD, Chhabra R (1997) Rheology of polymeric systems. Chapter 8. Hanser Publishers, Munich

Contents

Symbols and Nomenclature

a	Constant in damping function equation (Eq. 4.2)
$a_P(P)$	Horizontal shift factor for effect of pressure on viscosity (Eq. 2.18)
$a_T(T)$	Horizontal temperature shift factor for time/strain rate (Eqs. 3.60, 3.61)
$b_T(T)$	Vertical shift factor for stress (Eq. 3.65)
A	Area of sample in contact with fluid (Eq. 6.29)
A_0	Area of sample in contact with fluid (Eq. 6.29)
A_0	Area of die exit (Eq. 8.26)
B	Swell ratio for capillary extrudate (Eq. 8.16)
B_D	Diameter swell at annular die exit (Eq. 8.24)
B_H	Thickness swell at annular die exit (Eq. 8.25)
B_W	Weight swell for annular extrudate (Eq. 8.26)
B_A	Area swell for annular extrudate (Eq. 8.27)
$\mathbf{B}$	Finger tensor (Appendix B)
$\mathbf{C}$	Cauchy tensor (Appendix B)
C	Constant in transition region (Eqs. 3.82, 3.83)
D	Diameter of swollen extrudate (Eq. 8.16)
$D(t)$	Extensional creep compliance (Eq. 3.34)
De	Deborah number (Eq. 4.9)
D_0	Diameter of capillary (Eq. 8.17)
e	End correction for capillary rheometer (Eq. 6.20)
F	Force (Sect. 1.5.3)
$F(\tau)$	Continuous relaxation spectrum (Eq. 3.22)
G	Shear modulus of Hookeian solid (Eq. 1.11)
G_c	Crossover modulus (where $G' = G''$) (Eq. 3.71)
$G(t)$	Relaxation modulus in linear viscoelasticity (Eq. 4.1)
$G(t, \gamma)$	Relaxation modulus in nonlinear viscoelasticity (Eq. 4.1)
G_N^0	Plateau modulus (Sect. 5.3.1)
$G^*(\omega)$	Complex modulus (Eq. 4.37)
$G'(\omega)$	Storage modulus (Eq. 4.36a)
$G''(\omega)$	Loss modulus (Eq. 4.36b)
h	Gap between plates in rheometer (Eqs. 2.1, 6.8)
$h(t)$	Damping function for shear (Eq. 4.1)
$H(\tau)$	Relaxation spectrum function (Eq. 3.23)

I2	Flow rate (melt index) for load of 2 kg (Sect. 6.3.3)
I10	Flow rate (melt index) for load of 10 kg (Sect. 6.3.3)
$I_{m/n}$	(= σ_m/σ_n) Amplitude ratio in LAOS (Eq. 4.17)
I_1	First invariant of the finger tensor (Eq. 5.38)
I_2	Second invariant of the finger tensor (Eq. 5.39)
J	Compliance of a Hookean solid (Eq. 1.12)
$J(t)$	Creep compliance in linear viscoelasticity (Eq. 3.24)
J_e	Equilibrium compliance of a rubber
J_r	Recoverable compliance (Eq. 3.28)
J_s^o	Steady-state compliance of a melt (Eq. 4.22)
J_N^o	Plateau compliance
k, K	Constant in power-law for viscosity (Eq. 2.6)
$L(t)$	Length of spesiman for extensional test (Eq. 4.19)
$L(\tau)$	Retardation spectrum (Eq. 3.24c)
m	Constant in cross viscosity model (Eq. 2.9)
m(s)	Memory function (Eq. 5.43b)
MI	Melt index
M	Molecular weight of polymer (see note* at bottom of list)
M_0	Torque amplitude in rotational rheometer (Eqs. 6.6, 6.7, 6.11)
M_b	Molecular weight per backbone bond (Eq. 2.25)
M_C	Critical molecular weight for viscosity (Eq. 2.21)
M'_C	Critical molecular weight for compliance (Sect. 7.1.3)
M_e	Molecular weight between entanglements (Ferry) (Eq. 3.72)
M_n	Number average molecular weight
M_o	Monomer molecular weight (Eq. 3.76)
M_w	Weight average molecular weight
M_e^G	Molecular weight between entanglements (not recommended)
n	Exponent in power-law model for viscosity (Eq. 2.6)
N_1	First normal stress difference (Eq. 2.26)
N_1^+	First normal stress growth coefficient (Sect. 4.3.2)
N_2	Second normal stress difference (Eq. 2.27)
p	Constant in Carreau viscosity model (Eq. 2.10)
P	Degree of polymerization (M/M_0) (Eq. 3.74)
P_0	Reference pressure (Eq. 2.17)
P_d	Driving pressure, in capillary rheometer (Eq. 6.18)
ΔP_{ent}	Entrance pressure drop (Eq. 6.18)
ΔP_{cap}	Pressure drop for flow through a capillary (6.18)
PI	Polydispersity index defined as M_w/M_n
$Q(\omega, \gamma_0)$	Nonlinearity parameter for LAOS (Eq. 4.17)
Q	Volumetric flow rate in capillary rheometer (Eq. 6.17)
Q_{ij}	Component of Doi-Edwards strain tensor (Eq. 5.56)
R	Universal gas constant
$\langle R^2 \rangle_o$	Mean square radius of an unperturbed molecule (Eq. 3.74)
$\langle R_g^2 \rangle_o$	Mean square radius of gyration of an unperturbed molecule (Eq. 1.24)

s	Time elapsing between t' and t, i.e., t-t'
S	Engineering strain (Eq. 1.1)
t	In constitutive equations, the time at which the stress is to be evaluated
t_r	Time at which strain is stopped for creep recovery test (t-t_0) (Eq. 3.28)
t'	A time prior to the time t (Eq. 3.7)
T	Temperature
T_A	Empirical constant with units of absolute temperature (Eq. 8.33)
T_o	Reference temperature (Eq. 3.61)
$\boldsymbol{u}$	Displacement vector (Eq. 5.12)
v	Velocity of fluid at a point
V	Speed of rheometer plate
w_i	Weight fraction of component i (Eq. 2.20)
Wi	Weissenberg number (Eq. 4.10)
W	Width of slit (Eq. 8.6)
Z	Degree of entanglement (M/M_e) (Sect. 3.18.2)

Greek Letters

α	Exponent in relationship for zero-shear viscosity (Eqs. 2.20 and 2.21)
β	Average number of branch points per molecule (Eq. 7.7)
β	Pressure coefficient in Barus equation (Eq. 2.17)
β	Cone angle of rotational rheometer fixture (Eq. 6.1)
γ	Shear strain amplitude in step shear (Eq. 1.8)
γ_∞	Recoverable strain (ultimate recoil) (Eq. 3.27)
$\dot{\gamma}$	Shear rate (Eq. 1.9)
$\dot{\gamma}_A$	Apparent wall shear rate in capillary rheometer (Eq. 6.17)
$\dot{\gamma}_n$	Nominal shear rate (V/h) in sliding plate rheometer (Eq. 6.26)
$\dot{\gamma}_W$	Wall shear rate in capillary rheometer (Eq. 6.23)
$\dot{\gamma}_{ij}$	Component of rate of deformation tensor (Eq. 5.13)
δ	Loss angle (Eq. 3.39)
ε	Hencky strain (Eq. 1.4)
$\dot{\varepsilon}$	Hencky strain rate (Eq. 1.5)
ζ	Monomeric friction coefficient (Eq. 3.74)
η	Viscosity (Eq. 1.14)
η_0	Zero-shear viscosity (Sect. 2.4.1)
η_p	Plastic viscosity (Eq. 1.23)
$\eta(\dot{\gamma})$	Shear-rate dependent viscosity (Sect. 2.4.1)
$\eta^+(t)$	Shear stress growth coefficient-linear viscoelasticity (Eq. 3.10)
$\eta^+(t,\dot{\gamma})$	Shear stress growth coefficient-nonlinear behavior (Eq. 4.11)
η^*	Complex viscosity (Eq. 3.51)
$\eta_E^+(t,\dot{\varepsilon})$	Tensile stress growth coefficient (Eq. 4.21)
$\eta_E(\dot{\varepsilon})$	Extensional viscosity (Eq. 4.23)
λ	Time constant in viscosity models (Eqs. 2.8–2.11)
λ	Characteristic time for non-Newtonian behavior (Eq. 4.10)

λ	Avg number of branch points per 1000 carbon atoms (Eq. 7.7)
ρ, ρ_M	Melt density
ρ_S	Solid density (Eq. 6.29)
ρ_o	Density at reference temperature (Eq. 3.65)
σ	Shear stress in viscometric flow
σ_o	Stress amplitude in oscillatory shear (Eq. 3.37)
σ_A	Apparent wall shear stress in capillary rheometer (Eq. 6.16)
σ_E	Net tensile stress (Sect. 3.10.1)
$\boldsymbol{\sigma}_{ij}$	Components of stress tensor (Eq. 5.1)
σ_W	Wall shear stress in capillary (Eq. 6.19)
σ_y	Yield stress (Eq. 1.23)
τ	Time constant in Maxwell model (Eq. 3.18)
τ	Variable of integration with units of time (Eq. 3.22)
τ_d	Reptation (disengagement) time-Doi-Edwards model (Eq. 5.55)
τ_r	Characteristic relaxation time of polymer (Eq. 4.9)
τ_R	Longest Rouse stress relaxation time (Eq. 3.75)
τ_k	Time after which relaxation modulus curves superpose (Sect. 4.3.1)
φ	Angular displacement in rotational rheometer (Eqs. 6.1, 6.8)
φ_0	Angular displacement amplitude in rotational rheometer (Eqs. 6.5–6.7)
$\Psi_1(t, \gamma)$	First normal stress relaxation coefficient (Eq. 2.28)
$\Psi_{1,0}$	Limiting zero-shear-rate value of $\Psi_1(\dot{\gamma})$ (Eq. 5.49)
$\Psi_1(\dot{\gamma})$	First normal stress diff. coefficient in steady shear (Eq. 2.28)
$\Psi_2(\dot{\gamma})$	Second normal stress diff. coefficient in steady shear (Eq. 2.29)
$\Psi(t, \gamma)$	Normal stress relaxation ratio (Eq. 4.8)
ω	Frequency of oscillation (Eq. 3.37)
ω_c	Crossover frequency (where $G' = G''$) (Eq. 3.71)

*The International Union of Pure and Applied Chemistry (IUPAC) recommends the term *molar mass* (MM), which has SI units of g/mol. But *molecular weight* (MW) is widely used, and the American Chemical Society accepts both terms. However, MW is a dimensionless ratio that depends on the acceleration of gravity and is numerically very close to MM (g/mol), and one cannot change its units. The number often called "molecular weight (kg/mol)" is actually MW/1000 (no units). This number can also be called molar mass (kg/mol).

Chapter 1
Introduction to Rheology

Abstract Basic concepts of rheology are presented: what it is and how it is useful in applied polymer science and plastics engineering. Stress and strain are defined, and their use to describe material behavior is described. Rheological phenomena and the properties used to describe them are presented, including elasticity, viscous flow, viscoelasticity and plasticity. Finally the molecular structures that give rise to the complexity of melt rheology are discussed.

1.1 What is Rheology?

Some readers may have little previous knowledge of rheology but wish to find out how it can be useful in solving practical problems involving the flow of molten plastics. For this reason, the authors have tried to supply sufficient basic information about rheology to enable these readers to understand and make use of the methods described. With this in mind, we begin with the definition of rheology.

Rheology is the science that deals with the way materials deform when forces are applied to them. The term is most commonly applied to the study of liquids and liquid-like materials such as paint, ketchup, oil well drilling mud, blood, polymer solutions and molten plastics, i.e., materials that flow, although rheology also includes the study of elastic materials that do not flow, particularly cross-linked elastomers.

The two key words in the above definition of rheology are *deformation* and *force*. To learn anything about the rheological properties of a material, we must either measure the deformation resulting from a given force or measure the force required to produce a given deformation. For example, let us say you wish to compare the s*oftnesses* of several foam rubber pillows. Instinctively, one would squeeze (deform) the various products, noting the force required to deform them. A pillow that required a high force to compress it would be considered *hard*, and

J. M. Dealy and J. Wang, *Melt Rheology and its Applications in the Plastics Industry*, Engineering Materials and Processes, DOI: 10.1007/978-94-007-6395-1_1,

you probably wouldn't buy it, because it would be painful to sleep on. On the other hand, if it required too little force (too soft) it would not provide adequate support for your weary head. Foam rubber is a lightly cross-linked elastomer that is full of gas bubbles, and in squeezing it you are making your evaluation on the basis of its resistance to deformation, i.e., its elastic modulus, which is a rheological property.

Now let us say that you are looking for lubricating oil for a household application and that your local hardware store has a display in which samples of several oils are contained in glass bottles. You can pick up the bottles and invert them to make the oil flow from one end to the other. The "light oils" flow very rapidly and splash as they hit the end of the bottle. The "heavy oils" creep slowly in response to your test and take several seconds to accumulate at the end of the bottle. This is a rheological experiment. You have used the earth's gravitational pull to supply the force, and the speed with which the oil flows is a rough measure of its rate of deformation. Lubricating oils are viscous liquids, and in tipping the bottles you are evaluating them on the basis of their viscosity, i.e., their resistance to flow, and this is another example of a rheological property.

There are two principal aspects of rheology. One involves the development of quantitative relationships between deformation and force for a material of interest. The information for the development of such a relationship is obtained from experimental measurements. For example, for the foam pillow it might be observed that the force required to compress it a certain distance is proportional to the distance. In the case of the lubricating oil, it might be found that the speed with which it flows through a small hole in the bottom of a can is proportional to the height of the oil remaining in the can.

For a linear elastic rubber or a Newtonian fluid, such simple observations are sufficient to establish a general equation describing how the material will respond to any type of deformation. Such an equation is called a *constitutive equation* or a *rheological equation of state*. However, for more complex materials such as molten plastics, the development of a constitutive equation is a much more complex task that requires data from several types of experiment.

The second aspect of rheology is the development of relationships that show how rheological behavior is influenced by the structure and composition of the material and temperature and pressure. Ideally, one would like to know how these parameters affect the constitutive equation, but this is still a work in progress at the present time, except for linear, unbranched polymers. In the case of more complex materials, one can at least develop relationships showing how specific rheological properties such as the viscosity and the relaxation modulus are influenced by molecular structure, composition, temperature and pressure.

Molten plastics are rheologically complex materials that can exhibit both viscous flow and elastic recoil. A truly general constitutive equation has not been developed for these materials, and our present knowledge of their rheological behavior is largely empirical. This complicates the description and measurement of their rheological properties but makes polymer rheology an interesting and challenging field of study.

1.2 Why Rheological Properties are Important

Many materials of major commercial importance are examples of *soft matter*, a category of materials that includes molten polymers, colloids, foams and gels. When being deformed, these materials exhibit complex mechanical behavior, i.e., complex rheological behavior. Their rheological properties are of central importance in their industrial processing behavior and/or their end-use applications. The syrup we put on our pancakes has a high viscosity, but it is a Newtonian fluid, which means it has very simple rheological behavior. But many other everyday materials exhibit complex flow behavior. These include toothpaste, ketchup, shampoo and processed foods such as pudding and ice cream. Other materials are solids in their final application but require processing at high temperatures in liquid form. Examples are paving asphalt and molten plastics. There are also liquids such as lubricating oils, greases and oil-well drilling-fluids that experience high temperatures during normal use.

Polymers that melt and are processed as liquids are called thermoplastics, and these are nearly always converted to end-use products by means of melt forming operations, including profile extrusion and several molding techniques. In all these processes, the rheological properties of the molten polymer are of crucial importance and must be taken into account in the design of processing equipment and selection of operating conditions. And for resin manufacturers, the ease with which a polymer can be processed is of crucial importance in the development of new products. In an extruder, the dependence of viscosity on shear rate and temperature is of most interest, but once a melt leaves a die, its viscoelastic properties govern its behavior. For example, the swelling that occurs on exiting a die is of central importance in profile extrusion and blow molding. A polyethylene milk container is produced by blow molding, and the amount of swelling that the melt undergoes as it exits the die before being expanded into the mold is of crucial importance if a reliable and economically optimal product is to be obtained. And various flow instabilities that can cause havoc in melt processing are strongly affected by rheological properties. In summary, because rheological properties govern melt flow behavior, they have important applications in resin development, quality control, and the numerical simulation of forming processes.

Rheological data are also useful for polymer characterization, i.e., to determine molecular structure. This is because the viscoelastic behavior of a melt is very sensitive to molecular weight distribution and branching structure. In fact, melt rheology can often provide a more accurate picture of structure than gel permeation chromatography.

1.3 Stress: A Measure of Force

1.3.1 Foam Rubber Revisited

It was emphasized in Sect. 1.1 that force and deformation are the key words in the definition of rheology. In order to describe the rheological behavior of a material in a quantitative way, i.e., to define rheological material constants (such as the viscosity of a Newtonian fluid) or material functions (such as the relaxation modulus of a rubber), it is necessary to establish clearly defined and quantitative measures of force and deformation. Furthermore, it is necessary to define these measures in such a way that they describe the state of the material of interest without detailed reference to the procedure used to make the measurement. For example, in the case of the foam-rubber pillows mentioned in Sect. 1.1, one way to quantify the test results would be to place the pillow on a table, place a flat board on top, and measure the distance between the board and the table both before and after a weight of a certain mass was placed on top of the board, as shown in Fig. 1.1. If our objective is simply to compare several pillows of the same size, it would be sufficient to simply list the amount of compression, in centimeters, caused by a weight having a mass of say 1 kg.

However, if our objective is to make a quantitative determination of a property of the foam rubber, this way of reporting of test results is awkward at best. To make use of the data, one would have to know the size and shape of the sample (the pillow), the mass of the weight applied, the size and weight of the board used, and the amount of compression. If one wishes to compare the behavior of this foam with that of another, the second material would have to be tested in exactly the same way as the first. It would be advantageous to be able to describe the elastic behavior of the foam rubber using characteristics or properties of the *material*, without reference to the details of the test procedure. To do this, we will need to use a normalized measure of force, which is called stress.

1.3.2 Stress: A Normalized Measure of Force

Two types of force can act on a fluid element. A *body force* acts directly on the mass of the element as the result of a force field. Usually only gravity need be

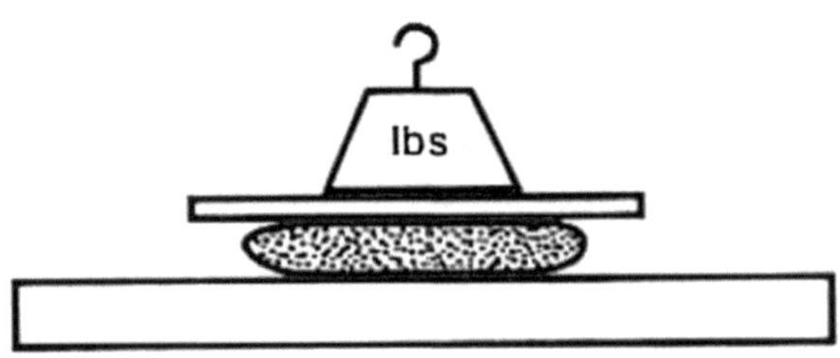

Fig. 1.1 Testing the softness of a pillow

considered, but a magnetic field can also generate a body force in a magnetic material. A *surface force* results from contact of the fluid element with a solid wall or with surrounding fluid elements. While gravity may play a role in a melt processing operation, for example in rotational and blow molding, when describing the rheological behavior of a tiny element of fluid, the force due to gravity is negligible compared to surface forces.

Surface forces are of two types; shear forces act tangentially to the surface of a fluid element, while normal forces act normal to the surface. And a normal force can be tensile or compressive. Both shear and normal forces cause deformation and are thus of importance in rheology. Now we will look at how we might determine experimentally the relationship between the deforming force and the deformation of the foam rubber mentioned above. Notice that we now focus our attention not on the pillow but on the foam of which it is made. If we could establish a relationship between force and deformation for the foam, we could calculate the deformation of any pillow made of this foam caused by a given force.

Placing a weight having a mass of 1 kg on a small sample of the foam will cause more compression than placing it on a larger sample. From the point of view of the material, it is obviously not the total force that is important. In fact, since the deforming force acts on the upper and lower surfaces of the sample, if this force is divided by the area of the surface we obtain a quantity suitable for describing the properties of the material. We call this quantity *stress*. In general, then, the stress is calculated by dividing the force driving a deformation by the area over which it acts. In the case of a test like the one with the foam rubber, which involves squeezing, we have a *compressive* stress. A more common type of test method for elastic materials involves stretching rather than compressing, as shown in Fig. 1.2. Again, the stress is the force divided by the cross-sectional area of the sample, but in this case it is a *tensile* stress.

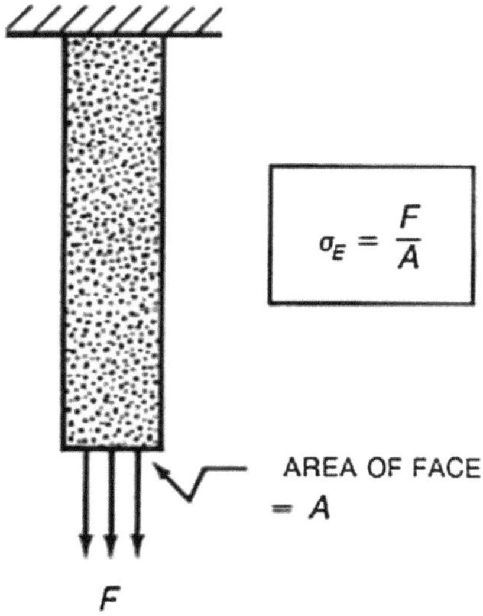

Fig. 1.2 Uniaxial (simple) extension

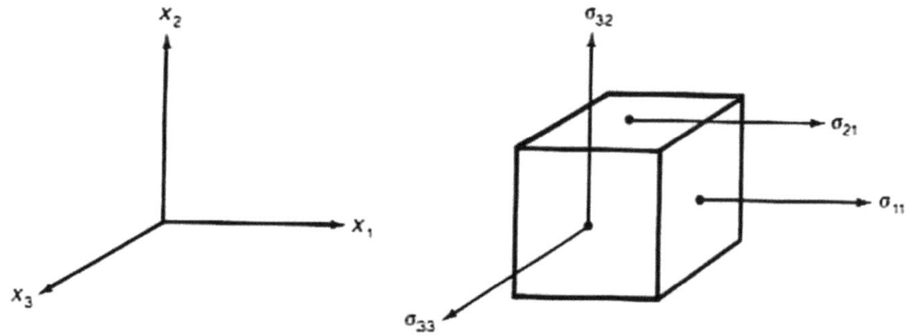

Fig. 1.3 Shear and normal stress components

Compressive and tensile stresses are the two types of *normal stress*, so called because the direction of the force is normal (perpendicular) to the surface on which it acts. In addition to normal stresses, there are *shear stresses*, which act in a direction tangential to the surface on which it acts, as shown in Fig. 1.3. This figure shows the deformation called *simple shear*, in which the sample is contained between two flat plates with a fixed gap h between them. The upper plate moves in a direction parallel to itself while the lower plate is stationary. The shear stress is the shear force divided by the area of the plate wetted by the fluid of interest. We will use the symbol σ with no subscript, to refer to the shear stress in a simple shear deformation.

The representation of stress components acting on a fluid element during deformations more complex than simple shear or simple compression requires two subscripts. This is because stress is a tensor–valued quantity rather than a vector-valued one that requires only one index. In this book, tensor notation is only used in Chap. 4, which deals with the theory of nonlinear viscoelasticity. We will use it here only to help illustrated the types of surface force that act on a fluid element. Figure 1.3 shows a fluid element along with several typical stress components. Note that when the two subscripts are the same, a normal stress is indicated, while when the two subscripts are different, it is a shear stress.

1.3.3 Role of Normal Stresses in Rheology

While all materials are compressible to some extent, in the case of molten plastics, quite high pressures are required to produce a significant change in volume. Thus, for many practical purposes these materials can be considered incompressible in the molten state. Now consider what happens when we subject an incompressible material to a compressive stress that is equal in all directions, i.e., an *isotropic* stress, and we call this stress the pressure P. However, pressure is positive when acting to compress a material, while a tensile stress is positive when it acts to stretch the material, so the normal stress in the fluid is $-P$. An incompressible fluid at rest will not be deformed by an isotropic stress, i.e., it will not change its size or shape. Thus, an isotropic stress field is of no rheological significance. Only when there are shear stresses acting, as in simple shear flow, or when the normal stress components are *different from each other*, will deformation occur in an incompressible material. This means that for an incompressible material a single normal component of stress has no rheological significance. Only *differences* between normal components are of rheological significance, because only differences can result in deformation. We will see in Chap. 2 that for shear flows the two rheologically meaningful normal stress differences are called *the first and second normal stress differences*.

In the case of simple extension, there is only one significant feature of the stress field, because there are no shear stresses acting, and there is axial symmetry. This is the principal stretching stress, σ_E, which is a normal stress difference, the difference between the tensile stress and the ambient pressure surrounding the sample. Because of the relatively high resistance to deformation of molten, high-molecular-weight polymers, the tensile stress driving the deformation is normally much larger than the ambient pressure. Thus, for practical purposes, $\sigma_E = \sigma_{tensile}$.

1.4 Strain: A Measure of Deformation

In the previous section, we said that shear stresses and normal stresses are useful measures of the forces that deform a material. Now we need a quantitative measure of *deformation* that is rheologically significant. There are many general measures of strain that can be used to describe, large, complex deformations, but for the purposes of this book we make use only of measures that are needed to describe the deformations used for most rheological measurements, namely simple shear and simple extension.

To define a measure of strain we need to consider two states of a material element, as deformation involves a change in shape and must refer to two states, one before and one after the deformation. In other words, to specify the strain of a material element at a time t, we must compare its shape at that time with its shape in some reference state. In the case of an elastic material that cannot flow, for example a cross-linked rubber, the selection of the reference state is straightforward, because there is a unique, unstrained state to which a material element always returns when deforming stresses are removed.

For materials that flow, i.e., fluids, such a unique reference state does not exist. In the case of a well-controlled experiment, however, in which a simple homogeneous deformation is imposed on a sample initially at rest and free of all deforming stresses, this initial condition provides a useful reference state with respect to which strain can be defined. We will make use of this fact in the next two sections to define strain measures for simple extension and simple shear.

1.4.1 Strain Measures for Simple (Uniaxial) Extension

Consider the simple stretching test illustrated in Fig. 1.4. Let L_0 be the length of the sample prior to the application of a tensile stress and L the length after deformation has occurred. A simple measure of the deformation is the quantity $(L-L_0)$. However, this quantity depends on the initial length of the sample, whereas we desire a measure of deformation that describes the state of a material element. We can easily form such a quantity by dividing the length difference by the initial length to obtain the *engineering* or *linear strain* for simple or uniaxial extension.

$$S = (L - L_0)/L_0 \tag{1.1}$$

For a uniform deformation, every material element of the sample experiences this same strain. For example, if the initial length at time t_0 of a material element measured in the direction of stretching is $X_1(t_0)$, and its length at a later time t after deformation has occurred is $X_1(t)$, the linear strain of the material element is:

$$S = [\delta X_1(t) - \delta X_1(t_0)]/\delta X_1(t_0) \tag{1.2}$$

This measure of deformation has some useful features. It is independent of sample size, and it is zero in the unstressed, initial state.

However, it is not the only measure of deformation that has these properties. Another is the Hencky strain, which is defined as follows in terms of the length of a material element.

$$\varepsilon = \ln[\delta X_1(t)/\delta X_1(t_0)] \tag{1.3}$$

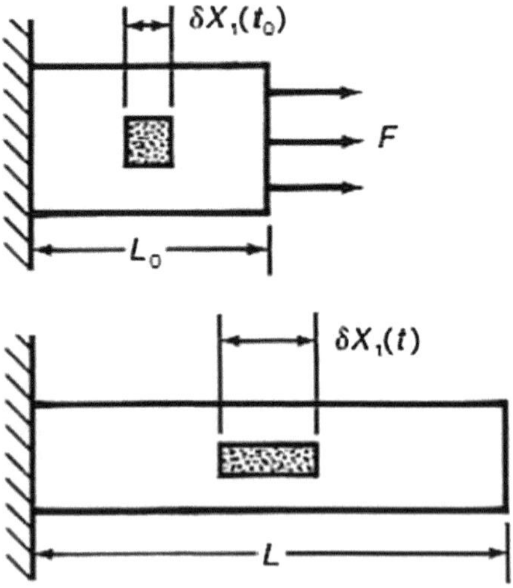

Fig. 1.4 Deformation of a fluid element in simple extension

For a sample with initial length L, undergoing uniform strain the Hencky strain is given by Eq. (1.4).

$$\varepsilon = \ln(L/L_0) \tag{1.4}$$

Note that this measure is also initially zero and independent of sample size.

For materials that flow, e.g., molten plastics, this quantity is more useful than the linear strain.[1] The Hencky strain *rate* is also useful for describing rheological behavior. This is defined by Eq. (1.5) in terms of the current length L of the sample.

$$\dot{\varepsilon} = \mathrm{d}[\ln(L)]/\mathrm{d}t \tag{1.5}$$

We note that the initial length does not enter into the Hencky strain rate but does enter into the linear strain rate dS/dr. Because an elastic liquid does not maintain any memory of its initial state we use the Hencky strain to describe the stretching of molten plastics.

1.4.2 A Strain Measure for Shear

Now consider simple shear, which is the type of deformation most often used to make rheological measurements on fluids. Referring to Fig. 2.1, an obvious choice of a strain measure is the displacement of the moving plate, ΔX, divided by the distance between the plates, h.

$$\gamma = \Delta X/h \tag{1.6}$$

Referring to the two material particles shown in Fig. 1.5 rather than to the entire sample, we can define the *shear strain* γ for the fluid element located at (x_1, x_2, x_3) as

$$\gamma(x_1, x_2, x_3) = \Delta X_1/\Delta x_2 \tag{1.7}$$

where ΔX_1 is the displacement in the x_1 direction of one particle relative to a neighboring particle separated from it by a distance ΔX_2 in the x_2 direction. In the absence of edge effects, i.e., for a uniform deformation, every fluid element undergoes the same strain, and the local shear strain at every location within the sample is equal to the overall strain of the sample:

$$\gamma(x_1, x_2, x_3) = \Delta X/h \tag{1.8}$$

[1] In the limit of very small strains the linear and Hencky strains become equivalent. This can be demonstrated for simple extension by noting that $\varepsilon = \ln(1 + S)$ and that the first term of the series expansion of $\ln(1 + S)$ is S.

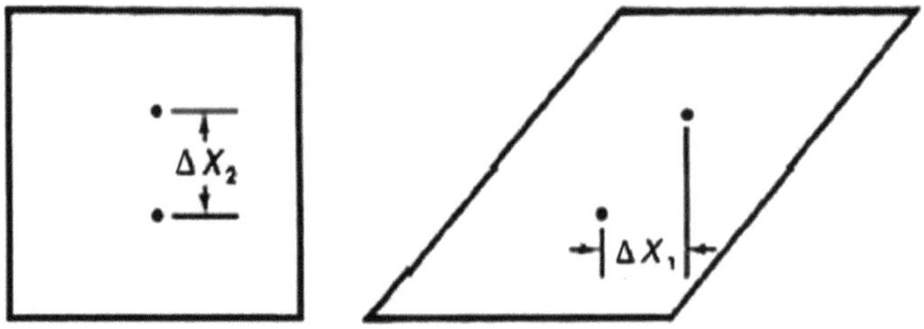

Fig. 1.5 Displacement of two material particles in simple shear

And the shear rate is simply the rate of change of the shear strain with time:

$$\dot{\gamma} = \frac{1}{h}\frac{\mathrm{d}X}{\mathrm{d}t} = \frac{V}{h} \tag{1.9}$$

where V is the velocity of the moving plate.

1.5 Rheological Phenomena

In this section we examine the general types of rheological behavior that are exhibited by materials. These are elasticity, viscosity, viscoelasticity, structural time dependency, and plasticity. Although we will use simple extension and simple shear behavior to describe rheological phenomena, it is important to remember that for rheologically complex materials such as polymeric liquids, the behavior observed in these simple deformations cannot be used to predict how such materials would behave in more complex flows.

1.5.1 Elasticity: Hooke's Law

Elasticity is a type of behavior in which a deformed material returns to its original shape whenever a deforming stress is removed. This implies that a deforming stress is necessary to produce and maintain any deviation in shape from the original (unstressed) shape, i.e., to produce strain. The simplest type of elastic behavior is that in which the stress required to produce a given amount of deformation is directly proportional to the strain associated with that deformation. For example, in simple extension this can be expressed as follows.

$$\sigma_E = E(L - L_0)/L_0 = ES \tag{1.10}$$

The constant of proportionality E is called Young's modulus, and this relationship is called Hooke's law. The corresponding form of Hooke's law for simple shear is:

$$\sigma = G\gamma \tag{1.11}$$

where G is the shear modulus or modulus of rigidity. We note that in a purely elastic material like this, all the work done to deform the material is stored as elastic energy and can be recovered when the material returns to its equilibrium configuration.

Another way of describing elastic behavior is to specify the *strain* that results from the application of a specific *stress*. For a Hookean material we have, for simple shear:

$$\gamma = J\sigma \tag{1.12}$$

where J is the shear compliance. Obviously, for a material following Hooke's law:

$$J = 1/G \tag{1.13}$$

1.5.2 Viscosity: The Newtonian Fluid

Viscosity is the property of a material that describes its resistance to continuous deformation. Unlike elasticity, the stress is not related to the *amount* of deformation but to the *rate* of deformation. Thus, it is a property peculiar to materials that flow. We will consider first the simplest type of rheological behavior for a material that can flow. For simple shear this type of behavior is described by a linear relationship between the shear stress and the shear rate:

$$\sigma = \eta\dot{\gamma} \tag{1.14}$$

where η is the viscosity. A material that obeys this equation, with the viscosity independent of shear rate, is called a Newtonian fluid.

The quantity measured in uniaxial extension is the net tensile stress σ_{E}, and in Chap. 4 it is shown that for a Newtonian fluid, this is simply related to the viscosity:

$$\sigma_{\mathrm{E}} = 3\eta\dot{\varepsilon} \tag{1.15}$$

We will see later that the *extensional viscosity* is defined as $\sigma_E/\dot{\varepsilon}$, so that for a Newtonian fluid, the extensional viscosity is just three times the (shear) viscosity.

For a Newtonian fluid, the viscosity is a *material constant*, in that it does not depend on the rate or amount of strain (But it does depend on temperature and pressure.). Single phase liquids containing only low molecular weight compounds are Newtonian for all practical purposes. For multiphase systems, for example suspensions and emulsions, and for polymeric liquids, the relationship between stress and strain rate is no longer linear and cannot be described in terms of a single constant. It is still convenient, however, to present the results of steady simple shear experiments in terms of $\sigma/\dot{\gamma}$, which becomes the *viscosity function* $\eta(\dot{\gamma})$.

$$\eta(\dot{\gamma}) \equiv \sigma/\dot{\gamma} \tag{1.16}$$

where σ is the shear stress and $\dot{\gamma}$ is the shear rate. A typical viscosity-shear rate curve for a polymer melt or important details of low-rate behavior that would disappear into the left axis on a linear scale (See Fig. 2.3). The important features of this curve are noted below. A detailed discussion of the viscosity function is provided in Chap. 2.

1. At sufficiently low shear rates, the viscosity approaches a limiting constant value η_0 called the zero-shear viscosity.
2. The viscosity decreases monotonically as the shear rate is increased. This type of behavior is called *shear thinning* (An older terminology was *pseudoplastic*).
3. At sufficiently high shear rates the viscosity might be expected to level off again, or even increase, although such behavior is not observed in melts, because viscous heating, wall slip, flow instabilities, or polymer degradation make it impossible to carry out experiments at sufficiently high shear rates.

1.5.3 Viscoelasticity

Polymeric materials, including solutions, melts, and cross-linked elastomers, exhibit both viscous resistance to deformation and elasticity. In the case of a vulcanized (cross-linked) rubber, flow is not possible, and the material has a unique configuration that it returns to in the absence of deforming stresses. However, the viscous resistance to deformation makes itself felt by delaying the response of the rubber to a change in stress. To illustrate this point, consider the phenomenological analog of a viscoelastic rubber shown in Fig. 1.6. This mechanical assembly consists of a spring in parallel with a dashpot. In this simple picture, the force in the spring is proportional to its elongation, and the force in the dashpot is proportional to its rate of elongation. Thus, the spring is a linear elastic element, in which the force is proportional to the extension, X, and the dashpot is a linear viscous element, in which the force is proportional to the rate of change of X. Note that this assembly will always return to a unique length, the rest length of the spring, when no force is acting on it. This assembly, called a Voigt body, is not intended to be a physical or quantitative model for a rubber. However, the qualitative characteristics of its response to changes in force are similar in some ways to those exhibited by rubbers.

Consider how this assembly responds to a sudden application of a tensile load F. The total force F is the sum of the force in the spring, K_sX, and that in the dashpot, $K_d(\mathrm{d}X/\mathrm{d}t)$. Thus:

$$F = K_sX + K_d(\mathrm{d}X/\mathrm{d}t) \tag{1.17}$$

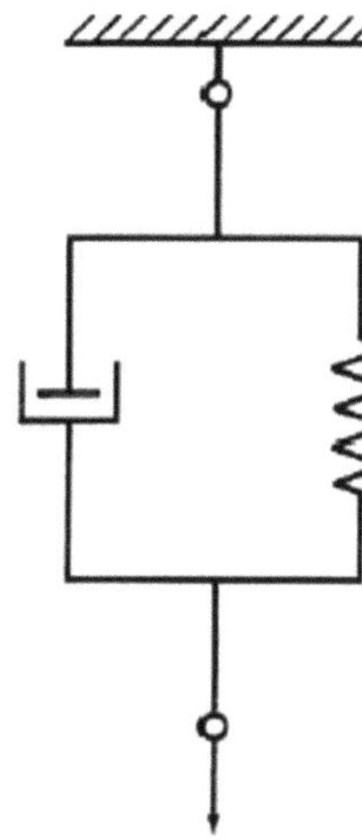

Fig. 1.6 Voigt body mechanical analog of a viscoelastic solid

We note that some of the work put into the assembly to deform it is dissipated in the dashpot, while the remainder is stored elastically in the spring. If X is initially zero, and the force F is suddenly applied at time $t = 0$, the solution to this ordinary differential equation is given below.

$$X(t) = F/K_s[1 - \exp(-K_s t/K_d)] \tag{1.18}$$

The important point to note is that the viscous resistance to elongation introduces a time dependency into the response of the assembly and that this time dependency is governed by the ratio (K_d/K_s), which has units of time. If we take the force to be analogous to the deforming stress in a viscoelastic material, and the elongation to be analogous to strain, we see that a viscoelastic rubber has a time constant and cannot respond instantaneously to changes in stress. This is called *retarded elasticity*. And as the time constant approaches zero, the behavior becomes purely elastic.

Now we turn to the case of an elastic liquid. To illustrate certain qualitative features of the rheological behavior of such a material, consider the mechanical assembly shown in Fig. 1.7. This assembly, called a Maxwell element, consists of a linear spring *in series* with a linear dashpot. Note first that unlike the Voigt body, this assembly has no unique equilibrium length and will deform forever under the influence of an applied force (assuming the dashpot is infinite in length). This is analogous to the behavior of an uncross-linked polymeric material above its glass transition and melting temperatures.

Now we examine the force on a Maxwell element that is subjected to sudden stretching by an amount X_o. The force, but not the displacement, is the same in both the spring and dashpot. The force is thus as follows.

$$F = K_s X_s = K_d(\mathrm{d}X_d/\mathrm{d}t) \tag{1.19}$$

Fig. 1.7 Maxwell mechanical analog of a viscoelastic liquid

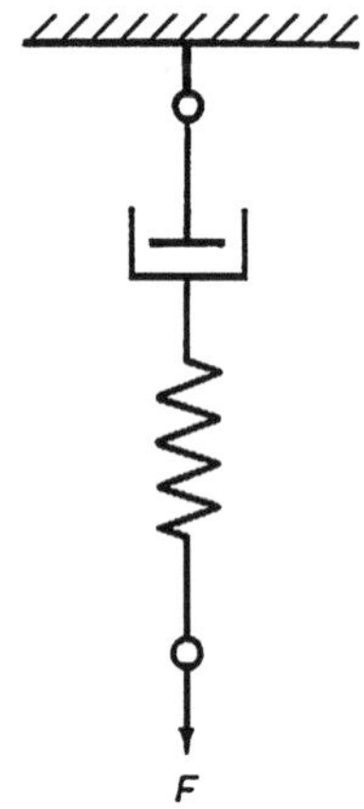

Again we note that some of the work done is dissipated in the dashpot and the remainder is stored in the spring. The total displacement of the assembly X_0 is the sum of X_s and X_d:

$$X_s + X_d = X_0 \tag{1.20}$$

Thus:

$$F(t) = K_s(X_0 - X_d) = K_d(\mathrm{d}X_d/\mathrm{d}t) \tag{1.21}$$

The solution of this ordinary differential equation is an exponential function:

$$F(t) = K_s X_0[\exp(-K_s t/K_d)] \tag{1.22}$$

Note once again that (K_d/K_s) is a constant with units of time. The force thus decays or relaxes exponentially. If we take the force to be analogous to the deforming stress in an elastic liquid and the elongation to be analogous to the strain, this process is analogous to a stress relaxation experiment. As in the case of the viscoelastic rubber, we note that the combination of viscous and elastic properties endows the material with a characteristic time and makes its behavior time-dependent. As this *relaxation time* becomes shorter and shorter, however, its behavior will approach that of a purely viscous material.

When we examine the rheological behavior of actual polymeric materials, we find creep and relaxation behavior that is qualitatively like those described above. In particular, the response to a sudden change in stress or strain is always time-dependent, never instantaneous, and there is both elastic storage of energy and viscous dissipation. However, for real polymeric materials the creep and relaxation curves cannot be described by a single exponential function involving a single characteristic time.[2] As will be explained in Chap. 3, however, practical use can

[2] Other deficiencies of these simple analogs are that the Voigt body cannot be stretched instantaneously, and the Maxwell element does not exhibit retarded creep.

still be made of the concept of a relaxation time by describing the viscoelastic behavior of real materials in terms of a *spectrum* of relaxation times.

1.5.4 Structural Time Dependency

In our discussion of the viscosity function, we took the shear stress to be independent of time at constant shear rate. For a Newtonian fluid this is appropriate, because the stress responds instantaneously to the imposition of a shear rate. However, some inelastic non-Newtonian fluids do not respond instantaneously, so that when shearing begins, there is a transient period during which the shear stress varies with time, first rising instantly to one value and later reaching a steady state value that can be used to calculate a viscosity by use of Eq. (1.16). The origin of such time dependency is most often a flow-induced change in the structure of the fluid, as in the case of a concentrated suspension of solid particles. For example, the state of aggregation of suspended particles can be changed significantly by shearing. This *structural time dependency* is entirely dissipative, that is to say there is no elastic energy storage, and all the work done to deform the material is converted immediately into heat, or more accurately, into internal energy. Time dependency can also arise from changes in the conformation and orientation of molecules, as in the case of polymer solutions, but this type of time dependency is a viscoelastic effect rather than a structural time dependency.

Restricting attention here to inelastic materials such as suspensions of rigid particles in Newtonian fluids, in which no elastic energy storage occurs, the most common type of structural time dependency is that shown in Fig. 1.8, where the shear stress decreases with time after the start up of steady shear flow. This type of behavior, in which the viscosity decreases with time, is called *thixotropy*. If the shearing is stopped, the viscosity will gradually increase as the particles respond to inter-particle forces in the absence of a shear stress, but the time required for the complete reformation of the equilibrium structure can be very long. There are

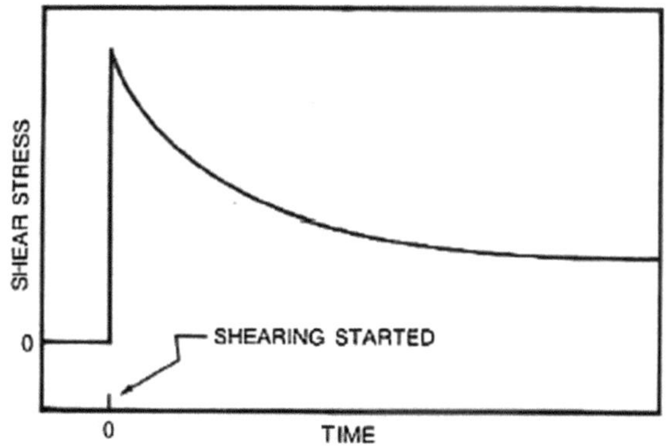

Fig. 1.8 Start-up of simple shear for a material with time-dependent structure

some materials, such as corn starch suspensions, that exhibit the opposite effect, viz., the stress increases with time after the start-up of steady shear. This type *of* behavior is called *rheopexy* or *antithixotropy*. In the case of thixotropy, the shear tends to break down structure that has been established in the unsheared material, while in the case of a rheopectic material, shearing promotes the buildup of structure.

1.5.5 Plasticity and Yield Stress

The structure in a concentrated suspension can be sufficiently rigid that it permits the material to withstand a certain level of deforming stress without flowing. The maximum stress that can be sustained without flow is called the *yield stress*, and this type of behavior is called *plasticity*. Polycrystalline metals generally exhibit plasticity, as do semicrystalline polymers at temperatures between their melting and glass transition temperatures. Highly filled melts are also thought to have a yield stress, although its precise measurement is difficult.

The simplest model of plastic behavior is that in which the excess stress, above the yield stress, is proportional to the shear rate. For simple shear flow, this type of behavior is described by Eq. (1.23).

$$\sigma = \sigma_y + \eta_p \dot{\gamma} \quad (\sigma \geq \sigma_y) \tag{1.23}$$

Here σ_y is the yield stress, and η_p is a *plastic viscosity*. A material that behaves this way is called a *Bingham plastic*. This is an idealized type of behavior that is not precisely followed by any real material, but it is sometimes a useful approximation to real behavior. Note that Eq. (1.23) is not a constitutive equation, as it does not

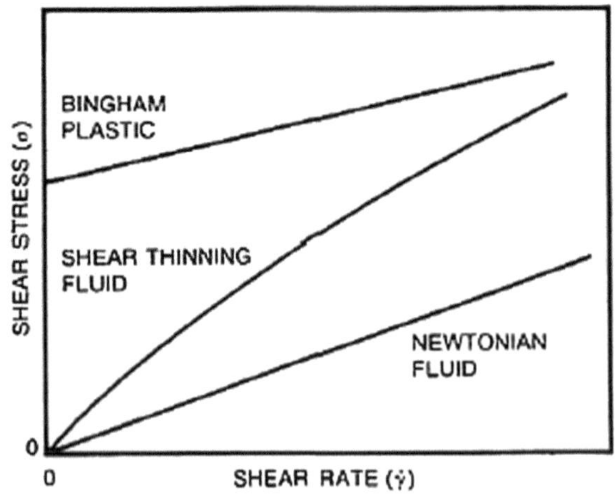

Fig. 1.9 Shear stress versus shear rate for three types of material

describe all the components of the stress tensor in any type of deformation but only the shear stress in simple shear. Figure 1.9 compares shear stress versus shear rate curves for a Bingham plastic, a Newtonian fluid and a shear thinning fluid.

1.6 Why Polymeric Liquids are Non-Newtonian and Elastic

It is important in applied polymer science to be able to relate physical properties, including rheological properties, to molecular structure. This subject is taken up in some detail in Chap. 7. We mention here only the general mechanisms by which polymeric molecules endow liquids with complex rheological behavior.

First, we consider why polymer molecules are elastic. A useful picture of a polymer molecule is that of a long chain with many freely jointed links allowing relative rotation of adjacent links. The presence of the large number of joints makes the chain quite flexible and allows it to have many different configurations. At temperatures above the glass transition temperature a molecule continually changes its configuration due to Brownian motion, but we can describe the states of a single molecule over a period of time or a large number of molecules at a given time in terms of statistical averages. For example, at a given temperature there will be a unique average value of the end-to-end distance, R, for the molecules of a polymeric liquid that has been at rest for a sufficient length of time that it is in its equilibrium state. Deforming the liquid will alter this average length, but if the deformation is stopped, Brownian motion will tend to return the average value of R to its equilibrium value. This is the molecular origin of the elastic and relaxation phenomena that occur in polymeric liquids. This mechanism, by which a molecule always tend to return to its equilibrium, isotropic configuration when free of deforming stresses, is sometimes referred to as an entropic spring.

Some types of deformation have a greater ability to change R and orient molecules than others. Deformations that can generate a high degree of stretching and orientation are said to be *strong flows*, while those that cannot are said to be *weak*. We have seen how the deformation of a polymeric liquid can alter the configuration and orientation of the molecules. Thus, the physical nature of the fluid is altered by the deformation, and this, in turn, alters the mechanical (rheological) properties of the fluid. At high shear rates, the shape of a polymer molecule is different from that at low shear rates, and this alters its resistance to flow and thus its viscosity. Furthermore, the partial orientation of the molecules, i.e., the shift of the average orientation of the R vectors away from zero, introduces a strain-induced anisotropy that is responsible for differences in the normal stresses in steady simple shear. Finally, the tendency of Brownian motion to return the system of molecules to its most probable, i.e., equilibrium, configuration distribution explains why the behavior of these fluids is time-dependent.

The short-time response to rapid deformations of high-molecular-weight, molten polymers is observed to be very similar to that of cross-linked rubbers. This has inspired the concept of a *temporary network* that exists in the melt and acts like a rubbery network at shorter times but whose junctions can slip over longer periods of time to permit flow. This network is said to arise from *entanglements* in the melt. The modern view is that the rubbery behavior of melts is due not to an actual looping or knotting of molecules around each other but simply to the constraints on their motion resulting from the fact that molecules cannot cross over each other, i.e., to their *uncrossability*. However, the word entanglements is still universally used to describe the important limitation on the motion of one molecule by its neighbors.

Entanglement is related to the high degree of spatial overlap of molecules in concentrated solutions and melts. The existence of overlap is readily demonstrated by considering the size of a polymer coil. One measure of molecular size is the radius of gyration R_g. For linear polyethylene, R_g depends on the molecular weight M as follows:

$$R_g(cm) = 4 \times 10^{-9} M^{1/2} \tag{1.24}$$

For a polyethylene with a molecular weight of 10^6 g/mol, the volume of the sphere occupied by one molecule is therefore about 2.6×10^{-9} cm^3. The mass of this coil is 10^6 divided by Avogadro's number, which is 1.7×10^{-18} g. The density of the coil based on its occupied volume is thus less than 0.01 g/cm^3. But the bulk melt density is about 0.76 g/cm^3 at 190 °C, and this implies that parts of many other coils are invading the volume explored by a given molecule. In other words, molecules penetrate each others space to a very large extent and thus strongly impede each others' movements.

A similar calculation for much shorter molecules, e.g., a polymer with a molecular weight of 10^4 g/mol, gives a density of 0.1 g/cm^3, which is considerably closer to the measured bulk density. This shows that the degree of coil overlap, and therefore the entanglement density, increases sharply with molecular weight.

Rubbery behavior occurs in a melt when the molecular weight is above a characteristic value that varies from one polymer to another. Above this molecular weight the number of entanglements is sufficient to produce strong rubberlike effects. The macroscopic effects of entanglements include high viscosity and high elastic recoil. At the same time, the nature of this strong interaction can be altered temporarily by deformation, so that high molecular weight melts have highly nonlinear properties; for example, the viscosity is a very strong function of shear rate.

Polymeric materials are said to have a memory, in the sense that when deforming stresses are eliminated, they tend to return to a previous configuration. Cross-linked polymers have perfect memory since their network consists of permanent chemical cross-links. Thus, a sample always returns to a unique equilibrium shape when deforming stresses are removed, whereas a molten polymer is said to have a fading memory, since the so-called entanglement network is not permanent and is altered by flow and relaxation processes. Therefore, only part of the deformation is recovered when deforming stresses are removed.

Chapter 2
Viscosity and Normal Stress Differences

Abstract Viscosity is the property most used with molten plastics. It relates the shear stress to the shear rate in steady simple shear flow, which is the deformation generated between two parallel plates, one of which undergoes linear displacement. For viscoelastic fluids, two other quantities are needed for a complete description of the stress field, and these are the *first and second normal stress differences*. The viscosity and the two normal stress differences are functions of shear rate that are called the *viscometric functions*, and flows governed by these are called *viscometric flows*. In addition to simple shear, other viscometric flows include flow in straight channels and rotational flows between concentric cylinders, between a cone and plate and between two disks. Flow in an extruder is dominated by the viscometric functions, mainly the viscosity. This chapter describes the dependence of viscosity on shear rate, temperature, molecular weight and its distribution, tacticity, comonomer content, and long-chain branching.

2.1 Simple Shear and Steady Simple Shear

Simple shear is of central importance in applied rheology for two reasons. First, it is the flow that is the easiest to generate in the laboratory, and melt data most often reported are based on flows that are rheologically equivalent to simple shear. Secondly, a number of processes of industrial importance, particularly extrusion and flow in many types of die, approximate steady simple shear flow. Finally, the flow behavior of melts in these flows is governed mainly by viscosity, and elasticity is not an important factor. It is thus appropriate to devote a chapter to the properties that govern simple shear flow.

Simple shear was introduced in Sect. 1.4.2, and we will review its essential features here. Referring to Fig. 2.1, we show this flow being generated by the rectilinear motion of one flat plate relative to another, where the two plates are parallel and the gap between them h is constant with time. This flow is completely

J. M. Dealy and J. Wang, *Melt Rheology and its Applications in the Plastics Industry*, Engineering Materials and Processes, DOI: 10.1007/978-94-007-6395-1_2,

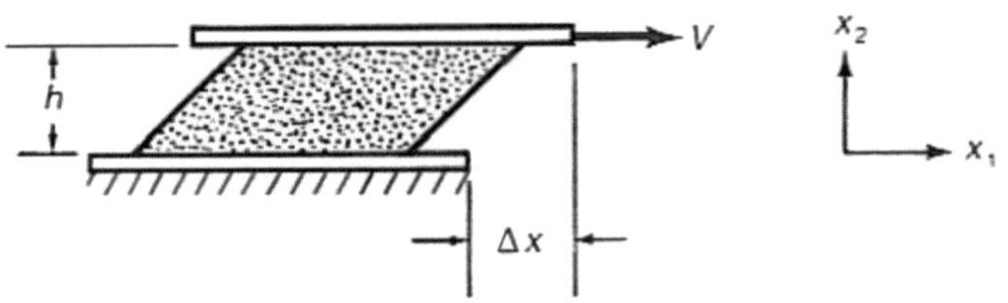

Fig. 2.1 Simple shear flow in x_1 direction

described by giving the shear strain γ as a function of time, where γ was defined by Eq. (1.8), which is repeated below as (2.1). The shear rate $\dot{\gamma}$ is the derivative of this quantity with respect to time and is thus the velocity of the moving plate divided by the gap, as was shown by Eq. (1.9), repeated below as (2.2). In *steady simple shear* the shear rate is constant with time.

$$\gamma = \Delta X/h \tag{2.1}$$

$$\dot{\gamma} = \frac{1}{h}\frac{\mathrm{d}X}{\mathrm{d}t} = \frac{V}{h} \tag{2.2}$$

According to a universally used convention, the velocity components in simple shear are defined as follows in terms of the coordinate system shown in the Fig. 2.1.

V is the velocity of the upper plate.
v_1 is the velocity of the fluid in the x_1 direction.
The velocities in the x_2 and x_3 directions are zero.
The velocity gradient is in the x_2 direction.

Simple shear is a uniform deformation, i.e., each fluid element undergoes the same deformation, and the strain and strain rate are independent of position in space; and for steady simple shear the strain rate is constant with time. Thus, the strain rate of every fluid element is the same and given by:

$$\dot{\gamma} = \mathrm{d}v_1/\mathrm{d}x_2$$

As a result, the rheologically meaningful stresses are also independent of position in space.

2.2 Viscometric Flow

We will define a viscometric flow as one that, from the point of view of a fluid element, is indistinguishable from steady simple shear. A more comprehensive mathematical definition can be found in the book by Bird et al. [1]. Thus, while various fluid elements in the field of flow may be subject to different shear rates, the shear rate experienced by any particular fluid element is constant with time. Steady simple shear is the simplest possible viscometric flow.

Viscometric flows are of practical interest, not only because of their wide use in experimental rheology, but also because many flows of practical importance are viscometric flows or close approximations thereto. Some examples are listed below:

1. Steady tube flow (Poiseuille flow)

This flow is the one most often used to determine the viscosity of molten plastics. It also occurs whenever a melt is transported by means of circular channel. The magnitudes of the shear stress and shear rate vary from zero on the axis to a maximum at the wall.

2. Steady slit flow

Sometimes called plane Poiseuille flow, slit flow can also be used to measure viscosity. The shear stress and shear rate vary from zero on the plane of symmetry to a maximum at the walls.

3. Annular pressure flow

This flow occurs in the axial direction in the space between two concentric cylinders as the result of a pressure gradient. If the ratio of the two diameters is close to one, the velocity distribution approaches that for slit flow.

4. Steady concentric cylinder (Couette) Flow

This is a *drag flow* generated by the rotation of either the inner or outer cylinder of a concentric cylinder apparatus. It is widely used for the measurement of viscosity in Newtonian fluids, but is not convenient for use with melts. If the ratio of the two diameters is close to one, the shear rate becomes nearly uniform in the annular gap containing the fluid and thus approximates that in steady simple shear flow.

5. Steady parallel disk flow

This involves torsional flow of fluid between two parallel discs generated by the rotation of one of the disks. Also called plate–plate flow

6. Steady cone and plate flow

This flow is approximately viscometric and is of special interest, because if the cone angle is very small, the shear rate and shear stress are very nearly uniform, and the viscosity and first normal stress difference can be readily measured. The use of this flow, however, is limited to quite low shear rates because of the substantial departures from viscometric flow that occur at higher rates.

7. Steady sliding cylinder flow

This is a drag flow of fluid between concentric cylinders generated by the linear displacement of the inner cylinder. If the ratio of the diameters of the two cylinders is close to one, the shear rate is nearly uniform in the gap.

8. Steady helical flow

This is a combination of flows 4 and 7. The fluid is contained in the annular space between two concentric cylinders; one of these rotates at a constant speed while either the same or the other cylinder is displaced along its own axis at a constant speed.

9. Combined drag and pressure flow

If we combine steady simple shear with pressure flow in a slit, there are two forces driving fluid flow. Drag flow results from the motion of one wall, while pressure flow results from the pressure gradient. If the direction of the pressure gradient is the same as the direction of the motion of the moving wall, the velocity profile is of one of the types shown in Fig. 2.2 [1]. Depending on the sign of the pressure gradient, pressure can work with or against the drag flow, and four situations are illustrated in the figure. Note that a negative pressure gradient ($\mathrm{d}p/\mathrm{d}x < 0$) promotes flow from left to right, while a positive gradient works against the drag flow. If the direction of the drag flow is at an angle to the direction of the pressure gradient, the resulting deformation is similar to that in the channel of a single screw extruder. A similar combined flow is Couette (concentric cylinder) flow with an axial pressure gradient. Here the drag flow is perpendicular to the pressure flow direction. When the gap is very small, this flow becomes equivalent to plane Couette flow with perpendicular pressure flow.

All of the analyses presented or cited above are based on the *no-slip* assumption that the fluid adheres to any wall with which it is in contact and that if this wall moves, the fluid in contact with it moves at the same speed. However, this assumption is not always valid for molten plastics. For certain combinations of shear stress and shear strain, the melt undergoes some type of fracture at or near the wall and subsequently undergoes slip. In the case of linear polyethylene this occurs at shear stresses in the neighborhood of 0.1 MPa unless the experiment is terminated while the total shear strain is still quite small. Once slip flow begins, the velocity of the sliding polymer surface relative to the wall is not known a priori. This complicates the interpretation of data to determine the viscometric functions. Wall slip is discussed in further detail in Chap. 6.

Another assumption that is made in the classical analyses of viscometric flows is that the deformation is homogeneous. In reality, an experimental apparatus is

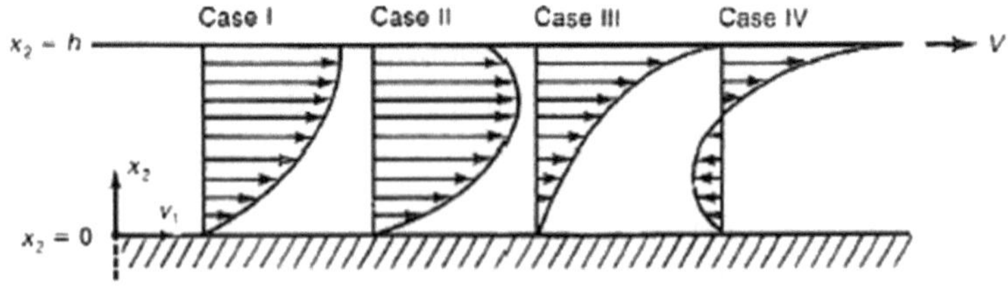

Fig. 2.2 Velocity profiles for combined drag and pressure flows. Cases I and II: $\mathrm{d}p/\mathrm{d}x < 0$; Cases III and IV: $\mathrm{d}p/\mathrm{d}x > 0$. Adapted from Ref. [1]

always finite in size, and the sample has one or more exposed free surfaces. Examples are the sample edge in a cone and plate or parallel disk rheometer. In addition, there may be zones in the field of flow where the deformation differs significantly from the assumed viscometric flow. Examples are the zones below the inner cylinder of a Couette viscometer and at the entrance of a capillary rheometer. These end and edge effects can be sources of error in viscometric measurements and are considered in some detail in Chap. 6.

2.3 The Viscometric Functions

As was pointed out in Chap. 1, for an incompressible material a normal stress by itself has no rheological significance, since if the normal stresses are the same in all directions, i.e. the stress is isotropic, there will be no deformation. Only normal stress differences can cause deformation, for example stretching and compression. There are two, independent, normal stress differences, and these are called the *first and second normal stress differences*. These, along with the viscosity, are functions of shear rate, and are called the *viscometric functions*.

$$\eta(\dot{\gamma}) \equiv \sigma/\dot{\gamma} \tag{2.3}$$

$$N_1(\dot{\gamma}) \equiv \sigma_{11} - \sigma_{22} \tag{2.4}$$

$$N_2(\dot{\gamma}) \equiv \sigma_{22} - \sigma_{33} \tag{2.5}$$

For any viscometric flow, the three viscometric functions completely describe the rheological behavior of a fluid. In other words, these constitute all the rheological information that can be obtained from measuring the stress components.

2.4 The Viscosity

Viscosity is the rheological property most often used to characterize molten plastics, because it is relatively easy to measure, provides some information about molecular structure, and plays an important role in melt processing. Like all rheological properties the viscosity (and normal stress differences) of a polymer depend on the following factors:

1. Flow conditions

   (a) Shear rate
   (b) Temperature
   (c) Pressure

2. Resin composition and molecular structure
   (a) Chemical nature of polymer
   (b) Molecular weight distribution
   (c) Presence of long chain branches
   (d) Nature and concentration of additives, fillers, etc.

The viscosity at high shear rates is determined by use of a capillary rheometer, while that at low rates it is measured using a rotational rheometer with cone-plate fixtures. The actual range of shear rates accessible using either instrument depends on the properties of the melt, as is explained in Chap. 6. These instruments are able to generate the steady shear flow that is required to measure viscosity. The entire viscosity curve is rarely used for routine quality control. The device most used for this application is the *melt indexer*, more properly called an *extrusion plastometer*. While this very simple flow tester involves flow through a short capillary, it does not provide a reliable value of viscosity. Experimental methods are described in detail in Chap. 6.

2.4.1 Effect of Shear Rate on Viscosity

As in the case of Newtonian fluids, the viscosity of a polymer depends on temperature and pressure, but for polymeric fluids it also depends on shear rate, and this dependency is quite sensitive to molecular structure. In particular, comprehensive and precise data of viscosity versus shear rate can be used to infer the molecular weight distribution of a linear polymer. And it can sometimes tell us something about the level of long-chain branching. This curve is also of central importance in plastics processing, where it is directly related to the torque and energy required to extrude a melt and is useful in the design of extruders and dies.

The viscosity of molten thermoplastics decreases sharply as the shear rate increases. Typical behavior is sketched in Fig. 2.3. Note that linear rather than logarithmic scales are used here. At very low shear rates, the viscosity normally becomes independent of shear rate, as shown in the magnified inset of Fig. 2.3. The constant viscosity that prevails at these low shear rates is called the *zero-shear viscosity* and has the symbol η_0. The zero shear viscosity is an important scaling parameter, as is shown below, but for many commercial resins, particularly those with very broad molecular weight distributions or a high degree of long chain branching, it is difficult to measure using controlled strain rotational rheometers. This is because the shear rate at which $\eta(\dot{\gamma})$ levels out to its limiting value is too low to be generated in these instruments. It is often necessary to resort to a long-duration creep experiment to determine η_0. More will be said about this in Chap. 6.

In order to show clearly the approach of the viscosity to its limiting, low shear-rate value, while also showing high shear-rate behavior, it is customary to display viscosity versus shear rate behavior as a plot of $\log(\eta)$ versus $\log(\dot{\gamma})$. An example of such a plot is shown in Fig. 2.4, which shows the data of Meissner [2] for a

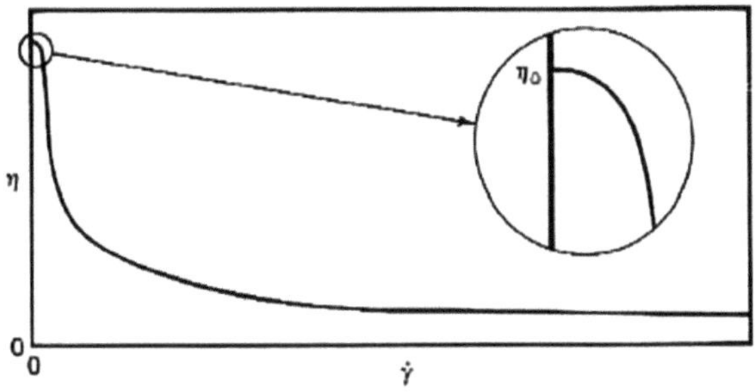

Fig. 2.3 Shape of viscosity versus shear rate curve for a molten polymer. Using linear scales important low shear rate features are crowded into the left-hand axis. The inset is an expansion of this region

low-density polyethylene at several temperatures. This polymer has a broad molecular weight distribution and a high level of long-chain branching. As a result the decrease in viscosity from its zero-shear value to a power-law region extends over many decades of shear rate. To obtain the data at very low shear rates, it was necessary to make major modifications to a commercial rotational rheometer, and the measurements required great skill and very long measurement times. At the lowest shear rates we see a clearly-defined "Newtonian" region over which the viscosity is constant. And at the highest shear rates, the curves tend to approach each other, appearing to move onto straight lines, suggesting power-law behavior.

For linear polymers with narrow molecular weight distributions, the viscosity curve (on a log–log plot) has a distinct region of essentially constant viscosity as well as a well-defined power law region, and the transition between the two occurs over about one decade of shear rate. This is illustrated in Fig. 2.5, where data for several, narrow molecular weight distribution polystyrenes are shown [3]. For these specially-polymerized samples the range of shear rates between the low-rate Newtonian region and a well-defined power law is much narrower than in the case of the LDPE data shown in Fig. 2.4. Also, we see that the curves for samples with

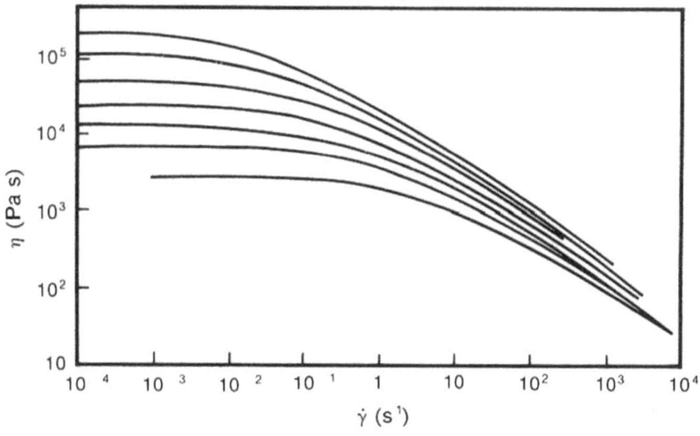

Fig. 2.4 Viscosity versus shear rate of a LDPE at several temperatures; from *top* to *bottom*: T(°C) = 115, 130, 150, 170, 190, 210, and 240. Data of Meissner [2]

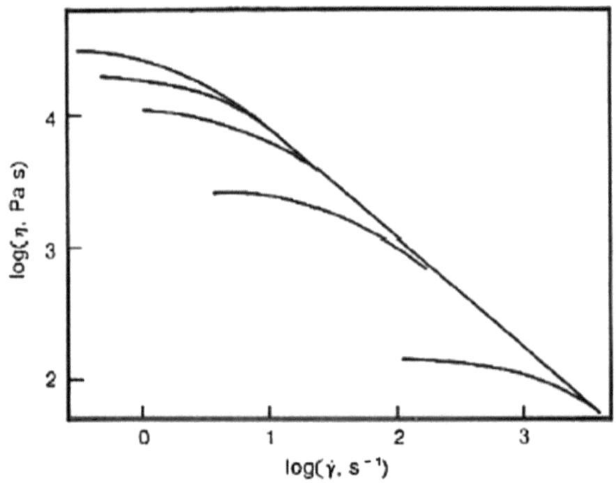

Fig. 2.5 Viscosity versus shear rate for narrow MWD polystyrenes; from *top to bottom* $M_w = 4.9 \times 10^4$, 12×10^4, 18×10^4, 22×10^4, and 24×10^4. From Stratton [3]

various molecular weights converge onto a single power-law line at shear rates that decrease with molecular weight. This power law is described by Eq. (2.6).

$$\sigma = k\dot{\gamma}^n \tag{2.6}$$

In terms of viscosity this becomes Eq. (2.7).

$$\eta = k(\dot{\gamma})^{n-1} \tag{2.7}$$

Obviously, a Newtonian fluid is a special case for which $n = 1$ and K is equal to the viscosity.

There are several undesirable features of the power law expressed by Eqs. (2.6) and (2.7).

1. The units of K depend on the value of n, and unless n is unity, they involve time to a power that is not an integer.
2. If the shear rate is negative, the equation does not yield a value for the viscosity (unless n is an integer).
3. The zero shear viscosity does not appear as a parameter.
4. The equation is only valid at high shear rates.

The first three undesired features can be eliminated by use of the following form:

$$\eta = \eta_0|\lambda\dot{\gamma}|^{n-1} \tag{2.8}$$

where λ is a material constant with units of time, i.e., a characteristic time of the material. Specifically, it is the reciprocal of the shear rate at which η becomes equal to η_0. A variation of (2.8) that is occasionally used is:

$$\eta = \eta_1|\dot{\gamma}|^{n-1}$$

where η_1 is numerically equal to the viscosity at a shear rate of 1 s^{-1}.

While Eq. (2.8) still cannot describe the low-shear-rate portion of the curve, where the viscosity approaches a constant value, for many polymers Eq. (2.6) [and Eqs. (2.7) or (2.8)] holds reasonably well over the high-shear-rate range of interest for processing. It has been widely used in the modeling of melt flow because of its mathematical simplicity, and this makes it possible to derive analytical expressions describing flows in extruders and channels. However, because of the wide availability of powerful computational facilities, the power law no longer offers an important advantage, and more realistic equations are now being used in most process simulations. These allow for the transition to Newtonian behavior over a range of shear rates.

As suggested by Eq. (2.8), the variation of η with $\dot{\gamma}$ implies the existence of at least one material property with units of time, and in that model this parameter, λ, is the reciprocal of the shear rate at which the power-law line reaches η_0. Models that can describe the approach to η_0 thus involve at lease one characteristic time. Examples of modified power laws whose parameters are η_0, a characteristic time, and an exponent, include the Cross equation [4] and the Carreau equation [5], shown below as Eqs. (2.9) and (2.10) respectively.

$$\eta(\dot{\gamma}) = \eta_0[1 + (\lambda|\dot{\gamma}|)^m]^{-1} \tag{2.9}$$

$$\eta(\dot{\gamma}) = \eta_0\left[1 + (\lambda\dot{\gamma})^2\right]^{-p} \tag{2.10}$$

These models approach power-law behavior at high shear rates, and the dimensionless material constants m and p are simply related to the power law exponent, i.e., $m = 1-n$ and $p = (1-n)/2$. Hieber and Chiang [6] compared the ability of these two models to fit data for a variety of commercial polymers for purposes of flow simulation. They found that the Cross equation provided a better fit for the polymers they considered.

For more flexibility in fitting data, Yasuda et al. [7] generalized Eq. (2.10) by adding an additional parameter as shown in Eq. (2.11) in order to adjust the curvature in the transition region.

$$\eta(\dot{\gamma}) = \eta_0[1 + (\lambda\dot{\gamma})^a]^{(n-1)/a} \tag{2.11}$$

This is often called the Carreau-Yasuda equation.

Elberli and Shaw [8] reported that time constants obtained by fitting data to two-parameter viscosity models were less sensitive to experimental error than those based on models with more parameters. Data at low shear rates and around the reciprocal of the time constant are most essential to obtain useful values of the parameters, while the high shear rate data are less important.

Plumley [9] evaluated the ability of the above models to fit data for linear and sparsely-branched metallocene polyethylenes. They found that the Cross equation gave a good fit to the data and that adding parameters did not lead to a significant improvement. On the other hand, for a linear polymer with a bimodal molecular weight distribution or polymers with long-chain branching, such as LDPE, three-parameter models are not able to fit data over a very broad range of frequencies. Wang [10] proposed a "double-Cross" model to describe the behavior of such materials. It is the sum of two Cross model terms (Eq. 2.9), each with its own set of three parameters. Wang found that this model could fit both the low-shear-rate approach to the zero-shear viscosity as well as the high-shear-rate approach to a power law with great precision.

Equations such as those presented above are sometimes used to extrapolate low-shear-rate data to estimate the zero-shear viscosity when data do not reach the Newtonian region. But this is not a reliable procedure, since there is no theoretical basis for it, and it can yield values that are 50 % or more away from the correct value.

In Chap. 7 we will see that the molecular weight distribution of a linear polymer can be inferred from viscosity data obtained over a broad range of shear rates.

2.4.2 The Cox-Merz "Rule"

In Chap. 3 it is shown that small amplitude oscillatory shear using a rotational rheometer is the most widely used technique for the determination of linear viscoelastic properties. Using a small sample and a single instrument, it is possible to obtain data over a wide range of frequencies. One way of presenting the resulting data is a plot of magnitude of the complex viscosity $|\eta^*(\omega)|$ versus frequency. This is often referred to simply as the "complex viscosity" or the "dynamic viscosity." This is much less trouble than measuring the viscosity over a wide range of shear rates using at least two rheometers. It would thus be useful to be able to relate the complex viscosity to the (steady-shear) viscosity. Based on data for two polystyrenes, Cox and Merz [11] reported that the curve of *apparent viscosity* versus shear rate measured using a capillary viscometer was very similar to the curve of complex viscosity versus frequency. We will see in Chap. 6 that the apparent viscosity calculated from capillary viscometer data, i.e., the wall shear stress divided by the apparent shear rate, is substantially different from the actual viscosity, but it is possible that for the two polymers Cox and Merz studied the apparent viscosity was closer to the true value than is generally the case. What they observed was that the curve traced by their $\eta_A(\dot{\gamma}_A)$ data was very similar to that traced by $|\eta^*(\omega)|$ data. If the two curves were in fact identical, it would imply that:

$$\eta_A \equiv \frac{\sigma_w}{\dot{\gamma}_A} \cong |\eta^*|(\omega) \quad \text{with } \dot{\gamma}_A = \omega. \tag{2.12}$$

Over time, however, these details have been forgotten, and one refers now to a "Cox-Merz rule" which is expressed by Eq. (2.13).

$$\eta(\dot{\gamma} = \omega) = |\eta^*|(\omega) \tag{2.13}$$

Thus, the original observation, based on the apparent viscosity, has come to be replaced by a quantitative relationship. However, it has often been reported that the relationship expressed by Eq. (2.13) is obeyed by experimental data, although this conclusion is sometimes based on the use of a small log–log graph on which significant deviations can easily hide. The Cox-Merz rule has been examined critically by Utracki and Gendron [12] and by Venkatraman et al. [13]. The latter authors reported that Eq. (2.13) works fairly well for LDPE, but not for HDPE. In any event, we will see below that many viscosity-structure relationships originally developed for use with viscosity data are now often applied to complex viscosity data. One can look at this in two ways. On the one hand, it is simply an application of Eq. (2.13). Or, on the other hand, one can say that the relationship itself applies to the complex viscosity.

2.4.3 *Effect of Temperature on Viscosity*

In Chap. 3, the procedure called time–temperature superposition is explained in detail. Here we make use of this technique to show how the shear rate range over which viscosity information can be obtained can be extended beyond the range actually accessible using a given viscometer. The procedure is based on the idea that increasing temperature has an equivalent effect on viscosity as decreasing shear rate. This trend can be seen by inspection of Fig. 2.4, noting that increasing the temperature moves data to lower curves. If this effect is quantitatively the same at all shear rates, this implies that a single *horizontal shift factor* can be used to shift data on a log–log plot taken at several temperatures along the shear rate axis to coincide with those measured at a *reference temperature* T_0. This idea is expressed by Eq. (2.14). As explained in Sect. 3.12, for greater precision, a second shift factor, $b_T(T)$, normally equal to $(T_0\rho_0/T\rho)$, should also be applied to the viscosity, but it is usually close to unity and is often neglected.

$$\eta(\dot{\gamma}, T_0) = [1/\alpha_T(T)]\eta(\dot{\gamma}a_T, T) \tag{2.14}$$

Such a representation is called a *master curve* and is a plot of *reduced viscosity* $\eta(T)/a_T(T)$ versus reduced shear rate $\dot{\gamma}a_T(T)$. Note that the horizontal shift factor $a_T(T)$ is applied to both axes, because the definition of the viscosity involves the shear rate. By use of this procedure, the range of reduced shear rate range over which a master curve can be prepared is broader than the range of shear rates over which measurements can be made.

For a polymer well above its glass transition temperature, it is often found that the zero-shear viscosity obeys the well-known Arrhenius relationship shown by Eq. (2.15).

$$\eta_0(T) = \eta_0(T_0)\exp\left[\frac{E_a}{R}\left(\frac{1}{T} - \frac{1}{T_0}\right)\right] \tag{2.15}$$

The constant E_a is called the activation energy for flow. But this is the special case of Eq. (2.14) for shear rates where the viscosity equals η_0, which implies that the shift factor is give by Eq. (2.16).

$$a_T(T) = \frac{\eta_0(T)}{\eta_0(T_0)} = \exp\left[\frac{E_a}{R}\left(\frac{1}{T} - \frac{1}{T_0}\right)\right] \tag{2.16}$$

Since $\eta_0(T_0)$ is a constant for a given master curve, the shift factor is proportional to $\eta_0(T)$. Thus, a viscosity master curve can be prepared by plotting $\eta(T)/\eta_0(T)$ versus $\dot{\gamma}\eta_0(T)$.

It should be mentioned that time–temperature superposition is of limited utility with crystallizable polymers, because the range of temperatures over which measurements can be made is limited to that between the melting point and the temperature at which the polymer starts to decompose.

Time–temperature superposition is not useful for long-chain branched systems, although such materials are sometimes characterized in terms of specially-defined activation energies. This subject is discussed in detail in Chap. 3.

2.4.4 Effects of Pressure and Dissolved Gas on Viscosity

Whereas increasing temperature decreases the viscosity of melts, increasing pressure increases it, because compression of the melt decreases free volume. Pressure shift factors can be used to generate master curves just as temperature shift factors are used in time–temperature superposition. The Barus equation is often found to describe the pressure dependence of viscosity. This is shown as Eq. (2.17).

$$\ln\left[\frac{\eta_0(P)}{\eta_0(P_0)}\right] = \beta(P - P_0) \tag{2.17}$$

This implies that the pressure shift factor a_P is given by:

$$\ln[a_P(P)] = \beta(P - P_0) \tag{2.18}$$

Figure 2.6 shows the effect of pressure on the viscosity of a high-density polyethylene at 180 °C, and Fig. 2.7 is a maser curve based on the same data [14]. The horizontal shift factor is $a'_P(P)$, the prime indicating that the vertical shift factor was neglected, i.e., set equal to unity. The Barus Eq. (2.18) was found to fit the entire viscosity curve very well with b_P set at unity. Increasing the pressure from atmospheric 0.1 to 69 MPa ($\approx$10,000 psi) increases the viscosity by a factor of about two.

The HDPE for which data are shown in Figs. 2.6 and 2.7 was saturated with carbon dioxide at several pressures, and viscosity data are shown in Fig. 2.8; the corresponding master curve is shown in Fig. 2.9. The horizontal shift factor $a_{P,C}$

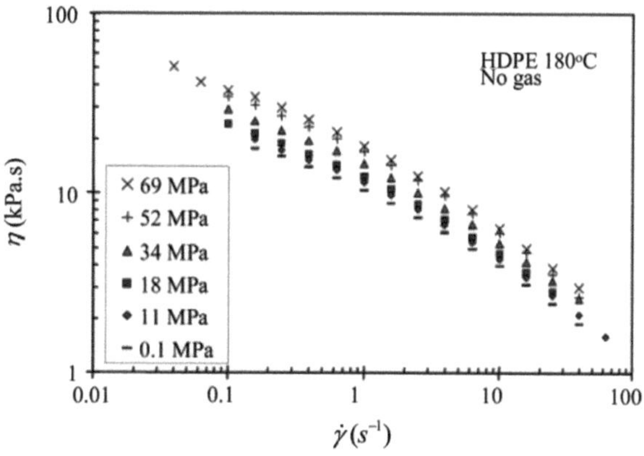

Fig. 2.6 Effect of pressure on the viscosity versus shear rate curve of HDPE. From Park and Dealy [14]

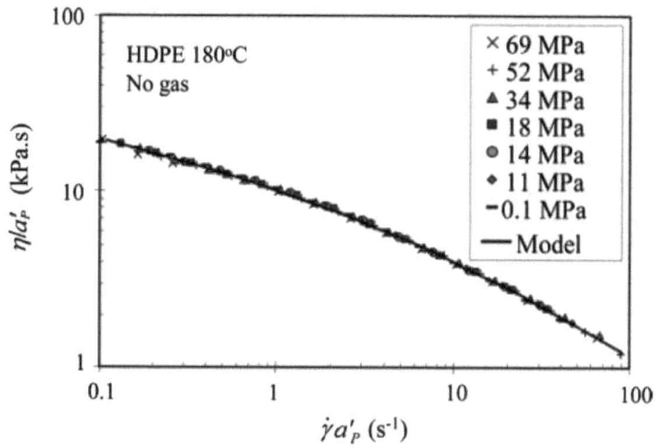

Fig. 2.7 Shifted viscosity curve taking b_P to be unity; reference pressure is 0.1 MPa. From Park and Dealy [14]

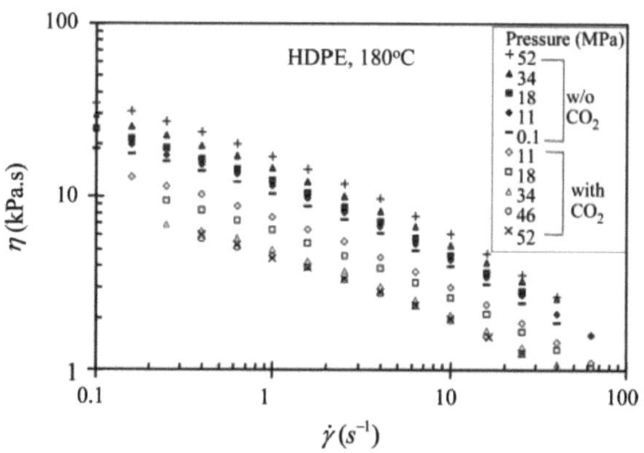

Fig. 2.8 Effects of pressure and dissolved CO_2 on viscosity. From Park and Dealy [14]

accounts for both pressure and gas concentration. The CO_2 reduced the viscosity by increasing free volume, thus counteracting the effect of pressure. The carbon dioxide not only neutralized the pressure effect but reduced the viscosity below its value at atmospheric pressure. At 69 MPa the viscosity is reduced by a factor of

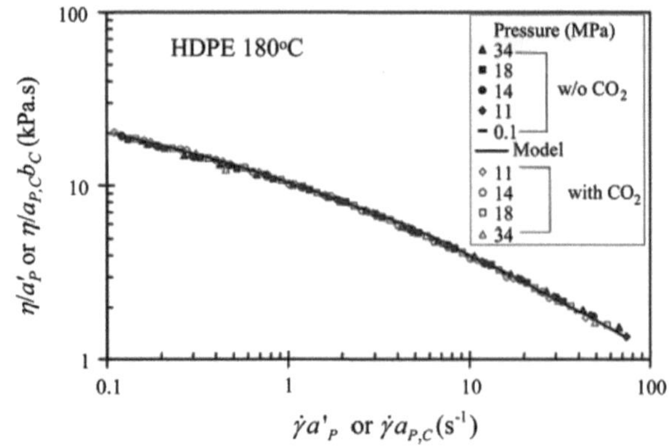

Fig. 2.9 Data of Fig. 2.8 shifted vertically and horizontally to make pressure-composition master curve. Vertical shift factor takes into account effect of concentration but not pressure. From Park and Dealy [14]

about 2.5. Park et al. [15] reported on the effect of long-chain branching on pressure shift factors for polystyrene and polyethylene.

2.4.5 *Effect of Molecular Weight on the Zero-Shear Viscosity*

Small molecules in the liquid state interact primarily through intermolecular forces that give rise at the microscopic level to friction and at the macroscopic level to viscosity. The viscosity of such a liquid is independent of shear rate. A polymeric liquid with a low molecular weight behaves in this way, and its viscosity increases linearly with molecular weight. For example, for linear polyethylene this behavior obtains up to a molecular weight around 3,500. But over a fairly narrow range of molecular weights the viscosity starts to decrease with shear rate and the increase of η_0 with molecular weight becomes much stronger than linear. In the same range of rates, the viscosity depends increasingly on shear rate.

Plots of $\log(\eta_0)$ versus $\log(M)$ for several linear, monodisperse polymers are shown in Fig. 2.10 [16]. At low molecular weights the viscosity is proportional to molecular weight and varies little with shear rate over a wide range of shear rates.

$$\eta \propto M \tag{2.19}$$

As the molecular weight increase, η_0 starts to increase much more rapidly with M, and the viscosity starts to depend strongly on shear rate. Over a fairly narrow range of M, data on a log–log plot approach a line with a slope between 3.4 and 3.6. In other words for linear, monodisperse polymers having sufficiently high molecular weight the relationship between $\log(\eta_0)$ and $\log(M)$ is given by Eq. (2.20).

$$\eta_0 = KM^{\alpha} \tag{2.20}$$

where α is usually the range of 3.5 ± 0.2

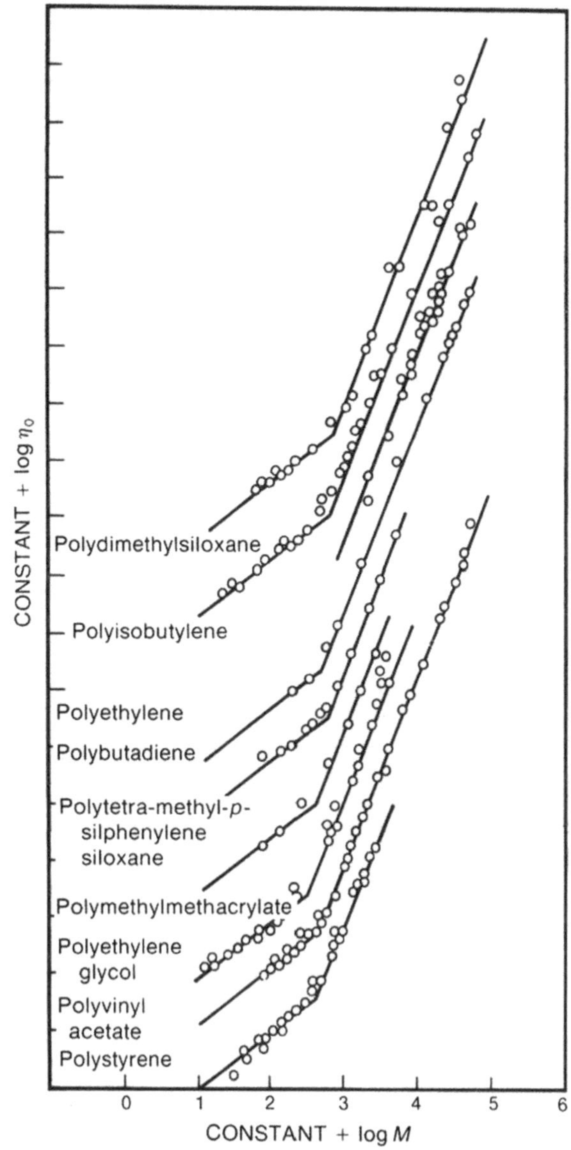

Fig. 2.10 Zero-shear viscosity versus molecular weight (logarithmic scales) for several polymers. The axes have been shifted to avoid crowding. The *low-MW lines* correspond to unentangled samples and have slopes of unity, while the *high-MW lines* correspond to entangled polymers and are fitted to lines having slopes of 3.4. From Berry and Fox [16]

The value of M where the lines described by Eqs. (2.19) and (2.20) intersect for a given polymer, M_C, is called the *critical molecular weight for entanglement.* Values of M_C for a number of polymers are given in Appendix A. This is not to be confused with two other rheologically meaningful critical molecular weights M'_C and M_e, which will be introduced later.

For polydisperse materials, it is found that Eq. (2.20) continues to be valid if M is simply replaced by the weight-average molecular weight, as long as there are very few unentangled molecules present, i.e. those with $M < M_C$.

$$\eta_0 = KM_w^\alpha \tag{2.21}$$

This relationship leads directly to a blending law for viscosity. For example, in a binary blend of two monodisperse samples of the same polymer having molecular weights M_1 and M_2, the weight-average molecular weight of the blend is given by:

$$M_{wb} = w_1 M_1 + w_2 M_2 \tag{2.22}$$

where w_1, and w_2, are the weight fractions of the blend components.

Using Eqs. (2.20) and (2.21) to eliminate the molecular weights, we have:

$$\eta_{0,b} = KM_w^\alpha = \left(w_1 \eta_{0,1}^{1/\alpha} + w_2 \eta_{0,2}^{1/\alpha}\right)^\alpha \tag{2.23}$$

This equation has been tested for blends of monodisperse [17, 18] and polydisperse [19] materials.

2.4.6 *Effect of Molecular Weight Distribution on Viscosity*

The effect of molecular weight distribution, MWD, is somewhat more subtle but still very important. In general, commercial polymers have a rather broad molecular weight distribution, although materials produced using metallocene catalysts can have polydispersities (M_w/M_n) as low as two. Figure 2.11 is a sketch of viscosity curves for two polymers having the same weight average molecular

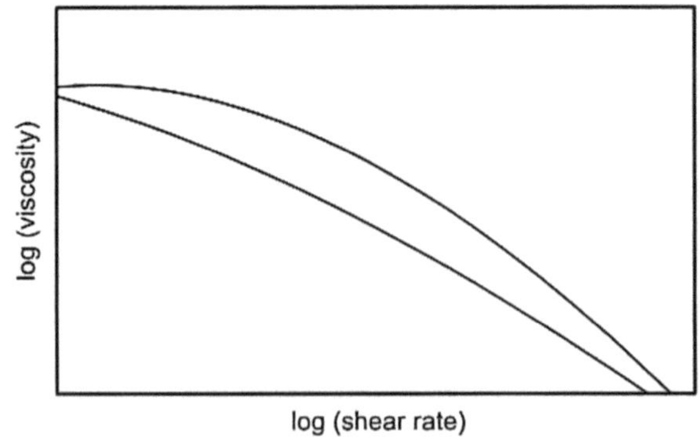

Fig. 2.11 Shapes of viscosity curves for two samples having the same M_w but with narrow (*upper curve*) and broad (*lower*) molecular weight distributions. The narrow MWD sample moves from a well-defined Newtonian region to power-law behavior over a narrower range of shear rates

weight but different molecular weight distributions. The upper curve is for a nearly monodisperse sample, while the lower one is for a sample with a moderately broad MWD. The broadening of the distribution stretches out the range of shear rates over which the transition from the zero-shear viscosity to the power law region occurs. Chapter 7 describes methods for using viscosity data to infer the MWD of a linear polymer, although it is to be noted that this requires data of high accuracy. In the plastics industry it is often desired to estimate polydispersity from easily measured quantities. Shroff and Mavridis [20] have compared several empirical correlations that have been proposed to do this.

2.4.7 *Effect of Tacticity on Viscosity*

The effect of tacticity on viscoelastic behavior is discussed in Chap. 3, but little has been published about its effect on viscosity. Fuchs et al. [21] studied a series of PMMAs that were 78 and 81 % syndiotactic as well as more conventional materials that were 59 % syndiotactic. They found that the zero-shear viscosities obeyed Eq. (2.20) with α equal to about 3.4. Figure 2.12 shows their results for two series of samples that were 59 and 81 % syndiotactic.

Figures 2.13 and 2.14 show data of Huang et al. [22] for iso-, stereo- and a tactic polystyrenes (iPS, sPS, aPS) having similar M_w values (about 2.5×10^5 g/mol) and polydispersities (2.3–2.5). The inset shows that dividing M_w by M_e made it possible to bring data for the three samples together on one line. M_e is the *molecular weight between entanglements* defined in Chap. 3. It is calculated from the plateau modulus and is one measure of the molecular weight at which

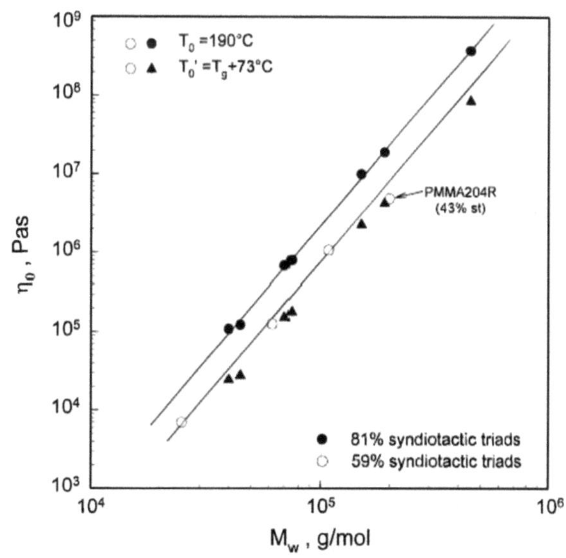

Fig. 2.12 Effect of tacticity on viscosity—Dependence of zero-shear viscosity at 190 °C on M_w for polypropylenes of varying tacticities. From Fuchs et al. [21]

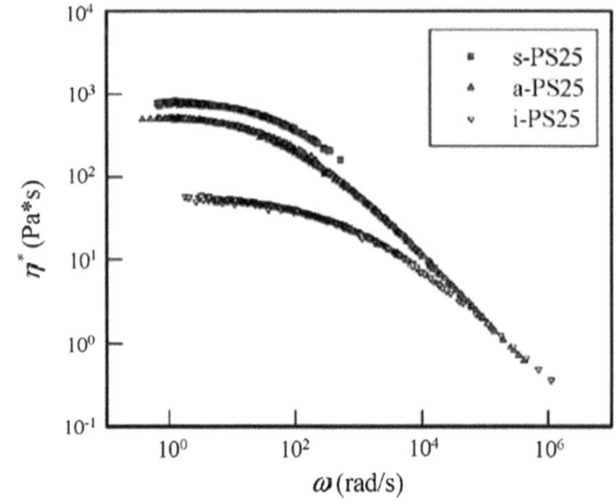

Fig. 2.13 Complex viscosity versus frequency of atactic, syndiotactic and isotactic polystyrenes at 280 °C. From Huang et al. [22]

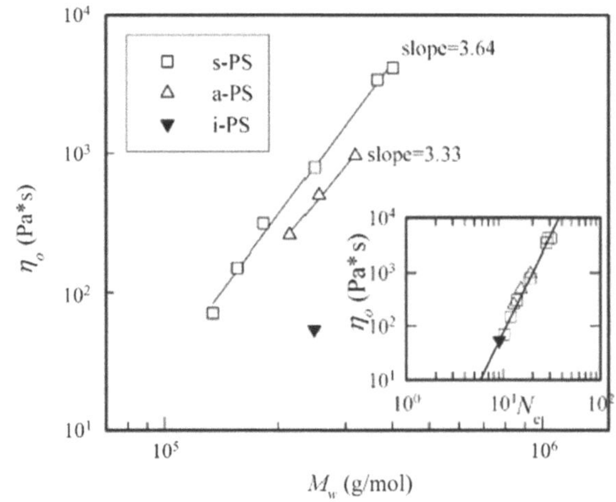

Fig. 2.14 Zero-shear viscosities of polystyrenes of Fig. 2.13 versus M_w. *The inset* shows viscosity versus "entanglement number" $N_e \equiv M_w/M_e$. From Huang et al. [22]

entanglement effects become prominent. The equation of the line in the inset of Fig. 2.14 is:

$$\eta_0 = 2.92 \times 10^{-2}(M_w/M_e)^{3.6} \tag{2.24}$$

2.4.8 *Viscosity of Ethylene/Alpha-Olefin Copolymers*

An important class of commercial polymers are copolymers of ethylene and an alpha-olefin, known as linear low density polyethylenes (LLDPE). The use of a copolymer introduces short-chain side branches onto the polyethylene backbone, and the details of the molecular structure depend on the method of polymerization. If a heterogeneous, Ziegler catalyst is used, the side-chains tend to be distributed in blocks rather than randomly along the backbone. If the copolymer is prepared using a single-site (metallocene) catalyst, the short-chain branching distribution is expected to be random.

Wood-Adams et al. [23] studied the effect of comonomer content in three ethylene-butene copolymers prepared using a single-site catalyst, in which the butene level ranged from 11 to 21 %. While the three materials studied had polydispersities of about 2.0, there was a modest variation in average molecular weight, and this was accounted for by dividing the complex viscosity by the zero-shear viscosity. While the resulting master plots for the three materials were not precisely identical, the authors concluded that there was no significant effect of comonomer content on η_0. Wood-Adams and Costeux [24] found that the copolymers used in this study were thermorheologically simple, i.e., that they obeyed time–temperature superposition, and that the activation energy was insensitive to butene content at levels up to 7 wt %.

Garcia-Franco et al. [25] later studied this issue using a set of twelve ethylene/butene copolymers and also looked at data from several studies that involved other comonomers. They were able to correlate all their data using, in place of M_w, the average molecular weight per backbone bond m_b, and Fig. 2.15 shows their correlation. The line corresponds to Eq. (2.25), and while there is considerable scatter on this highly compressed log–log plot, a general trend is suggested.

$$\eta_0 = 4.73 \times 10^{-10} \left(\frac{M_w}{M_b}\right)^{3.33} \tag{2.25}$$

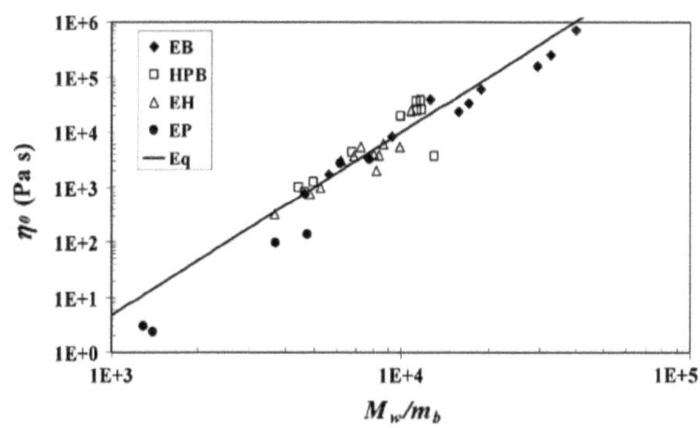

Fig. 2.15 Zero-shear viscosity at 190 °C for various polyolefins versus M_w/M_b, where M_b is the average molecular weight per backbone bond. *EB* Ethylene/butane copolymers; *EP* Ethylene/propylene; *EH* Ethylene/hexane; *HPB* Hydrogenated polybutadiene. From Garcia-Franco et al. [25]

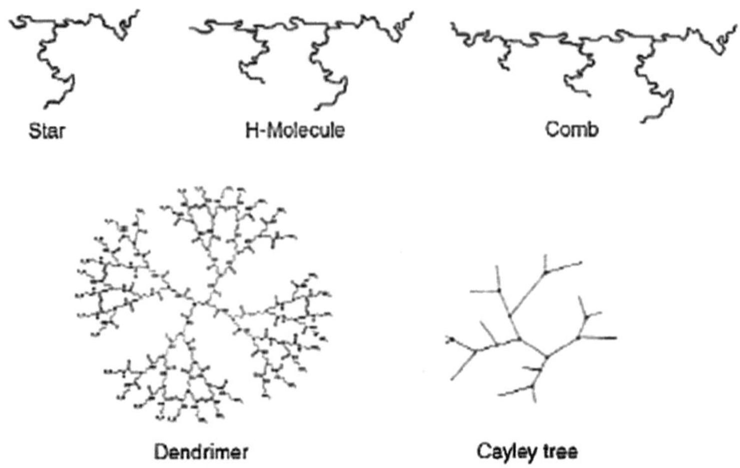

Fig. 2.16 Sketches of several branching structures: *star*, *H-shaped*, *comb*, *dendrimer* and *cayley tree* (branch-on-branch)

2.4.9 Effect of Long-Chain Branching on Viscosity

Branches are "long" from our point of view if they are sufficiently entangled to affect rheological behavior. For viscosity we might expect the molecular weight for the onset of branching effects to be M_C, which is about $3M_e$. The effect of branching also depends very much on the branching structure, i.e., lengths of branches, distance between branch points, and level of branching (branches on branches). Figure 2.16 shows several types of branching structure that will be referred to in the following discussion. It is not possible to infer branching structure from viscosity data unless something is known about the way a sample was polymerized. The effect of branching structure on rheological behavior is dealt with at length by Dealy and Larson [26]. Levels of long-chain branching too low to be detected using GPC can have a significant effect on viscosity, so rheology is a valuable tool for characterizing LCB.

To study the effects of specific types of LCB, it is necessary to work with polymers that have well-known branching structures. Small samples of such materials can be prepared by means of anionic polymerization, and this technique has been widely used in rheological studies. There is still a gap, however, between what is known about such materials and our understanding of structure-rheology relationships for highly-branched, commercial polymers.

2.4.9.1 Zero-Shear Viscosity of Monodisperse, Star-Shaped Polymers

For symmetric star polymers of a given molecular weight whose arms are too short to entangle, as the number of arms is increased η_0 falls progressively further below that of a linear polymer with the same molecular weight. This is because the size of a molecule decreases as the number of branches increases at constant M. For a given number of arms, if the arm length is increased, η_0 rises above that of the linear polymer having the same M. When the arms become long enough to be well-entangled, however, i.e. when the arm molecular weight M_a reaches two or three times the molecular weight between entanglements M_e (defined in Chap. 3) η_0 increases approximately exponentially with molecular weight [27]. This phenomenon is illustrated in Fig. 2.17 which shows data of Kraus and Gruver [28] for linear, three-arm and four-arm polybutadiene stars. At moderate molecular weights the data for the stars lie below, but parallel to, the line for linear, entangled polymer with $\alpha = 3.4$, in accord with Eq. (2.20), but when the branch length reaches about $3M_e$ for three-armed stars, and about $4M_e$ for four-arm stars, the star data rise sharply and cross the line for linear polymers described by Eq. (2.20). Somewhat surprisingly, the viscosity of symmetric stars with entangled branches depends only on branch length and not on the polymer type or number of arms [27, 29], up at least 33 arms. The sharp increase in viscosity with M_a when arms are highly entangled is interpreted in terms of a molecular model in Chap. 4.

The picture is more complicated in the case of asymmetric stars. Gell et al. [30] studied a series of asymmetric stars made by adding arms of varying length at the midpoint of a very long backbone ($M/M_e \approx 40$). They found that even a short arm

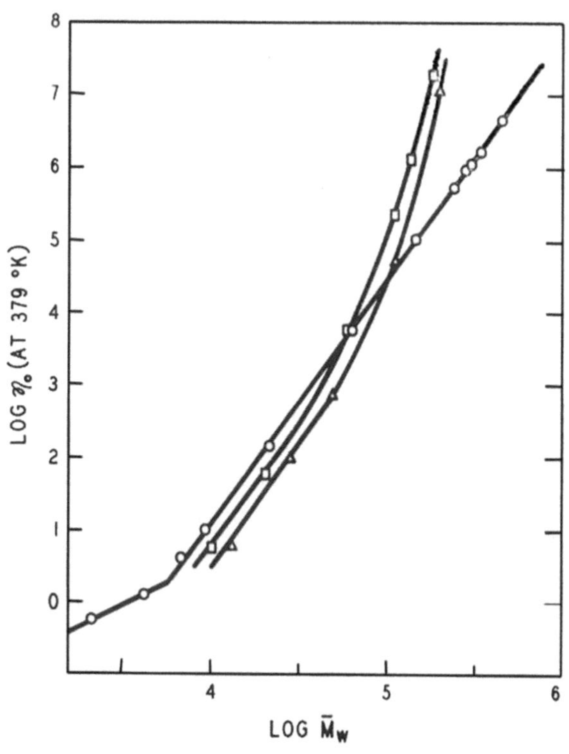

Fig. 2.17 Zero-shear viscosity at 379 °C versus M_w for polybutadienes having various structures: linear (*circles*), three-arm stars (*squares*), four-arm stars (*trangles*). At low Mw branched systems fall below the line for linear molecules, and at higher Mw their values rise above the line and approach an exponential behavior. From Kraus and Gruver [28]

with ($M_a/M_e = 0.5$ had the effect of tripling η_0, and for $M_a/M_e = 2.4$, η_0 was increased by a factor of ten. This illustrates the difficulty of inferring structural details from viscosity data unless one knows the type of structure present.

It is explained in Chap. 7 that the presence of chain segments with branch points at both ends, as in all the structures shown in Fig. 2.16 except the star, has a strong effect on extensional flow behavior.

While the use of specially synthesized polymers has advanced our knowledge of the effect of specific branching structures, relating rheological behavior to the structure of commercial branched polymers is a considerably more complex task. But the branching structure of commercial polymers is of great importance because of the very strong effect of certain branching structures on the processing behavior of thermoplastics.

2.4.9.2 Branched Metallocene Polymers

Polyolefins made using metallocene or related catalyst systems were the first commercial commodity polymers whose molecular structures were well-controlled and reproducible. It was thus possible to establish quantitative relationships between rheological behavior and structure. The linear homopolymers and ethylene-alpha-olefin copolymers have polydispersity indexes (M_w/M_n) very close to two. And *constrained geometry catalysts* [31, 32] can produce polymers with well-controlled, low levels of long-chain branching. These materials are sometimes said to be "substantially linear." Moreover, the branching structure of these polymers is well described by theory [33–37]. The first branches to be formed have one branch point, i.e., they are three-arm stars. As the branching level increases, molecules with two branch points start to appear; these are like the H-molecule shown in Fig. 2.16. A more detailed description of these polymers can be found in Sect. 7.2.3.

Wood-Adams et al. [23] reported the rheological properties of a series of polyethylenes made using such a catalyst. The branching levels were very low, ranging from 0.1 to 0.8 branches per 10,000 carbon atoms, and the polydispersities were very close to 2.0. The zero-shear viscosity increased very strongly with branching level, reaching nearly 70 times that of a linear polymer with the same molecular weight at the highest branching level, as shown in Fig. 2.18. Wood-Adams and Costeux [24] studied the effect of comonomer on branched polymers of this type. They found that all the long-chain branched polymers were thermorheologically complex, i.e., they did not obey time–temperature superposition, and that the activation energy based on the zero-shear viscosity was much higher for branched copolymers than for comparable branched homopolymers. The behavior of these samples is discussed in more detail in Chap. 7 where a technique is described for inferring the branching level in metallocene polymers from viscosity data.

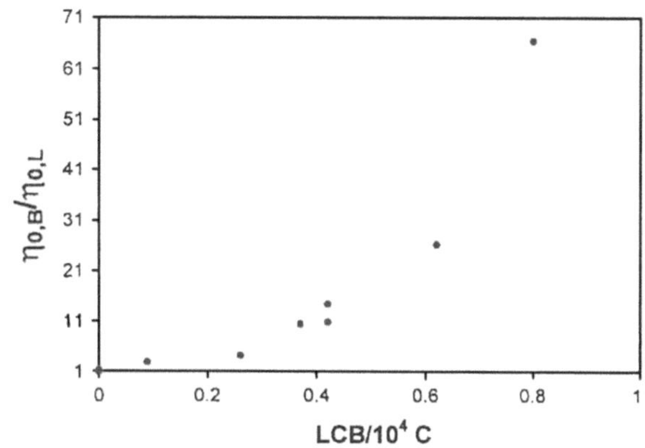

Fig. 2.18 Zero-shear viscosity of sparsely branched metallocene polyethylenes versus number of long-chain branches per 1,000 carbons. The viscosities are divided by that of linear PE having the same molecular weight. The first branched molecules to form at the *left* are stars, while as we move to the *right* there start to be H-molecules and very low levels of more highly-branched structures. From Wood-Adams et al. [23]

2.4.9.3 Viscosity of Randomly Branched Polymers and LDPE

We have already seen that the presence of long-chain branching, even at quite low levels, has a strong effect on the zero-shear viscosity and the shape of the viscosity curve. Random branching leads to a broad distribution of structures, making it difficult, if not impossible, to distinguish between the effects of branching and polydispersity. In fact, Wood-Adams and Dealy [38] demonstrated that one can, in principle, prescribe the molecular weight distribution of a linear polymer that would have the same complex viscosity as any given branched polymer. Low-density polyethylene is the commercial polyolefin with the most complex branching structure and poses the largest challenge in characterizing its structure. Laboratory studies of random branching usually make use of techniques in which increasing levels of branching are introduced into a linear precursor. While the results of these studies are instructive, their direct application to low-density polyethylene (LDPE) is limited.

Figure 2.19 shows complex viscosity as a function of frequency (similar to viscosity versus shear rate) for one linear and four branched ethylene/1-butene copolymers [39]. All five samples have nearly the same absolute molecular weight ($M_w \approx 155$ kg/mol), MWD ($M_w/M_n \approx 2$), and comonomer content. Sample A is linear, and the level of long-chain branching increases in the order B–C–D–E. We see that the zero-shear viscosity increases sharply with the level of branching but that all the data converge onto one curve at high shear rates.

Auhl et al. [40] studied the behavior of a series of polypropylenes that had been subjected to doses of electron beam radiation to generate various levels of long-chain branching. The ratio of the viscosity of a branched polymer η_0(br) to that of the linear precursor η_0(lin) is shown as a function of radiation dose d in Fig. 2.20. Assuming little chain scission, all these samples have the same molecular weight as

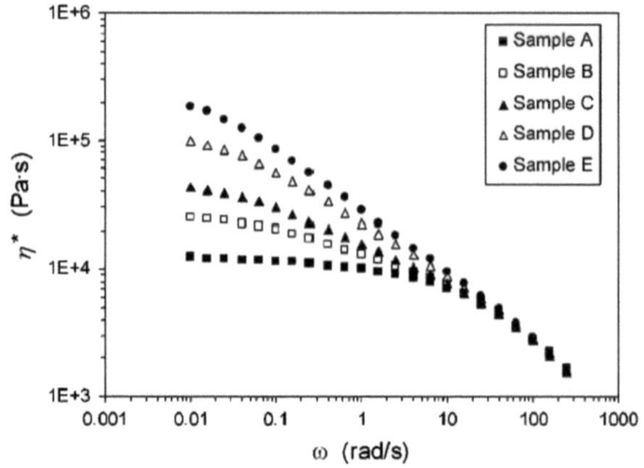

Fig. 2.19 Complex viscosity magnitude at 190 °C versus frequency for a linear (*Sample A*) and four branched polyethylene/1-butene copolymers. The branching level increases as we move through the alphabet. All samples have nearly the same absolute M_w (155 kg/mol) and polydispersity (2). Shear thinning begins at higher frequencies as we move from A to E, and at the highest frequencies all the curves come together. From Robertson et al. [39]

the linear precursor. Thus the point for $d = 0$ corresponds to $\eta_0(\text{br})/\eta_0(\text{lin}) = 1.0$. These data show the trend of increasing viscosity at low branching levels, reaching a peak and then decreasing. Comparing Figs. 2.18 and 2.20 makes it clear that the details of the branching structure have as important an effect on the zero-shear viscosity as the average degree of branching. The metallocene polyethylene (mPE)

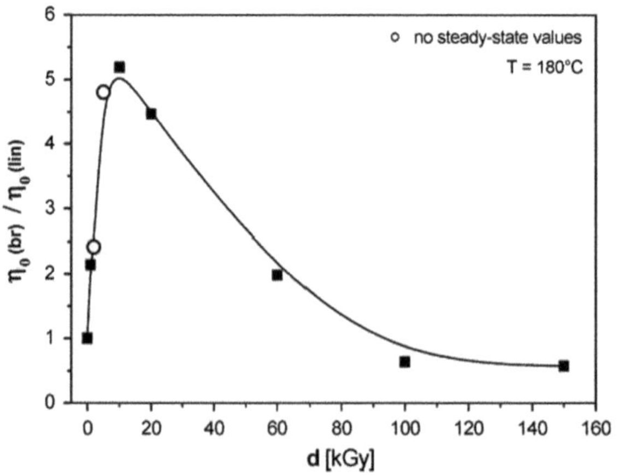

Fig. 2.20 Branching factor g, the ratio of η_0 of radiated polypropylene to that of linear PP precursor, versus radiation dose d. If little chain scission occurred all samples have the same molecular weight; i.e. $d = 0$ implies a branching factor of unity. Electron beam radiation has resulted in long-chain branching. The viscosity first increases sharply with branching level but then falls continuously to values below that of the linear precursor. From Auhl et al. [40]

(Fig. 2.18) always contains a large fraction of linear chains, and all the branch segments have the same size distribution as the linear chains. The irradiated polypropylene (PP) (Fig. 2.20), on the other hand, probably contains tree-like molecules with short branch segments. It is thus somewhat similar to a LDPE.

The most important commercial, branched polymer is low-density polyethylene (LDPE), which has a broad range of branching structures, with many short branches attached to tree- or comb-like molecules. In addition, LDPEs made in autoclaves have a distinctly different structure from those made in tubular reactors, as is explained in Chap. 7. All LDPEs are strongly shear-thinning, and we will see in Chap. 4 that they have a distinctive extensional flow behavior that is associated with high levels of long-chain branching.

The dependence of the zero-shear viscosity of LDPE [41–44] on average molecular weight is often compared with that of linear polymer having the same M_w, which is described by Eq. (2.20). However, two major questions arise in making such a comparison: obtaining a reliable value for η_0, and choosing an appropriate average molecular weight. Highly-branched, heterogeneous polymers such as LDPE have a very broad range of shear rates over which viscosity data make the transition from the Newtonian limiting value to a power-law region. It is generally not feasible to determine the zero-shear viscosity of LDPE using small-amplitude oscillatory shear, because the frequency required and the torque signal are extremely low. And extrapolation using an empirical viscosity model is highly unreliable. The LDPE viscosity data shown in Fig. 2.4 were obtained using a specially modified rotational rheometer, the operation of which required exceptional skill. The only reliable method is the long-time creep test described in Chap. 6. Such measurements can last several hours and require extra stabilization of the sample against thermo-oxidative degradation. Also, the activation energy defined by Eq. (2.30) has different values for linear PE and LDPE, so changing the temperature will alter the relationship between the viscosities of the two polymers.

The second issue regarding η_0 correlations for LDPE is the selection of a molecular weight average. Two averages have been used, one based on the size of molecules in solution, determined using GPC with universal calibration, and one based on mass, determined using light-scattering. Whichever one is selected, a collection of molecules having either similar sizes or similar masses will contain chains having a distribution of branching structures. Furthermore, LDPEs made in autoclaves have distinctly different branching structures than those made in tubular reactors. This issue is taken up in more detail in Sect. 7.2.5.

2.5 Normal Stress Differences

It was mentioned at the beginning of this chapter that there are two rheologically meaningful normal stress differences that can be measured in steady simple shear, the first and second normal stress differences. Based on the coordinate convention shown in Fig. 2.1, these are defined as:

$$N_1(\dot{\gamma}) \equiv \sigma_{11} - \sigma_{22} \tag{2.26}$$

$$N_2(\dot{\gamma}) \equiv \sigma_{22} - \sigma_{33} \tag{2.27}$$

These stress differences are associated with strain-induced anisotropy in a fluid, and in the case of polymeric liquids the anisotropy arises from the departure of molecules from their equilibrium, symmetrical average shape. In Chap. 8 we will find that the first normal stress difference is indirectly related to extrudate swell and that the second normal stress difference governs flow instabilities that occur in multi-layer profile coextrusion and flow through non-circular channels.

The first normal stress difference can be measured at low shear rates using a rotational rheometer equipped with cone-plate fixtures, although problems arise that are not encountered in the measurement of viscosity. The second normal stress difference is considerably more difficult to determine, and relatively few data have been reported. It has been observed that N_2 is negative with a magnitude about 1/3 that of N_1.

For a Newtonian fluid the shear stress is proportional to the shear rate, the normal stress differences are zero, and at sufficiently low shear rates viscoelastic materials approach Newtonian behavior. There are two simplified models that provide useful information about the first departures from Newtonian behavior as the shear rate is increased from zero. One of these is the second-order fluid model, and the other is the rubberlike-liquid model. Both of these models predict that the shear stress is still linear in shear rate (constant viscosity) but that the normal stress differences are quadratic in shear rate. This observation led to the definitions of the *first and second normal stress difference coefficients*, which these models predict are independent of shear rate.

$$\Psi_1(\dot{\gamma}) \equiv N_1(\dot{\gamma})/\dot{\gamma}^2 \tag{2.28}$$

Fig. 2.21 First normal stress difference versus shear rate for a polystyrene measured using a Lodge stressmeter and a sliding plate rheometer (SPR). There is scatter in the stressmeter data around 20 s^{-1}, but most of the data show good agreement. From Xu et al. [48]

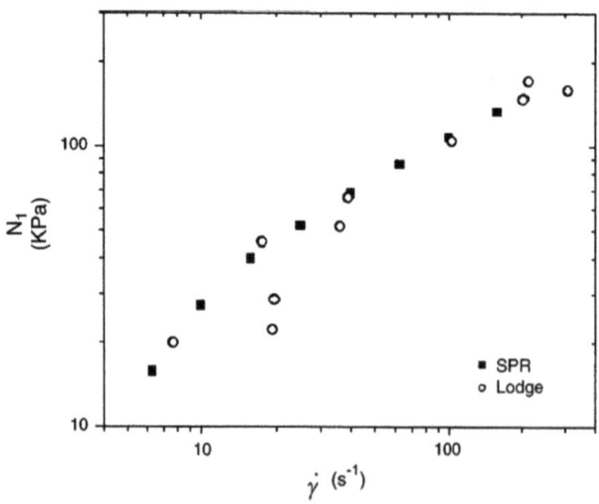

$$\Psi_2(\dot{\gamma}) \equiv N_2(\dot{\gamma})/\dot{\gamma}^2 \tag{2.29}$$

Data obtained at low shear rates have shown that Ψ_1 and Ψ_2 do indeed have limiting, non-zero values, which are assigned the symbols $\Psi_{1,0}$ and $\Psi_{2,0}$.

The second normal stress difference is usually reported in relation to the first normal stress difference by use of a shear-rate dependent normal stress ratio:

$$(\dot{\gamma}) \equiv \frac{-{}_2(\dot{\gamma})}{{}_1(\dot{\gamma})} = \frac{-N_2(\dot{\gamma})}{N_1(\dot{\gamma})} \tag{2.30}$$

This ratio has a non-zero limiting value as the shear rate approaches zero, and this value has been reported to be 0.24 for linear melts [45] and about 0.3 for stars [46].

As the shear rate increases, Ψ_1 decreases, but its value at high shear rates can not be determined using rotational rheometers due to flow disturbances. Data for a polybutadiene and a polystyrene at shear rates up to several hundred have been obtained using a Lodge Stressmeter [47] and a sliding plate rheometer (SPR), and polystyrene data obtained using both techniques are compared in Fig. 2.21 [48]. The data were found to be in good agreement with an empirical relationship proposed by Laun [49], shown below as Eq. (2.31)

$$\Psi_1(\dot{\gamma}) = 2\frac{G'}{\omega^2}\left[1 + \left(\frac{G'}{G''}\right)^2\right]^{0.7}. \tag{2.31}$$

The dimensionless quantity $N_1(\dot{\gamma})/\sigma(\dot{\gamma})$ is called the *stress ratio* SR and indicates the relative importance of orientation or stored elastic energy at a given shear rate. The ratio $N_1(\dot{\gamma})/2\sigma(\dot{\gamma})$, i.e., SR/2, is often called the *recoverable shear.* However, it is only equal to the actual strain recovered after sudden release of the shear stress during steady shear, at low shear rates.

References

1. Bird RB, Armstrong RC, Hassager O (1987) Dynamics of Polymeric Liquids, vol 1. Wiley, New York
2. Meissner J (1971) Deformationsverhalten der Kunststoffe im flüssigen und im festen Zustand. Kunststoffe 61:576–582
3. Stratton RA (1966) The dependence of non-Newtonian viscosity on molecular weight for "Monodisperse" polystyrene. J Colloid Interface Sci 22:517–530
4. Cross MM (1965) Rheology of non-Newtonian fluids: a new flow equation for pseudoplastic systems. J Coll Sci 20:417–437
5. Carreau PJ (1972) Rheological equations from molecular network theories. Trans Soc Rheol1 6:99–127
6. Hieber CA, Chiang HH (1992) Shear-rate-dependence modeling of polymer melt viscosity. Polym Eng Sci 14:931–938
7. Yasuda KY, Armstrong RC, Cohen RE (1981) Shear flow properties of concentrated solutions of linear and star-branched polystyrenes. Rheol Acta 20:163–178
8. Elberli B, Shaw MT (1978) Time constants from shear viscosity data. J Rheol 22:561–570

9. Plumley TA, Lai S, Betso SR, Knight GW (1994) Rheological modeling of Insite technology polymers. SPE ANTEC Tech Papers 40:1221–1224
10. Wang J (2010) Double cross model—A novel way to model viscosity curves, Society of Rheology. 82nd annual meeting, Santa Fe, NM
11. Cox WP, Merz EH (1958) Correlation of dynamic and steady flow viscosities. J Polym Sci 28:619–621
12. Utracki LA, Gendron R (1984) Pressure oscillation during extrusion of polyethylenes. J Rheol 28:601–623
13. Venkatraman S, Okano M, Nixon AA (1990) A comparison of torsional and capillary rheometry for polymer melts: the Cox-Merz rule revisited. Polym Eng Sci 30:308–313
14. Park HE, Dealy JM (2006) Effect of pressure and supercritical CO2 on the viscosity of Polyethylene. Macromolecules 39:5438–5452
15. Park HE, Dealy JM, Münstedt H (2006) Influence of long-chain branching on time-pressure and time-temperature shift factors for polystyrene and polyethylene. Rheol Acta 46:153–159
16. Berry GC, Fox TG (1968) The viscosity of polymers and their concentrated solutions. Adv Polym Sci 5:261–357
17. Bartels CR, Crist B, Fetters LJ, Graessley WW (1986) Self-diffusion in branched polymer melts. Macromol 19:785–793
18. Struglinski MJ, Graessley WW (1985) Effects of polydispersity on the linear viscoelastic properties of entangled polymers 1: experimental observations for binary mixtures of linear polybutadiene. Macromol 18:2630–2643
19. Kumar R, Khanna YP (1989) SPE Tech. Papers 35, 1675
20. Shroff AR, Mavridis H (1995) New measures of polydispersity from rheological data on polymer melts. J Appl Polym Sci 57:1605–1626
21. Fuchs K, Chr Friedrich, Weese J (1996) Viscoelastic properties of narrow-distribution poly(methylmethacrylates). Macromol 29:5893–5901
22. Huang CL, Chen YC, Hsiao TJ, Tsai JC, Wang C (2011) Effect of tacticity on viscoelastic properties of poplystyrene. Macromol 44:6155–6161
23. Wood-Adams P, Dealy JM, deGroot AW, Redwine OD (2000) Rheological properties of metallocene polyethylenes. Macromol 33:7489–7499
24. Wood-Adams P, Costeux S (2001) Thermorheological behavior of polyethylene: effects of microstructure and long chain branching. Macromol 34:6281–6290
25. Garcia-Franco CA, Harrington BA, Lohse DJ (2006) Effect of short-chain branching on the rheology of polyolefins. Macromol 39:2710–2717
26. Dealy JM, Larson RG (2006) Structure and rheology of molten polymers. Hanser Publishers, Munich
27. Pearson DS, Helfand E (1984) Viscoelastic properties of star-shaped polymers. Macromol 17:888–895
28. Kraus G, Gruver JT (1965) Rheological properties of multichain polybutadienes. J Polym Sci A 3:105–122
29. Graessley WW, Roovers J (1979) Melt Rheology of four-arm and six-arm star polystyrenes. Macromol 12:959–965
30. Gell CB, Graessley WW, Efstratiadis V, Pitsikalis M, Kadjichristidis N (1997) Viscoelasticity and self-diffusion in melts of entangled asymmetric star polymers. J Polym Sci, Part B: Polym Phys 35:1943–1954
31. Stevens JC (1994) INSITE™ catalyst structure/activity relationships for olefin polymerization. Stud Surf Sci Catal 89:277–284; Constrained geometry and other single site metallocene polyolefin catalysts: A revolution in olefin polymerization. *Ibid.* (1996) 101:11–20
32. Lai SY, Wilson JR, Knight JR, Stevens JC (1993) Elastic substantially linear olefin polymers. US Patent 5(380):810
33. Soares JBP, Hamielec AE (1996) Bivariate chain length and long chain branching distribution for copolymerization of olefins and polyolefin chains containing terminal double-bonds. Macromol Theory Simul 5:547–572

34. Soares JBP, Hamielec AE (1997) The chemical composition component of the distribution of chain length and long chain branching for copolymerization of olefins and polyolefin chains containing terminal double bonds. Macromol Theory Simul 6:591–596
35. Read DJ, McLeish TCB (2001) Molecular rheology and statistics of long chain branched metallocene-catalyzed polyolefins. Macromol 34:1928–1945
36. Costeux S, Wood-Adams P, Beigzadeh D (2002) Molecular structure of metallocene-catalyzed polyethylene: rheologically relevant representation of branching architecture in single catalyst and blended systems. Macromol 35:2514–2528
37. Soares JBP (2004) Polyolefins with long chain branches made with single-site coordination catalysts: a review of mathematical modeling techniques for polymer microstructure. Macromol Mater Eng 289:70–87
38. Wood-Adams PM, Dealy JM (2000) Using rheological data to determine the branching level in metallocene polyethylenes. Macromol 33:7481–7488
39. Robertson CG, García-Franco CA, Srinivas S (2004) Extent of branching from linear viscoelasticity of long-chain branched polymers. J Polym Sci, Part B: Polym Phys 42:1671–1684
40. Auhl D, Stange J, Münstedt H, Krause B, Voigt D, Lederer A, Lappan U, Lunkwitz K (2004) Long-chain branched polypropylenes by electron beam irradiation and their rheological properties. Macromol 37:9465–9472
41. Gabriel C, Münstedt H (2003) Strain hardening of various polyolefins in uniaxial elongational flow. J Rheol 47:619–630
42. Gabriel C, Kokko E, Löfgren B, Seppälä J, Münstedt H (2003) Analytical and rheological characterization of long-chain branched metallocene-catalyzed ethylene homopolymers. Polymer 43:6383–6390
43. Wang J, Mangnus M, Yau W, deGroot W, Karjala T, Demirors M (2008) Structure-property relationships of LDPE. SPE ANTEC Tech Papers, pp 878–881
44. Gabriel C, Lilge D (2006) Molecular mass dependence of the zero shear-rate viscosity of LDPE melts: evidence of an exponential behavior. Rheol Acta 45:995–1002
45. Schweizer T, van Meerveld J, Öttinger HC (2004) Nonlinear shear rheology of polystyrene melt with narrow molecular weight distribution—experiment and theory. J Rheol 48:1345–1363
46. Lee CS, Magda JJ, DeVries KL, Mays JW (1992) Measurements of the second normal stress difference for star polymers with highly entangled branches. Macromol 25:4744–4750
47. Lodge AS (1996) On-line measurement of elasticity and viscosity in flowing polymeric liquids. Rheol Acta 35:110–116
48. Xu J, Costeux S, Dealy JM, De Decker MN (2007) Use of a sliding plate rheometer to measure the first normal stress difference at high shear rates. Rheol Acta 46:815–824
49. Laun HM (1986) Prediction of elastic strains of polymers melts in shear and elongation. J Rheol 30:459–501

Chapter 3
Linear Viscoelasticity

Abstract Linear viscoelasticity is a type of behavior exhibited by molten polymers when the deformation is very small or very slow. Such behavior can be described completely by the relaxation modulus, which is determined by measuring the response of the melt to a sudden, small deformation. Alternatively it can be characterized in terms of the storage and loss moduli that are measured in small-amplitude oscillatory shear or the creep compliance measured by suddenly imposing a shear stress and tracking the deformation. A convenient mathematical form for the relaxation modulus is a sum of exponentials, which is called the generalized Maxwell model. The set of moduli and time constants involved in this model comprise a *discrete relaxation spectrum*. The dependence of these properties on temperature is described, and molecular models for their prediction are presented.

3.1 Introduction

The simplest type of viscoelastic behavior is linear viscoelasticity. This type of behavior is observed when the deformation is sufficiently mild that the molecules of a polymeric material are disturbed from their equilibrium configuration and entanglement state to a negligible extent. Obviously, very small deformations are in this category of deformation. This might be a deformation in which the total strain is very small, or the early stages of a larger deformation. For melts, which have a fading memory and can flow, linear behavior is also observed when a deformation occurs very slowly, as in steady simple shear or another viscometric flow at very low shear rates. This is because relaxation processes due to Brownian motion are always acting to return molecules to their equilibrium state, and if the deformation is tending to take them away from this state sufficiently slowly, the relaxation mechanism has plenty of time to keep up with this process, with the net result that no significant deviation from equilibrium occurs. One manifestation of this is that at very low shear rates, the viscosity of a polymeric liquid becomes independent of shear rate.

J. M. Dealy and J. Wang, *Melt Rheology and its Applications in the Plastics Industry*,
Engineering Materials and Processes, DOI: 10.1007/978-94-007-6395-1_3,

While nonlinear behavior is more directly relevant to processing, deformation in the linear regime is the most fruitful source of information about molecular structure of all shear flows. Linear data provide information about average molecular weight and molecular weight distribution. While some information about long-chain branching can also be obtained, we will see in Chap. 4 that stretching flows provide valuable additional information about branching. In this chapter, we present the theory of linear viscoelasticity along with formulas that are useful for the treatment of data. We then describe the linear behavior of typical molten polymers and the molecular mechanisms underlying this behavior. A more thorough treatment of this subject can be found in the excellent monograph by Ferry [1] on polymer viscoelasticity.

3.2 Stress Relaxation and the Relaxation Modulus

In Sect. 1.5.3 we described an experiment in which the Maxwell element shown in Fig. 1.7 was subjected to a sudden elongation, and the force was calculated as a function of time. The analogous rheological experiment is one in which a sample is suddenly deformed at time $t = 0$, and the resulting stress is measured as a function of time. This is called a *stress relaxation* experiment. In reporting the results of such an experiment, it is customary to divide the stress by the magnitude of the strain that is introduced to start the experiment. In the case of a sudden shear strain of magnitude γ_0, the quantity reported is the *shear relaxation modulus* $G(t,\gamma_0)$.

$$G(t, \gamma_0) \equiv \sigma(t)/\gamma_0 \tag{3.1}$$

In the case of an extensional strain of magnitude ε_0, the quantity reported is the tensile relaxation modulus $E(t, \varepsilon_0)$, which is the net tensile stress, σ_E, divided by ε_0.

$$E(t, \varepsilon_0) \equiv \sigma_E(t)/\varepsilon_0 \tag{3.2}$$

In general, the relaxation moduli defined above are functions of the strain magnitude, but when the strain is very small, they are independent of the strain. In this case the stress at any particular value of t is proportional to the strain.

$$\sigma(t) = G(t)\,\gamma_0 \tag{3.3}$$

It is because of this linear relationship that we call small-strain or small-rate behavior *linear viscoelasticity*.

3.3 The Boltzmann Superposition Principle

It is useful to have a general equation that describes all types of linear viscoelastic behavior. We will derive this equation starting from the assumption that the relaxation modulus is independent of strain amplitude, i.e., from Eq. (3.3).

Consider a sequence of small shear strains as shown in Fig. 3.1. The shear stress resulting from the strain $\delta\gamma(t_1)$ that occurs at time t_1 is:

$$\sigma(t) = G(t - t_1)\,\delta\gamma\,(t_1) \tag{3.4}$$

To calculate the stress resulting from the strain introduced at time t_2, if both $\delta\gamma(t_1)$ and $\delta\gamma(t_2)$ are very small we can assume that the incremental response of the material to the second step strain will be independent of the strain introduced at time t_1. Thus, we can simply add on or superpose the stress resulting from the strain at time t_2, as follows:

$$\sigma(t) = G(t - t_1)\delta\gamma(t_1) + G(t - t_2)\delta\gamma(t_2) \tag{3.5}$$

For a combination of N small strains, we can continue to add up the contributions to the stress:

$$\sigma(t) = \sum_{\mathrm{i}=1}^{N} G(t - t_\mathrm{i})\delta\gamma(t_\mathrm{i}) \tag{3.6}$$

For a smooth strain history not consisting of finite steps, we can make use of the definition of the integral to show that:

$$\sigma(t) = \int_{-\infty}^{t} G(t - t')d\gamma(t') \tag{3.7}$$

Noting that the strain that occurs during the time interval dt' is simply $\dot{\gamma}(t')dt'$, this can also be written as:

$$\sigma(t) = \int_{-\infty}^{t} G(t - t')\dot{\gamma}(t')dt' \tag{3.8}$$

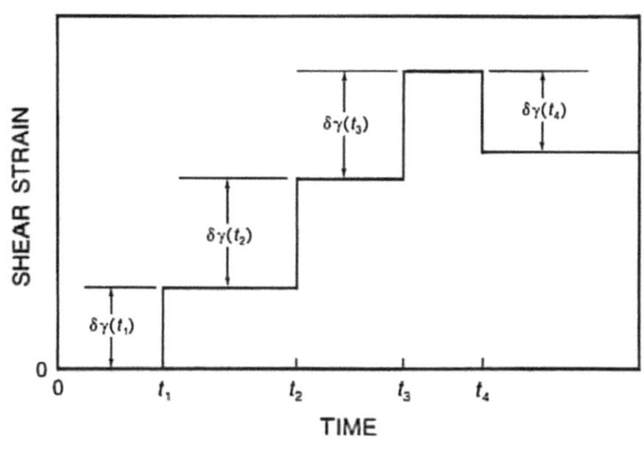

Fig. 3.1 A sequence of step strains. If the steps are sufficiently small, the Boltzmann superposition principle (Eq. 3.6) can be used to calculate the resulting stress

The use of the lower limit of minus infinity is a mathematical convenience; it implies that to calculate the stress at time t in the most general case one must know the strain history infinitely far into the past, i.e., at all times t' prior to t. But polymeric liquids have a fading memory, which means that as time passes the effect of a given deformation diminishes steadily, eventually falling to zero. In other words, the relaxation modulus approaches zero at long times.

$$\lim_{t\to\infty} G(t) = 0$$

And in an experiment, we generally start ($t = 0$) when the material is in a stress-free state. In this case, $\sigma(0) = 0$, and we can set the lower limit in 3.8 to zero.

$$\sigma(t) = \int_0^t G(t - t')d\gamma(t') \tag{3.9}$$

3.4 Start-Up of Steady Simple Shear

An example of using the relaxation modulus to calculate the response to any shear deformation that is very slow or very small is start-up of steady-simple-shear. With the sample in a stress-free state, a constant shear rate is suddenly imposed at $t = 0$. From Eq. (3.8) the shear stress is given by:

$$\sigma(t) = \dot{\gamma} \int_0^t G(t - t')dt'$$

Or, letting $s = t - t'$,

$$\sigma = \dot{\gamma} \int_0^t G(s)ds$$

The ratio $\sigma(t)/\dot{\gamma}$ is called the *shear stress growth coefficient* $\eta^+(t)$.

$$\eta^+(t) = \int_0^t G(s)ds \tag{3.10}$$

The viscosity is the steady-state (t $\to \infty$) value of $\eta^+(t)$, so we evaluate the above integral in 3.10 at $t = \infty$ to obtain:

$$\eta_0 = \int_0^\infty G(s)ds \tag{3.11}$$

In Chap. 5 we will see how the Boltzmann superposition principle can be written in tensorial form so that it can be used to calculate the response to any type of deformation, as long as it is small or slow. For example, for steady uniaxial extensional flow, the net tensile stress σ_{E} is simply three times the product of the viscosity and the Hencky strain rate.

$$\sigma_{\mathrm{E}} = 3\eta\dot{\varepsilon} \tag{3.12}$$

The extensional viscosity is defined as $\sigma_E/\dot{\varepsilon}$, so for a Newtonian fluid, the extensional viscosity is just three times the (shear) viscosity.

And for step Hencky strain, the tensile stress is:

$$\sigma_E \equiv \sigma_{11} - \sigma_{22} = 3\varepsilon_0 G(t) \tag{3.13}$$

In other words, Young's modulus $E(t)$ for a Newtonian fluid is just three times the shear relaxation modulus.

$$\sigma_E(t)/\varepsilon_0 \equiv E(t) = 3G(t) \tag{3.14}$$

3.5 Relaxation Moduli of Rubbers and Molten Polymers

Figure 3.2 is a sketch of $G(t)$ for a cross-linked elastomer. A logarithmic time scale is used to expand the short time behavior and show that there is a very short initial time period during which an elastic, glassy mechanism operates. And a logarithmic modulus scale is used to accommodate the very large difference between the glassy modulus G_{g} and the equilibrium modulus G_{e}. If the deformation occurs instantaneously, the only mechanism for deformation is chemical bond distortion, and this is the origin of the *glassy* behavior observed in most materials at extremely short times and or sufficiently low temperatures. Once the very fast, initial deformation is accommodated by the glassy mechanism there is time for changes in molecular conformation via Brownian motion within segments of a molecule, and the stress decays more rapidly. Since this is a cross-linked material, the modulus approaches an equilibrium value G_{e} at long times.

Turning now to uncross-linked melts, we find behavior similar to that of a rubber at very short times, but at longer times the melt flows instead of approaching an equilibrium configuration. Figure 3.3 shows sketches of $G(t)$, again using logarithmic scales, for three samples of a linear polymer. Samples A and B have narrow molecular weight distributions. One of these, B, has a molecular weight that is sufficiently high that there is a high degree of entanglement, while sample A has a molecular weight below that required for entanglement. The mean molecular weight of sample C is well above the entanglement threshold, but it has a broad molecular weight distribution and contains a significant number of unentangled molecules. We note that for all three materials, if the deformation is nearly instantaneous there is a very short time period in which

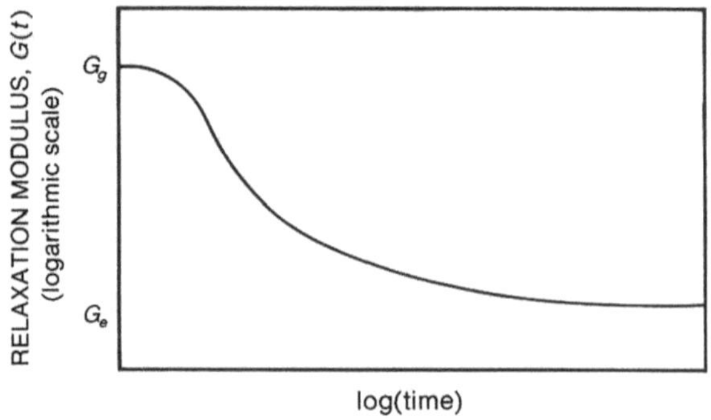

Fig. 3.2 Sketch of log–log plot of relaxation modulus versus time for a cured elastomer. This material has a glassy modulus G_g at very short times and an equilibrium modulus G_e at long times

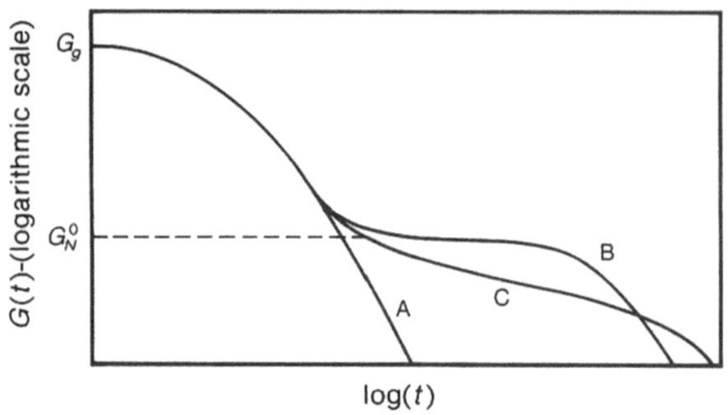

Fig. 3.3 Relaxation modulus curves on logarithmic axes for three linear polymers: *A* is monodisperse with $M < M_C$; *B* is monodisperse with $M \gg M_C$, and *C* is polydisperse with $M_w \gg M_C$

glassy behavior is observed, as in the case of the cross-linked polymer. At longer times, thermal motion occurs, accelerating relaxation. This is the transition region. All three materials exhibit the same behavior in this region, as only very small scale molecular motions are involved, and the molecular weight is not important. For the low molecular weight material A we move directly to a *terminal zone* where $G(t)$ falls off toward zero, as must happen eventually for any uncross-linked, unfilled polymer. For the high molecular weight, narrow distribution melt B the transition zone is followed by a plateau zone in which the modulus is nearly constant. The modulus in this region is called the *plateau modulus*, and has the symbol G_N^0. At longer times flow occurs, and the $G(t)$ curve moves into the terminal zone where the modulus relaxes eventually to zero.

The existence of a plateau implies that there are two relaxation mechanisms. One type of relaxation occurs at short times, and this governs the behavior in the transition zone. And there is another type that occurs at longer relaxation times, which governs behavior in the terminal zone. When the times over which these two relaxation mechanisms operate are widely separated, i.e., for a very high molecular weight, monodisperse polymer, there is an intervening range of times, the plateau region, in which little relaxation occurs, and the melt behaves much like the cross-linked material shown in Fig. 3.2.

The curve for sample C shows the behavior observed for a material with a broad molecular weight distribution. There is neither a well-defined plateau zone nor a clear transition to a well-defined terminal zone. It is not possible, therefore, to determine a reliable value for the plateau modulus. Long-chain branching has a similar effect on the relaxation modulus.

Molecular structure has significant effects on behavior only in the plateau and terminal zones, and melt processing involves only longer-time processes, so we will be interested in this book primarily with behavior in these zones.

3.6 The Maxwell Model for the Relaxation Modulus

It is useful to have a general functional form for the relaxation modulus $G(t)$ that can be used to describe experimental data. In Chap. 1, to illustrate the general nature of viscoelasticity we used a mechanical assembly consisting of a linear (Hookean) spring and a linear dash-pot, connected in series, which was shown in Fig. 1.7. This assembly was proposed by Maxwell as a model for the behavior of gases, and it is referred to as Maxwell element. We saw in Chap. 1 that if this element is suddenly stretched a distance X at time $t = 0$, the force will be transmitted through the spring and the dashpot and will thus be the same in both. This force is proportional to the spring extension X_s and also to the rate of extension of the dashpot dX_d/dt. The differential equation that describes the force as a function of time $F(t)$ was given as Eq. (1.19), which is repeated below as Eq. (3.15).

$$F(t) = K_s X_s = K_d \frac{dX_d}{dt} \tag{3.15}$$

The total displacement X is $X_s + X_d$, and for a step-strain experiment is given by a step function. The differential equation can then be solved as follows for $F(t)$:

$$F(t) = XK_s \exp(-tK_s/K_d) \tag{3.16}$$

We note that XK_s is the initial value of the force, since all of the extension must initially be absorbed by the spring and that K_d/K_s has units of time, which we will call the relaxation time of the Maxwell element τ_M.

$$F(t) = F_0 \exp(-t/\tau_M) \tag{3.17}$$

In general one can say that the dynamic behavior of a system that incorporates both energy storage and energy dissipation must have at least one characteristic time. Another example is an electrical circuit that includes both resistance and capacitance.

It will be useful in our discussion of linear viscoelastic behavior to use the relaxation modulus that is the viscoelastic analog of the Maxwell element. This model for melt elasticity in shear is obtained by taking the force F to be analogous to the shear stress σ, the spring constant K_s analogous to the initial shear modulus G_0 of a polymeric liquid, and τ_M analogous to the relaxation time τ. According to this simple model of viscoelasticity the response to a step shear strain is:

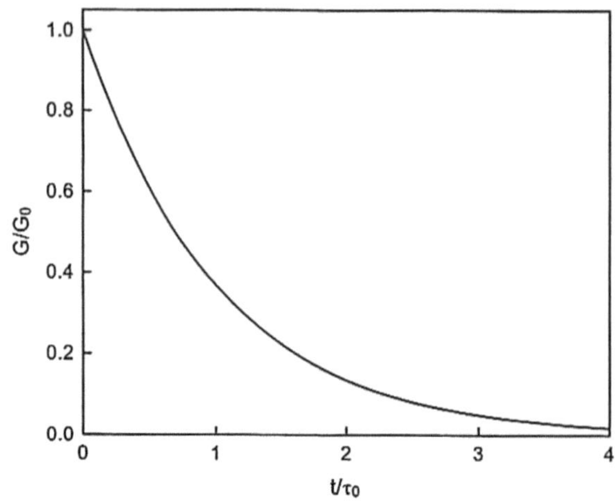

Fig. 3.4 Dimensionless relaxation modulus versus dimensionless time modeled as the exponential function shown as Eq. (3.19)

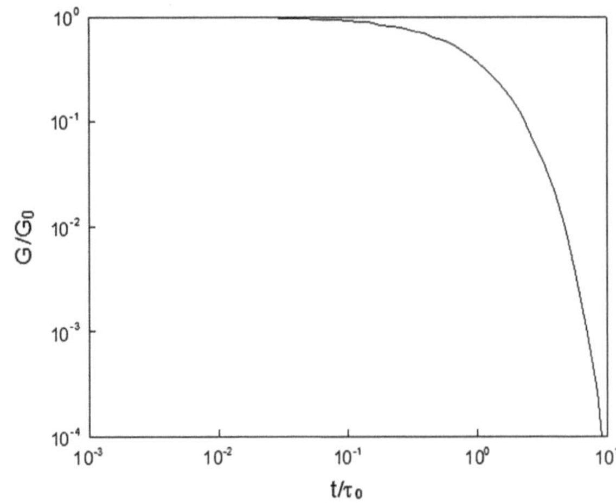

Fig. 3.5 The function shown in Fig. 3.4 plotted using logarithmic scales. Note the drastic change in the appearance of the curve. Logarithmic scales are very useful, but one must be careful in their interpretation

$$\sigma(t) = G_0\gamma_0[\exp(-1/\tau)] \tag{3.18}$$

And the shear relaxation modulus is:

$$G(t) = G_0[\exp(-1/\tau)] \tag{3.19}$$

This is called the Maxwell model and is plotted in Fig. 3.4 in dimensionless form, i.e., (G/G_0) versus (t/τ_o). We note the fast initial decay and the long-time, asymptotic approach to zero. In order to reveal details of the short-time behavior without losing the long-time data off the graph on the right, logarithmic scales are often used in plotting rheological material functions. For example, the curve shown in Fig. 3.4 is replotted using logarithmic scales in Fig. 3.5. This is a useful technique, but it is important to realize that the use of nonlinear scales greatly changes the shape of the curve. For example, one gets the impression from Fig. 3.5 that there is a short-time plateau in the behavior, but this is not so in this case.

3.7 The Generalized Maxwell Model and the Discrete Relaxation Spectrum

The exponential stress relaxation described by Eq. (3.18) is qualitatively similar to the behavior of polymeric liquids, but it does not describe the response of real materials However, it can be generalized by means of an assembly of Maxwell elements connected in parallel, as shown in Fig. 3.6. This leads to a multi-mode Maxwell model for linear viscoelastic behavior, which consists of the sum of exponentials shown by Eq. (3.20).

$$G(t) = \sum_{i=1}^{N} G_i[\exp(-t/\tau_i)] \tag{3.20}$$

The response to a small or slow deformation is described by substituting Eq. (3.20) into Eq. (3.7) and integrating. For example, Eq. (3.11) can be used to show that the zero-shear viscosity according to the multi-mode Maxwell model is:

$$\eta_0 = \sum_{i=1}^{N} G_i \tau_i \tag{3.21}$$

The set of parameters $[G_k, \tau_k]$ constitutes a *discrete relaxation spectrum*, and it is useful for the conversion of one response functional into another. If a sufficient number of terms are used, it is possible to fit the behavior of real materials to this model to a level of accuracy limited only by the precision and time or frequency range of the experimental data. Methods of determining the parameters of a discrete spectrum from experimental data are described in Sect. 3.12.

For a high-molecular-weight polymer with a narrow molecular weight distribution, the largest τ_i is well separated from the next largest one and is called the *terminal relaxation time* of the sample. This leads to classical terminal zone

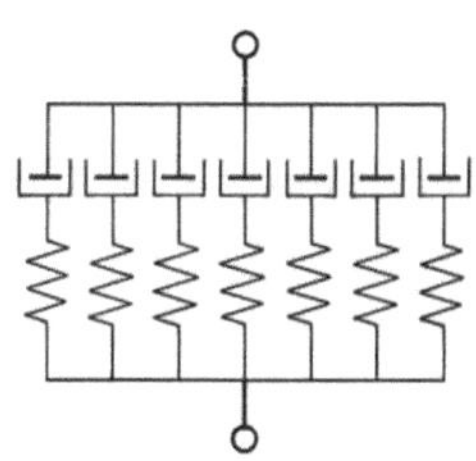

Fig. 3.6 Several maxwell elements in parallel used to model viscoelastic behavior that more complex than a single exponential, i.e., that involves a spectrum of relaxation times

behavior in which $G(t)$ decreases exponentially with time. This means that on a plot of $\log(G)$ versus $\log(t)$ such as Fig. 3.3, the data will approach a line having a slope of minus one at long times.

3.8 The Continuous Spectrum

In the computation of one linear viscoelastic function from another, it is convenient to make use of a continuous spectrum. This can be defined by letting the number of elements in the generalized Maxwell model increase without limit so that $G(t)$ can be represented in terms of a continuous spectrum $F(\tau)$, such that $F\mathrm{d}\tau$ is the contribution to the modulus of relaxation times between τ and $\tau + \mathrm{d}\tau$.

$$G(t) = \int_{0}^{\infty} F(\tau)[\exp(-t/\tau)]\mathrm{d}\tau \tag{3.22}$$

However, a logarithmic time scale is more convenient for representing data, and the spectrum $H(\tau)$ is generally used in place of $F(\tau)$, where $H = F\tau$, and $H\mathrm{d}[\ln(\tau)] = F\mathrm{d}\tau$. The relaxation modulus is related to $H(\tau)$ as follows.

$$G(t) = \int_{-\infty}^{\infty} H(\tau)[\exp(-t/\tau)]\mathrm{d}(\ln\tau) \tag{3.23}$$

Figure 3.7 [1, p. 371] shows typical relaxation spectra for: (1) a monodisperse, (2) a somewhat polydisperse, and (3) a polydisperse sample. The minimum occurs at about midway in the plateau zone. A basic assumption of polymer physics is that a homogeneous melt has a unique relaxation modulus $G(t)$ and a unique spectrum $H(\tau)$. However, using Eq. (3.23) to infer $H(\tau)$ from experimental data is an *ill-posed* problem. Since data are always subject to random error it is not possible to infer the theoretical unique spectrum from data, which are usually in the form of storage and loss moduli values determined at several frequencies $\left[G'_{\mathrm{k}}, G''_{\mathrm{k}}, \omega_{\mathrm{k}}\right]$. Dealing with this issue to achieve the most accurate representation of data is addressed in Sect. 3.12.

3.9 Creep and Creep Recovery: The Compliance

In a shear creep test, a material initially in its equilibrium state is subjected to a constant shear stress σ at time $t = 0$. The shear strain $\gamma(t)$ is then monitored as a function of time. This is often the preferred experimental technique when it is desired to study linear viscoelastic behavior at very long times. The results are presented in terms of the *shear creep compliance* defined as follows:

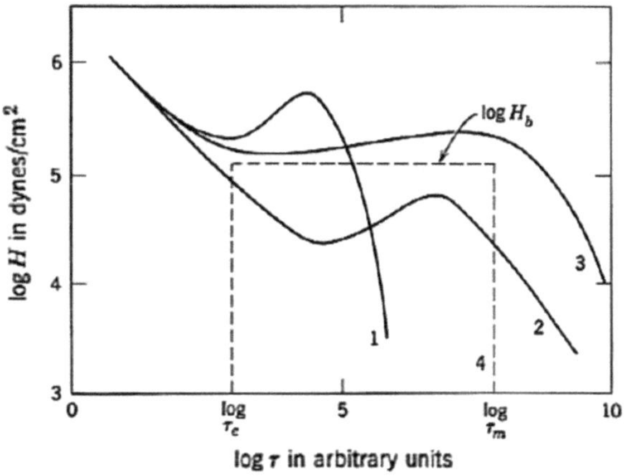

Fig. 3.7 Relaxation spetra $H(\tau)$ in the plateau zone of three materials: *1*—monodisperse polystyrene; *2*—poly(n-octyl methacrylate); *3*—broad MWD polyisobutylene. The *dashed lines 4*—outline the "box distribution," a primitive approximation of this type of behavior. From Ferry [1, p. 371]

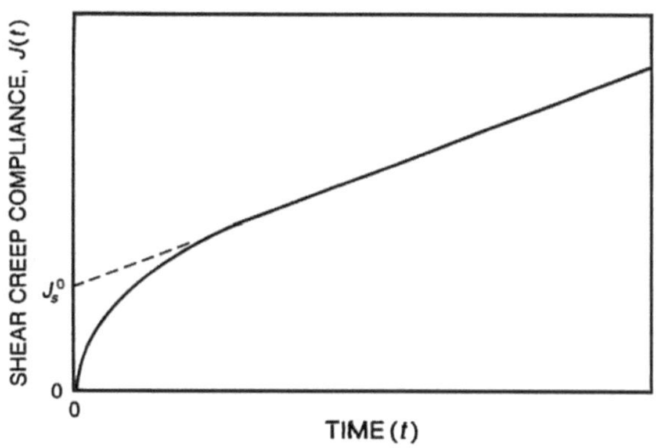

Fig. 3.8 The shape of the creep compliance curve for a typical molten polymer. If the stress is sufficiently small, the zero-time intersect of the extrapolation of the linear portion of the curve is the LVE steady-state compliance J_s^0

$$J(t) \equiv \gamma(t)/\sigma \tag{3.24}$$

For a cross-linked polymer, e.g. a cured elastomer or vulcanized rubber, for at a constant stress, the strain will reach an equilibrium value, and the creep will reach a steady value called the *equilibrium compliance* J_e. A melt, however, does not have an equilibrium compliance, and the strain will increase approaching linearity with time, and the slope of the line will be stress divided by the viscosity, so that the compliance approaches a line whose slope is the reciprocal of the viscosity.

A typical creep compliance curve is sketched in Fig. 3.8. In the case of linear behavior, the creep compliance is independent of σ. For a melt, the shear rate eventually becomes constant and equal to the shear stress divided by the viscosity. The intercept on the vertical J axis is called the *steady state compliance* J_s^0, and the compliance at long times is given by:

$$J(t) = J_s^0 + t/\eta_0 \tag{3.25}$$

For unentangled polymers, the steady-state compliance increases linearly with molecular weight. While it is independent of molecular weight for well-entangled, monodisperse polymers, it is pointed out in Chap. 7 that J_s^0 depends strongly on polydispersity.

In terms of the parameters of the multi-mode Maxwell model, the creep compliance of a melt is:

$$J(t) = \sum_{i=1}^{N} J_i[1 - \exp(t/\tau_i)] + t/\eta_0$$

And corresponding to the relaxation spectrum $H(\tau)$ there is a *retardation spectrum* $L(\tau)$ defined as follows:

$$J(t) = \int_{-\infty}^{\infty} L(\tau)[1 - \exp(t/\tau_i)]\mathrm{d}\ln\tau + t/\eta_0$$

If, at a time t_0 after the start of a creep experiment, the shear stress is suddenly eliminated, for example by removing a weight from a pulley mechanism, the material will spring back or recoil in a direction opposite that of the original applied force. The amount of recoil or *recovered shear strain* γ_r is a function of t and of the time $t - t_0$ that has elapsed since the elimination of the shear stress, but if the stress is removed only after steady state has been achieved, the recoil depends only on $t - t_0$. It is convenient to start a new time scale ($t_r = t - t_0$) that is zero at $t = t_0$, so that the recoverable strain is:

$$\gamma_r(t_r) \equiv \gamma(t_0) - \gamma(t) \tag{3.26}$$

Note that the recoverable strain is defined so that it is positive.

The ultimate recoil, or recoverable shear, when the deformation is halted, is:

$$\gamma_\infty \equiv \lim_{t_r \to \infty} [\gamma_r(t_r)] \tag{3.27}$$

The recoverable compliance $J_r(t_r)$ is defined as follows.

$$J_r(t_r) \equiv \gamma_r(t_r)/\sigma \tag{3.28}$$

And making use of the Boltzrnann superposition principle, it can be shown that:

$$J_r(t_r) = J(t) - t/\eta_0 \tag{3.29}$$

By comparison with Eq. (3.25), we see that the ultimate value of the recoverable compliance is equal to the steady state compliance:

$$\lim_{t_r \to \infty} J_r(t_r) = J_e^0 \tag{3.30}$$

For very polydisperse materials, particularly those with a small amount of high-molecular-weight polymer, the time required to reach steady state in a creep test can be extremely long because of the need to measure extremely slow deformations while maintaining the stress at a very low value to ensure that the behavior is linear. Also, if the measurement is made at an elevated temperature thermal stability may be a problem. Kraft et al. [2] proposed a technique for determining the creep compliance up to the steady-state without dealing with these issues. During a standard creep experiment at a stress σ_0 that would take the polymer into the nonlinear regime if maintained to steady state, they reduce the stress back to zero at time, t_1, well before steady-state is reached, and monitor the resulting recovery. This experiment can be analyzed by use of Eq. (3.31), which is an expression of the Boltzmann superposition principle where strain rather than stress is the dependent variable.

$$\gamma(t) = \int_{-\infty}^{t} J(t - t')d\sigma(t) \tag{3.31}$$

The unloading that takes place at t_1 starts a second creep experiment commencing at that time and driven by a negative stress of $-\sigma_0$. From Eq. (3.31) the resulting deformation $\gamma(t)$ is related to the creep compliance as follows:

$$\gamma(t) = J(t)\sigma_0 + J(t - t_1)(-\sigma_0) = \sigma_0[J(t) - J(t - t_1)] \tag{3.32}$$

The creep compliance is only measured directly from $t = 0$ to t_1, but solving (3.32) for $J(t)$, we we can use the recovery data to extend $J(t)$ up to $t = 2t_1$ as follows:

$$J(t) = \gamma(t)/\sigma_0 + J(t - t_1) \quad (t_1 < t < 2t_1) \tag{3.33}$$

Now that $J(t)$ has been determined up to $t = 2t_1$, this information can be used in combination with the next portion of the recoil curve to determine the compliance at times up to $3t_1$. This procedure is repeated until steady state is reached and the zero-shear viscosity can be calculated as the reciprocal of the slope of the creep curve.

There remains the problem of combining creep data with storage and loss modulus data to obtain a characterization valid over the broadest possible time range. This issue is examined in Sect. 3.13.

Figure 3.9 shows sketches of shear creep compliance curves for the three samples of linear polymer whose relaxation moduli were shown in Fig. 3.3. At very short times glassy behavior is observed. At times sufficient to allow motion within a molecule, all three samples exhibit a transition zone. For the low molecular weight sample A this transition leads directly to a steady flow or terminal region, while for sample B there is a well defined plateau and then a sharp transition to the terminal zone. The value of the compliance in the plateau zone is the plateau compliance J_N^0. The broad molecular weight sample C has neither a clearly defined plateau nor a sharp transition to the terminal zone.

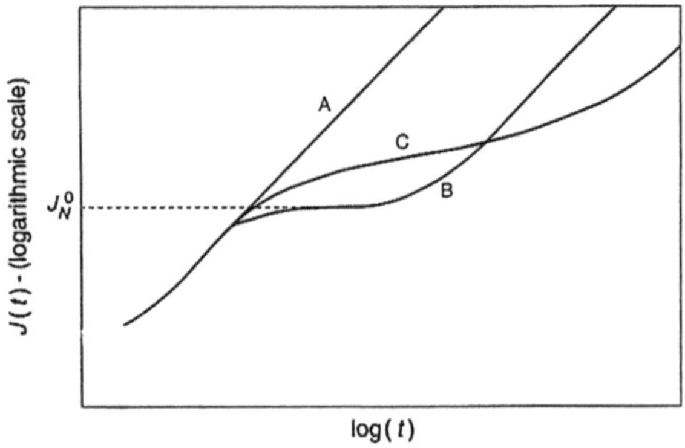

Fig. 3.9 Creep compliance curves using logarithmic axes for three linear, molten polymers. *A* is monodisperse with $M < M'_C$; *B* is monodisperse with M $\gg M'_C$, *C* is polydisperse with $M_w \gg M'_C$

3.10 Uniaxial (Simple) Extension

3.10.1 The Net Stretching Stress

We saw in Chap. 2 that a single normal stress component has no rheological significance for a constant density material, because only normal stress differences can cause deformation. The principal stress generating uniaxial extension is the tensile stress σ_{zz}, if z is the direction of stretch. But the sides of the sample normal to the stretch direction must be exposed to a lower normal stress if deformation is to occur. In most experiments, this stress is due to the ambient pressure, which is small compared to the stress driving the flow. The stress difference of rheological importance is the difference between the tensile stress σ_{zz} and the stress due to the ambient pressure, and we call this the *net stretching stress* σ_E. In nearly all melt measurements, it is very close to the tensile stress.

3.10.2 Extensional Creep

Analogous to the shear creep and creep recovery functions described above are the extensional creep and creep recovery functions. Rather than applying a constant shear stress, we apply a constant net stretching stress σ_E starting at time $t = 0$. The tensile creep compliance is the time-dependent Hencky strain divided by σ_E. By use of the Boltzmann superposition principle, it can be shown that $D(t) = J(t)/3$. Thus, at long times, when $D(t)$ becomes linear with time, we have:

$$D(t) = J_e^0/3 + t/3\eta_0 \tag{3.34}$$

3.10.3 Start-Up of Steady Simple Extension

Simple, i.e., uniaxial, extension is not used to determine the relaxation spectrum but when it is used to study nonlinear behavior it is useful to calculate the stress for the limiting case of very small strain or strain rate. In Sect. 3.4 we considered start-up of steady simple shear, and we now deal with the analogous deformation in which a sample initially in an equilibrium state is subjected to a constant Hencky strain rate starting at $t = 0$. Data are usually reported in terms of the *tensile stress growth coefficient* defined as follows:

$$\eta_{\mathrm{E}}^{+}(t) \equiv \sigma_{\mathrm{E}}(t)/\dot{\varepsilon} \tag{3.35}$$

The Boltzmann principle can be used to show that this is proportional to the corresponding material function for simple shear:

$$\eta_E^{+}(t) = 3\int_0^t G(s)ds = 3\eta^{+}(t) \tag{3.36}$$

When plotting tensile stress growth coefficient data obtained under conditions outside the linear regime, it is useful to show on the same graph the linear behavior given by Eq. (3.36), as it provides a test of the validity of the nonlinear data and is helpful in interpreting these data. This will be explored in detail in the following chapter.

At a sufficiently long time the deformation will approach steady simple extension, and $\eta_E^{+}(t)$ will approach the *extensional viscosity*, which is equal to three times the viscosity measured in shear.

$$\lim_{t\to\infty} \eta_{\mathrm{E}}^{+}(t) \equiv \eta_{\mathrm{E}} = 3\int_0^\infty G(s)ds = 3\eta_0$$

The relationship $\eta_{\mathrm{E}} = 3\eta$ also applies to Newtonian fluids, and was first discovered by Trouton.

3.11 Small-Amplitude Oscillatory Shear

While it is convenient to use the shear relaxation modulus $G(t)$ to introduce basic concepts, the experiment that is generally used to determine the linear viscoelastic *behavior* of polymeric liquids is small amplitude oscillatory shear. In this experiment, a thin sample of material is subjected to a simple shearing deformation such that the shear strain as a function of time is given by:

$$\gamma(t) = \gamma_0 \sin(\omega t) \tag{3.37}$$

where γ_0 is the strain amplitude and ω is the frequency. By differentiating, we find that the shear rate as a function of time is given by Eq. (3.38).

$$\dot{\gamma}(t) = \gamma_0\omega\cos(\omega t) = \dot{\gamma}_0\cos(\omega t) \tag{3.38}$$

where $\dot{\gamma}_0$ is the shear rate amplitude $\omega\gamma_0$.

The stress is then measured as a function of time, and if γ_0 is sufficiently small the stress will also be periodic.

$$\sigma(t) = \sigma_0\sin(\omega t + \delta) \tag{3.39}$$

where σ_0 is the stress amplitude and δ is the phase shift, which is called the *loss angle*. The amplitude ratio $G_d = \sigma_0/\gamma_0$ and the loss angle are functions of frequency but are independent of strain amplitude, as long as the deformation is sufficiently small that the behavior is in the linear regime. The maximum strain for linear behavior generally depends on frequency.

3.11.1 The Storage and Loss Moduli and the Complex Viscosity

The results of an oscillatory shear experiment could be presented by plots of amplitude ratio and phase shift versus frequency, but it is customary to make use of a trigonometric identity to write Eq. (3.39) in the form shown by Eq. (3.40).

$$\sigma(t) = \gamma_0[G'(\omega)\sin(\omega t) + G''(\omega t)\cos(\omega t)] \tag{3.40}$$

where $\gamma_0 G'$ is the amplitude of the in-phase component of the stress, and G' is the *storage modulus*; while $\gamma_0 G''$ is out-of phase amplitude, and G'' is the *loss modulus*. These two quantities are simply related to the amplitude ratio and the phase shift.

$$G' = G_d\cos\delta \tag{3.41}$$

$$G'' = G_d\sin\delta \tag{3.42}$$

To understand the physical significance of the storage and loss moduli, it is instructive to look at the behavior of a Newtonian (linear, purely viscous) fluid and a Hookean (linear, purely elastic) solid when subjected to oscillatory shear. For the linear solid with a shear modulus G, stress is proportional to strain, and for oscillatory shear we have:

$$\sigma = G\gamma = \gamma_0 G[\sin(\omega t)] \tag{3.43}$$

Thus, $G' = G$, $G'' = 0$, and the loss angle is zero. Thus we say that the in-phase component of Eq. (3.40) describes the *elastic* aspect of the response. For a

Newtonian fluid, on the other hand, stress is proportional to strain rate, and we have:

$$\sigma = \eta\dot{\gamma} = \gamma_0\eta[\cos(\omega t)] \tag{3.44}$$

For this material $G' = 0$, $G'' = \eta\omega$, and the loss angle is $\pi/2$, since $\sin(\omega t + \pi/2) = \cos(\omega t)$.

The out-of-phase component of 3.40 describes the *viscous* aspect of the response. It can easily be shown that the energy dissipation or work per cycle per unit volume W is proportional to G''.

$$W = \oint \sigma\dot{\gamma}\,dt = \pi\gamma_0^2 G'' \tag{3.45}$$

The ratio of the moduli is the tangent of the loss angle, tan δ.

$$G''/G' = \tan\delta$$

This useful function of frequency is a measure of the ratio of dissipated or lost energy to stored, elastic energy. At high frequencies, where energy storage predominates, it approaches zero, while at very low frequencies, where dissipation predominates, the loss angle approaches $\pi/2$, and its tangent grows without bound.

It is sometimes convenient to think of G' and G'' as the real and imaginary components, respectively, of a *complex modulus* $G^*(\omega)$,

$$G^*(\omega) \equiv G'(\omega) + iG''(\omega) \tag{3.46}$$

The amplitude ratio G_d is thus the magnitude of G^*.

$$G_d \equiv \sigma_0/\gamma_0 = |G^*| = \sqrt{(G')^2 + (G'')^2} \tag{3.47}$$

An alternative representation of the results of a small amplitude oscillatory shear test is based on material functions with units of viscosity.

$$\sigma(t) = \dot{\gamma}_0[\eta'(\omega)\cos(\omega t) + \eta''(\omega)\sin(\omega t)] \tag{3.48}$$

where

$$\eta' = (\sigma_0/\dot{\gamma}_0)\sin\delta = G''/\omega \tag{3.49}$$

$$\eta'' = (\sigma_0/\dot{\gamma}_0)\cos\delta = G'/\omega \tag{3.50}$$

The complex viscosity is:

$$\eta^* = \eta'(\omega) - i\eta''(\omega) \tag{3.51}$$

And its absolute value is

$$|\eta^*| = \sigma_0/\dot{\gamma}_0 = \sqrt{(\eta')^2 + (\eta'')^2} \tag{3.52}$$

In Chap. 2 it was pointed out that when this quantity is plotted versus frequency, the data often fall remarkably close to a plot of viscosity versus shear rate.

$$\eta(\dot{\gamma}) = |\eta^*|(\omega) \qquad \dot{\gamma} = \omega \tag{3.53}$$

This relationship is called the Cox-Merz rule. This so-called rule is based on an observation [3] that when apparent viscosity data for two molten, commercial polystyrenes obtained using a capillary rheometer were plotted versus shear rate, they fell close to data for $|\eta^*|$ versus ω. It is explained in Chap. 6 that the apparent viscosity measured in a capillary rheometer deviates from the true viscosity of a non-Newtonian fluid. Nonetheless, it has been reported that the Cox-Merz rule defined by Eq. (3.53) is often obeyed by linear, molten polymers [4–6], and since the measurement of the viscosity at low shear rates using a rotational rheometer is more difficult than oscillatory shear measurement $|\eta^*(\omega)|$ is often used in place of $\eta(\dot{\gamma})$ in the plastics industry.

All linear behavior is governed by the Boltzmann superposition principle, in which the material‘s viscoelasticity is described completely by the relaxation modulus $G(t)$. For example $G'(\omega)$ and $G''(\omega)$ can be shown to be the Fourier sine and cosine transforms, respectively, of the relaxation modulus. And if the relaxation modulus is represented in terms of the generalized Maxwell model (Eq. 3.20) we obtain:

$$G'(\omega) = \sum_{i=1}^{N} \frac{G_i(\omega\tau_i)^2}{\left[1 + (\omega\tau_i)^2\right]} \tag{3.54}$$

$$G''(\omega) = \sum_{i=1}^{N} \frac{G_i(\omega\tau_i)}{\left[1 + (\omega\tau_i)^2\right]} \tag{3.55}$$

Methods to infer discrete and continuous spectra from oscillatory shear data are summarized in Sect. 3.12.

3.11.2 The Storage and Loss Compliances

As an alternative to the use of the components of the complex modulus or the complex viscosity to report the results of a sinusoidal shear experiment, the storage and loss compliances J' and J'' can be used. Here, an oscillatory stress is considered to be the stimulus, with the sinusoidal strain reflecting the material response. In fact, the same data that are used to calculate G' and G can also be used to calculate J' and J''.

$$J' = (\gamma_0/\sigma_0)\cos(\delta) \tag{3.56}$$

$$J'' = (\gamma_0/\sigma_0)\sin(\delta) \tag{3.57}$$

We note that

$$|J^*| = (\gamma_0/\sigma_0) = 1/G_d \tag{3.58}$$

3.12 Inferring a Discrete Relaxation Spectrum from Storage and Loss Moduli

The information required to determine the parameter set $[G_i, \tau_i]$ for a specific polymer is usually obtained by means of oscillatory shear measurements that yield a set of storage and loss modulus data $\left[G'_k, G''_k, \omega_k\right]$. In other words, we want to determine N pairs of model parameters $[G_i, \tau_i]$ starting with m sets of experimental data $\left[G'_k, G''_k, \omega_k\right]$. As was mentioned in Sect. 3.8 this is an *ill-posed problem*, and there is no unique solution. As a result, the model parameters have no physical significance, but they are useful for calculating one rheological property from another.

In attempting to improve agreement between data and values recalculated from an inferred discrete spectrum by increasing the number of relaxation times τ_i, more and more relaxation strengths are found to be negative, and the standard error for the fit of the entire curve increases. This is because the model is trying to fit the noise in the data. Special techniques are required to achieve the most accurate representation of the data. Honerkamp and Weese [7] have demonstrated the use of Tikhonov regularization to solve this problem. A further advance is the use of regularization along with nonlinear regression [8, 9], with which it is possible to obtain a discrete spectrum that contains as much as possible of the information in the data.

Figure 3.10 compares discrete spectra obtained from oscillatory shear data using three methods: linear regression with regularization, nonlinear regression, and linear regression without regularization [10]. The use of linear regression with regularization yielded the spectrum that gave the most accurate re-calculation of the original data.

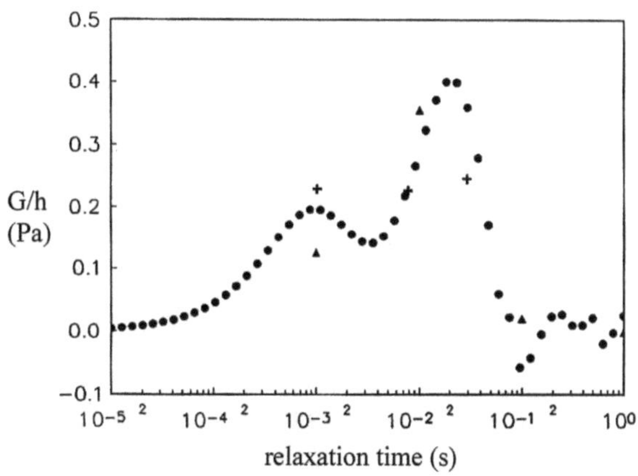

Fig. 3.10 Discrete relaxation spectra calculate by three methods: linear regression without regularization (LRwReg); nonlinear regression (NLR); linear regression without regularization. h is a normalization factor. ● LRwReegul, + MLR, ▲ LRwReegul from Orbey and Dealy [10]

3.13 Combining Creep and Oscillatory Shear Data

To relate rheological behavior with molecular structure, it is important to obtain data extending into the terminal zone. Oscillatory shear data are often limited to frequencies above the terminal zone, but creep data can be used to explore this region. One then wishes to display all the data on a single plot, and this is usually accomplished by converting creep compliance data to storage and loss moduli by use of a continuous or discrete spectrum. He et al. [11] recommend that one calculate the continuous retardation spectrum using both modulus and creep data and plot these together. The resulting graph shows clearly the zones in which each technique provides reliable data and the zone of overlap. They believe this procedure is better than using a discrete spectrum to combine data from the two types of experiment.

3.14 Time–Temperature Superposition

3.14.1 Basic Principle

In Chap. 2 we showed that viscosity data obtained at several temperatures can often be brought together on a single master plot by the use of a shift factor. Now we will extend this principal to deal with linear viscoelastic data. Time-temperature superposition is also useful in the presentation of nonlinear data, as is explained in Chap. 4.

A single rheometer operating at a single temperature can usually provide small-amplitude oscillatory shear data over only about three decades of frequency, and this is inadequate to track viscoelastic behavior from the high-frequency end of the plateau zone into the low-frequency terminal zone. By obtaining data at several temperatures, time–temperature superposition can be used to generate master curves showing the behavior at a *reference temperature* over many decades of time or frequency. A material to which this technique is applicable is said to be *thermorheologically simple*.

This also applies to material functions that depend on time instead of frequency, such as the relaxation modulus $G(t)$ and the creep compliance $J(t)$. The implication is that when material function data are shown on a double-logarithmic plot, data at various temperatures can be shifted horizontally, on the frequency or time axis, by a constant (independent of time or frequency) amount $\log(a_T)$ and vertically by another temperature-dependent distance identified as b_T to obtain a single master curve bringing all the data together. The quantity $b_T(T)$ is called the *vertical shift factor*, and $a_T(T)$ is the *horizontal shift factor*. The master curve displays data over a wide range of frequencies or times *reduced* to a common reference temperature T_0. A thorough discussion of time–temperature superposition can be found in the book by Ferry [1, Sect. 11-C], who also discusses the shifting of data obtained at several pressures.

Thermorheological simplicity obtains when the characteristic times of all relaxation mechanisms have the same temperature dependence and when stress magnitudes at all frequencies have the same temperature dependence. This is valid for describing behavior in the plateau and terminal zones and in the longest-time (lowest-frequency) portion of the transition zone. In this range of times or frequencies, behavior is dominated by fairly long chain segments. But at shorter times, relaxation modes involving shorter segments come into play, and as a result a value of a_T obtained using data in the terminal and plateau zones will not superpose data obtained at significantly shorter times (or higher frequencies). For polydisperse samples, there is a gradual transition from one zone to another, so one must examine the data carefully to detect the beginnings of thermorheological complexity as the frequency is increased. It is also important to note that short-time relaxation mechanisms will come into play at longer times as the temperature is lowered. One must therefore be on the alert for complexity not only at short times (high frequencies) but also at low temperatures.

As an example of this technique, a master curve of the frequency-dependent storage modulus, based on data obtained at various temperatures T can be obtained by plotting:

$$b_T G'(\omega, T) \text{ versus } \omega a_T \tag{3.59}$$

While it is not indicated explicitly, b_T and a_T are functions of temperature, although the b_T dependence is relatively weak. The quantity on the left is called the reduced storage modulus G'_r, while the reduced frequency is ω_r. Using these variables a master curve is obtained by plotting of G'_r versus ω_r. And the temperature dependence is described by a plot of $\log(a_T)$ versus T.

A master curve of the relaxation modulus is obtained by plotting

$$G_r(t) \text{ versus } t_r \tag{3.60}$$

However, the reader should be aware that an alternative definition of b_T has been used, particularly in the earlier publications of Graessley and his colleagues and in the recent book by Rubinstein and Colby [12]. In these publications the reduced modulus is $(1/b_T)G(t)$.

The concept of time–temperature superposition can be expressed in terms of an equation relating the property at the reference temperature T_0 to that measured at a temperature T and then shifted to T_0:

$$b_T G'(T, \omega a_T) = G'(T_0, \omega) \tag{3.61}$$

or:

$$G'_r(T, \omega_r) = G'(T_0, \omega) \tag{3.62}$$

Some authors write these equalities in a different way, which can cause confusion.

Viscosity, which involves both stress and time, requires the application of both shift factors. For example, a master curve of the absolute value of the complex viscosity is constructed by means of a double-logarithmic plot of

$$\frac{b_\mathrm{T}}{a_\mathrm{T}}|\eta^*|(T) \quad \text{versus} \quad \omega a_\mathrm{T} \tag{3.63}$$

where

$$|\eta^*| = \sqrt{G'^2 + G''^2}\Big/\omega$$

A special case is the zero-shear viscosity.

$$\frac{b_T(T)}{a_T(T)}\eta_0(T) = \eta_0(T_0) \tag{3.64}$$

If the vertical shift factor is taken to be unity, the horizontal shift factor can be inferred from the zero-shear viscosity.

$$a_T(T) = \frac{\eta_0(T)}{\eta_0(T_0)}$$

Since viscosity is then proportional to the shift factor, the viscosity ratio or even the viscosity itself $\eta_0(T)$ is often used as a horizontal shift factor.

Time-temperature superposition can be applied to all the material functions of linear viscoelasticity to generate master curves. For example to obtain a compliance master curve one plots $J\,(t,T)/b_\mathrm{T}(T)$ versus $t/a_\mathrm{T}(T)$.

In the case of crystallizable melts, the useful temperature range extends only from the crystallization temperature at the low end, to the degradation temperature, at the high end, and this limits the applicability of superposition to these polymers.

In general, establishing thermorheological complexity by inspection of a log–log plot of data can be misleading, and Wood-Adams and Costeux [13] propose the use of linear rather than logarithmic scales to detect thermorheological complexity. Figure 3.11 shows data for a metallocene polyethylene homopolymer with a very low level of long-chain branching. On this log–log plot the data appear to superpose well, but the linear plot shown in Fig. 3.12 shows clearly that this material is not thermorheologically simple. In this example, the detection of complexity is important, as it provides evidence of long-chain branching. Figure 3.13 shows master curves of the storage and loss moduli of a polybutadiene binary blend [14]. The actual frequency range of the rheometers used was only about three decades, but by making measurements at eight temperatures, the reduced frequency extends over nine decades. At the highest reduced frequency, i.e., the lowest temperature, the superposition is starting to fail, because the horizontal shift factor appropriate for the plateau and terminal zones is not appropriate for the Rouse modes that dominate the high-frequency behavior.

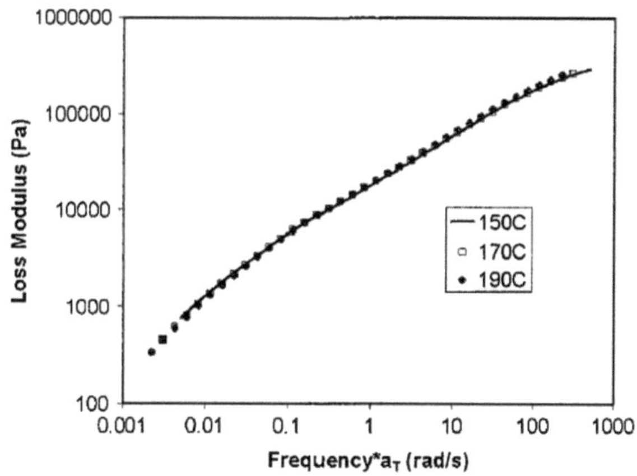

Fig. 3.11 Master curve of G″ versus $a_T\omega$ for an ethylene-butene copolymer with a *low level* of *long-chain* branching. Superposition was achieved by shifting the data to the reference curve (*solid curve*) such that the departures were averaged over the entire range of frequencies. However, we note that the points are a bit above the curve at high frequencies and a bit below at low frequencies. From Wood-Adams and Costeux [13]

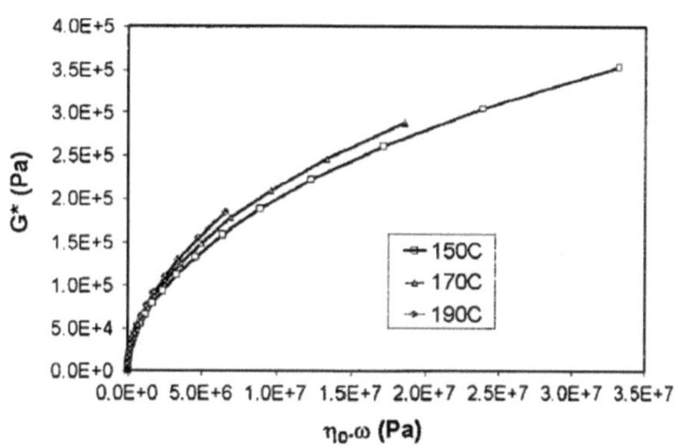

Fig. 3.12 Wood-Adams and Costeux [13] recommend the use of a plot of the magnitude of the complex compliance G^* versus reduced frequency using linear scales to reveal thermorheological complexity. We see here data for the same sample as in Fig. 3.12 but plotted using linear scales, and the lack of superposition is quite apparent

While some star polymers [15] and combs [16] have been found to be thermorheologically simple, more complex branching structures result in the failure of superposability. For example, metallocene polyethylenes containing even low levels of long-chain branching have been found to be thermorheologically complex [13, 17]. And low-density polyethylene data do not obey time–temperature superposition.

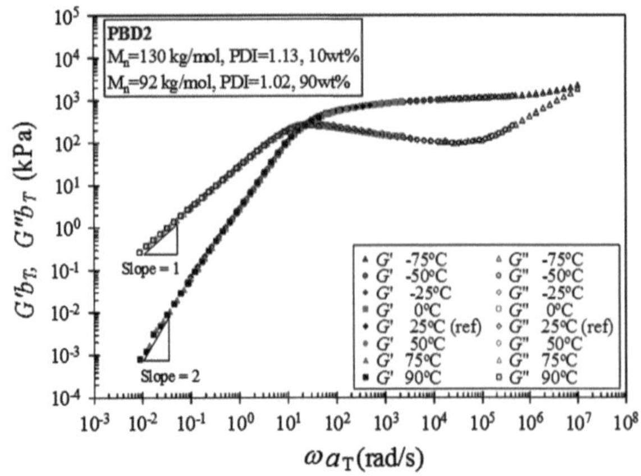

Fig. 3.13 Master curves of storage and loss moduli for a blend of two narrow MWD polybutadienes. The reference temperature is 25 °C, and the vertical shift factor was calculated using Eq. (3.65). From Li et al. [14]

3.14.2 The Vertical Shift Factor

In the Bueche-Rouse model of the linear viscoelasticity of unentangled polymer melts, which is presented in Sect. 3.17, the stress magnitudes, e.g. $G(t)$, $G'(\omega)$ and $G''(\omega)$, are proportional to the product of density and temperature. This implies that the vertical shift factor $b_T(T)$ is:

$$b_T = T_0\rho_0/T\rho \tag{3.65}$$

For entangled polymer systems, the molecular models most widely used predict that the magnitude of the relaxation modulus is proportional to the factor ρT that appears in the Rouse prediction multiplied by a factor that depends very weakly on temperature. While the vertical shift factor is thus predicted to be slightly different from the ρT ratio, data are rarely if ever precise enough to reveal this difference.

The vertical shift factor is sometimes used as a fitting parameter for an entire set of data, and the data are shifted both vertically and horizontally in an attempt to achieve a superposition that looks OK. This empirical procedure can disguise the failure of superposition by averaging the discrepancy over the entire frequency range. Systematic deviations of $b_T(T)$ from its expected behavior (Eq. 3.65) can also result from experimental error. Finally, if both shifts are carried out by fitting, neither factor has any physical significance. The best practice is to either use Eq. (3.65) or let b_T be unity.

3.14.3 The Horizontal Shift Factor

Since a_T cannot be predicted, it is found by shifting data horizontally on a log–log plot until they superpose. This should be done after the vertical shift factor has been applied, but one can determine the horizontal shift factor without making any assumption about $b_T(T)$ by plotting as a function of frequency a quantity that is not affected by b_T, such as the loss angle or its tangent, which is equal to G''/G'. Honerkamp and Weese [18] describe a procedure for determining horizontal shift factors in an objective manner, and software for this procedure is part of the IRIS software [19].

The empirical Arrhenius relationship has been found to describe the dependence of a_T on temperature in the plateau and terminal zones as long as the temperature is well above T_g.

$$a_T(T) = \exp\left[\frac{E_a}{R}\left(\frac{1}{T} - \frac{1}{T_0}\right)\right] \tag{3.66}$$

By analogy with reaction rate theory, the constant E_a is called the *activation energy for flow*. However, an activation energy is often used to characterize polymers even when rheological data do not superpose, and this results in ambiguity as to the meaning of E_a. For example, E_a is sometimes reported to vary with frequency or time or modulus. Various approaches have been proposed for identifying meaningful activation energies when data do not superpose [13, 20]. One value is based on η_0. In any event, when data do not superpose the use of an activation energy is questionable.

At temperatures within about 100 °C of T_g, the WLF equation [1, Sects. 11-B, C] is usually obeyed.

$$\log a_T = \frac{-c_1(T - T_0)}{[c_2 + (T - T_0)]} \tag{3.67}$$

where c_1 and c_2 are empirical constants.

An extensive tabulation of WLF parameters for many polymers can be found in Reference [21].

3.15 Cole–Cole and Related Plots of Linear Data

Several plotting techniques that involve plotting one material function versus another have been proposed for oscillatory shear data to reveal time–temperature superposability or structural features. These were inspired by a way of plotting data for the complex dielectric constant. Cole and Cole [22] found that when they plotted the imaginary versus real components of this, their dependence on frequency could be described by Eq. (3.68).

$$\varepsilon^* - \varepsilon_\infty = \frac{(\varepsilon_0 - \varepsilon_\infty)}{\left[1 + (i\omega\tau_0)^{1-\alpha}\right]} \tag{3.68}$$

Plots of this type are called Cole–Cole plots, and we will call this the Cole–Cole function. Equation (3.68) implies that a plot of imaginary versus real component of a complex variable is a circular arc with its origin below the real axis, as shown in Fig. 3.14, where the parameter α is related to θ by $\theta = \alpha\pi/2$, and is independent of temperature.

Marin and Graessley [23] used Cole–Cole plots together with Eq. (3.68) to interpret data for several polystyrenes prepared by anionic polymerization. They plotted the imaginary versus real components of the complex retardational compliance $J_r^*(\omega)$ defined as $[J_r^*(\omega) - 1/(i\omega\eta_0)]$. They found that for the sample with a molecular weight of about 37,000, which is near the critical molecular weight for viscosity M_c, a plot of $J_r''(\omega)$ versus $J_r'(\omega)$ took the form of a circular arc and could thus be fitted to Eq. (3.69) by analogy with Eq. (3.68).

$$J_r^*(\omega) = \frac{J_t}{\left[1 + (i\omega\tau_t)^{1-b}\right]} \tag{3.69}$$

The subscript t refers to transition zone behavior, while the glassy compliance is relatively negligible and has been omitted. The time constant, τ_t, has the same temperature dependence as the viscosity, and b is independent of temperature. For samples with molecular weights well above M_C, a second circular arc appeared at lower frequencies, reflecting the appearance of a plateau in the storage modulus. Montfort et al. [24] used this approach to interpret their data for narrow MWD polystyrenes. This technique has also been used to describe the behavior of star molecules [15] and combs [16]. Other applications of Cole–Cole plots have also been reported [25–30].

One should remember that frequency-dependence information is not revealed in these types of data presentation, so they are not complete descriptions of linear viscoelastic behavior.

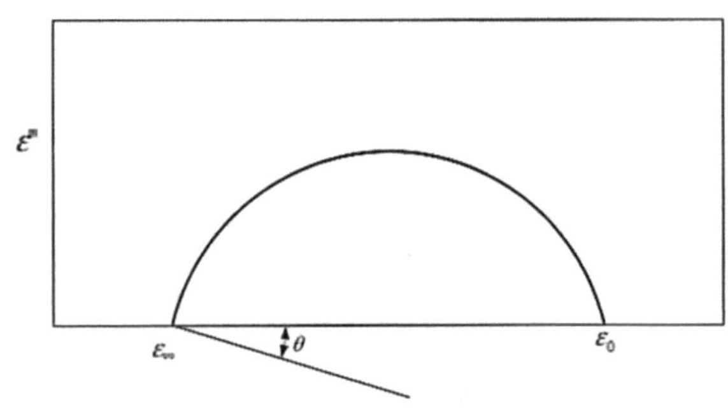

Fig. 3.14 Imaginary versus real components of the complex dielectric constant as described by the Cole–Cole function, Eq. (3.68). The significance of the angle θ is shown

3.16 Van Gurp-Palmen Plot of Loss Angle Versus Complex Modulus

Another plot of one linear viscoelastic property versus another, in which the frequency varies along the curve, was proposed by van Gurp and Palmen [31]. This is a plot of loss angle δ versus the logarithm of the magnitude of the complex modulus $|G^*|$. Such a plot is temperature invariant if the data obey time–temperature superposition. Thus, if data obtained at several temperatures all fall on the same curve on such a plot, this indicates that the time–temperature superposition is valid. Van Gurp and Palmen found that such a plot provides a more critical test of time–temperature superposability than data plotted in the usual manner, i.e., G' and G'' versus frequency. As mentioned above, frequency-dependence is lost in this representation. Finally, since the loss angle depends on the ratio of the loss to the storage modulus, it is not subject to calibration error, which might arise from sample trimming in a rotational rheometer.

This representation is also invariant with regard to average molecular weight but is very sensitive to polydispersity and long-chain branching. For example, Trinkle and Friedrich [32] used this type of plot to characterize polydispersity in linear polymers. They found that the value of $|G^*|$ where the loss angle δ is 60° correlated approximately with M_w/M_n. Trinkle et al. [33, 34] also suggested using this type of plot to reveal the presence of long chain branching in polyethylene. Lohse et al. [35] found that the use of a van Gurp-Palmen plot to display data for hydrogenated polybutadienes having well-defined structures was a useful way to bring out the distinctive effects of branching on linear viscoelastic behavior. Figure 5.15 shows such a plot of their data for linear, star and comb polymers, and we see clearly the distinctive features of the curves that are associated with each branching structure (Fig. 3.15).

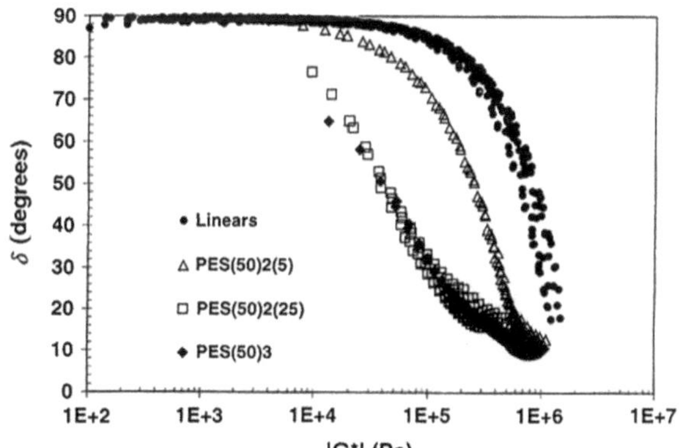

Fig. 3.15 van-Gurp Palmen plots of data for linear (*circles*), star (*triangles*) and comb (*squares*) molecules prepared by hydrogenation of polybutadienes to make samples similar to monodisperse polyethylene. From Lohse et al. [35]

3.17 Storage and Loss Moduli of Molten Linear Polymers

Figure 3.16 is a sketch showing typical $G'(\omega)$ curves for the three polymers, A, B and C, whose relaxation moduli were shown in Fig. 3.3. The general features of these curves are qualitatively similar to those seen in the curves of the relaxation modulus (Fig. 3.3) where the short time behavior of the $G(t)$ curve corresponds to the high frequency portion of the $G'(\omega)$ curve and vice versa. At the highest frequencies shown, glassy behavior is exhibited. At somewhat lower frequencies, rearrangement within molecules becomes possible during a cycle, and there is a transition zone to more fluid-like behavior. For the low molecular weight material A we move directly into a terminal zone, whereas for the high molecular weight material B, we have a plateau zone. In the terminal zone where behavior is governed by the longest relaxation time, according to Eq. (3.54) the storage modulus becoms proportional to ω^2, while according to Eq. (3.55) the loss modulus becomes proportional to ω. As in the case of the relaxation modulus, we note that for material C, which has a broad molecular weight distribution (MWD), the plateau modulus is not clearly defined, and there is no sharp transition to the terminal zone.

Samples for scientific study having very narrow molecular weight distributions and known molecular structures are generally made by anionic polymerization. Two that have been widely used are polystyrene and polybutadiene. These polymers have the added advantage that they do not crystallize, so time–temperature superposition can be used to great advantage. Figures 3.17 and 3.18 show master curves of data of Wang et al. [36] for four monodisperse polybutadienes having various molecular weights. The value of M_e for polybutadiene is about 1,800, so the degrees of entanglement Z of these samples range from about 20 to about 200. This means that all except the 44K sample are well entangled and have well-developed plateau zones, with the terminal zone moving to lower frequencies as the molecular weight increases. For this entangled polymer, G'' passes through a

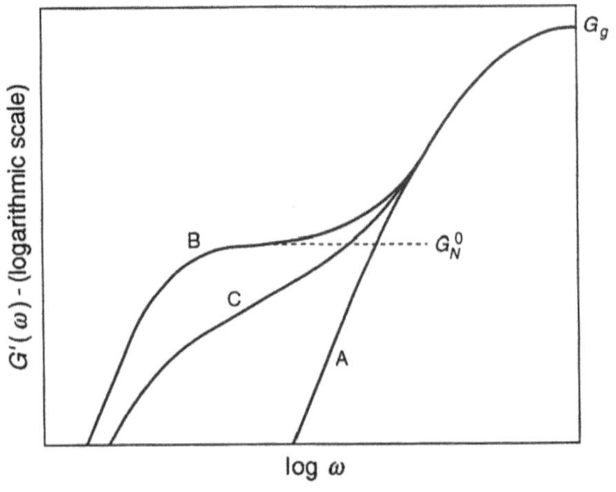

Fig. 3.16 Typical plots of storage moduli for the three linear polymers of Figs. 3.3 and 3.9. *A* is monodisperse with $M < M_C$; *B* is monodisperse with $M \gg M_C$, and *C* is polydisperse with $M_w \gg M_C$

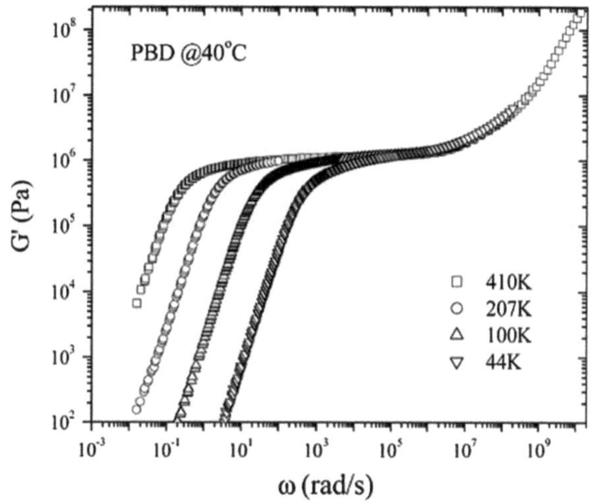

Fig. 3.17 Storage modulus master curves for monodisperse polybutadienes of four molecular weights. The degree of entanglement *(M/Me)* ranges from about 20 to 200 so that all except the 44K sample are well entangled and have well-developed plateau zones. From Wang et al. [36]

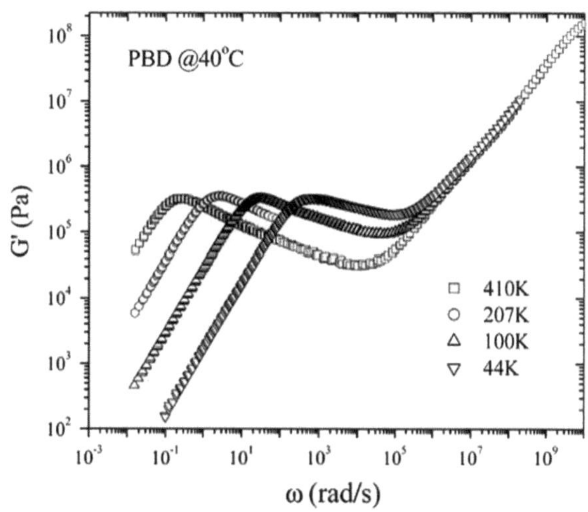

Fig. 3.18 Loss modulus master curves for the samples of Fig. 3.17. All except the least entangled (44K) sample have well-developed minima in their plateau zones. From Wang et al. [36]

minimum in the plateau zone reflecting the fact that little dissipation occurs in the region of rubbery behavior.

3.18 The Plateau Modulus and the Molecular Weight Between Entanglements

3.18.1 Methods for Estimating the Plateau Modulus

Except for very monodisperse samples having extremely high molecular weights, there is no true plateau in the storage modulus. Therefore, all methods for inferring a value for G_N^0 provide only estimates. But because this parameter plays a central role in the molecular modeling of polymer behavior it is important to determine a meaningful value based on experimental LVE data. Liu et al. describe a number of methods for doing this, and we provide here a brief summary.

The method with the best theoretical basis involves the integration of $G''\mathrm{d}(\ln\omega)$ over all frequencies in the neighborhood of its maximum. However, it is necessary to avoid the effect of high-frequency Rouse relaxation modes, and this requires an arbitary extrapolation of the storage modulus beyond its peak. For polymers that can be made by a polymerization method that yields highly-entangled (Z > 30), very nearly monodisperse samples, a definitive value can be identified as the storage modulus at the frequency where the loss modulus is a minimum.

$$G_N^0 = G'(\omega)_{G''=\min}$$

Figure 3.19 shows data for a monodisperse polybutadiene [36] with M_w/M_e equal to about 260, i.e., this is a very highly entangled polymer. There is a well-defined minimum in the loss modulus, which corresponds very well to a value that one might infer from the behavior of G'. This method has the advantages that it can be precisely determined directly from experimental data without the use of an empirical relationship and is free of interference from Rouse modes.

This value is very close to the value of G' where $\tan(\delta)$ is minimum. For somewhat polydisperse samples, when there is no minimum in G'' but there is one in the loss angle, Wu [38] suggested this as a useful estimate of G_N^0. There are also relationships based on the maximum value of G'' [1, 39]. The effect of broadening the molecular weight distribution is illustrated in Fig. 3.20, where we see master curves of the storage and loss moduli and tan δ for a polyisobutylene with $M_w = 8 \times 10^4$ and $M_w/M_n \approx 2$ and $M_w/M_e \approx 15$ [37]. Although the polydispersity is well below that of many commercial polymers there is no longer a distinct plateau in G', and the minimum in G'' has disappeared. For this case, and the tan $\delta_{\min}$ method was used to estimate the plateau modulus.

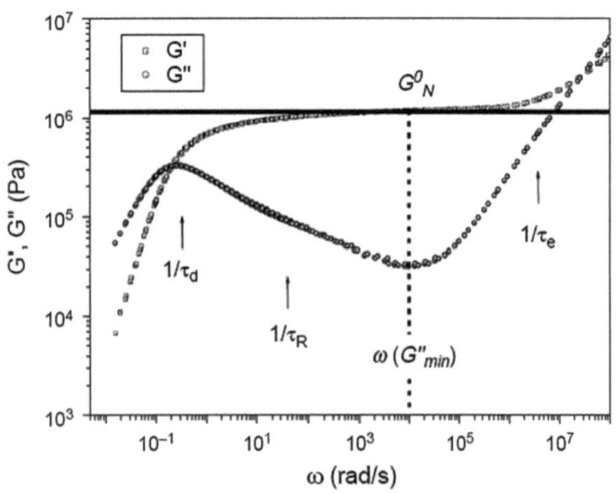

Fig. 3.19 Storage and loss moduli of a highly entangled, monodisperse poolybutadiene. The loss modulus has a very distinct minimum that points to an unambiguous value of the plateau modulus. Data of Wang et al. [36]

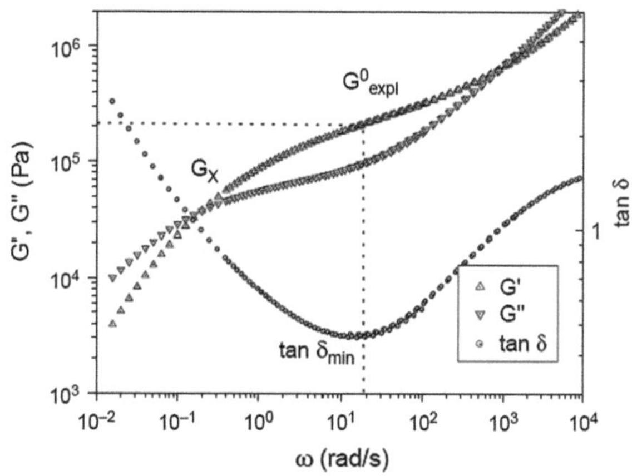

Fig. 3.20 Master curves of storage and loss moduli for a polyisobutylene with a broad MWD. There is no minimum in the storage modulus, but there is in the loss angle, which can be used to identify a plausible value for the plateau modulus on the plot of $G'(\omega)$. From Liu et al. [37]

In the worst case, when only the crossover modulus G_c, evaluated at the frequency where $G' = G''$, is known, Wu [39] proposed the following empirical equation that he said was applicable when the polydispersity is less than three for condensation polymers with limited M_w and some crystalline polymers.

$$\log\left(\frac{G_N^0}{G_c}\right) = 0.38 + \frac{2.63\log(M_w/M_n)}{1 + 2.45\log(M_w/M_n)} \tag{3.71}$$

All of the methods for use with polydisperse samples are empirical, and there is no way to know how much they may be in error, but they can be of use for comparing similar materials.

Polyolefins cannot be made by anionic polymerization, and for polyethylene several methods have been used to prepare narrow-distribution samples thought to be similar to polyethylene. These methods have yielded a wide range of values, in MPa: 1.58 @ 190C [40], 2.6 @ 140C, [41], and 3.5 @ 25 C [42].

The plateau moduli of several common polymers are tabulated in Appendix A, and a more extensive list can be found in Ref. [42].

3.18.2 The Molecular Weight Between Entanglements M_e

We have noted that the plateau in $G'(\omega)$ of a highly entangled melt imitates the behavior of a cross-linked polymer at low frequencies. This rubbery behavior implies the presence of strong interactions between molecules that mimic the effects of chemical cross-links over a certain period of time. These strong interactions between molecules are referred to as *entanglements*, as they are clearly different from the simple hydrodynamic frictional interactions that occur in low molecular weight liquids. Because these strong interactions result in behavior similar to that exhibited by a rubber, which has a permanent network, they are sometimes said to comprise a *temporary network*. The plateau modulus was used by Ferry [1] as the basis for defining an *average molecular weight between entanglements* M_e.

$$M_e \equiv \frac{\rho RT}{G_N^0} \tag{3.72}$$

It was later recognized that in stress relaxation, one fifth of the initial stress relaxes by fast Rouse modes, i.e., the motion of small chain segments between entanglements, before reaching the plateau region. This inspired a new definition of M_e as 4/5 of the value defined by Eq. (3.72); the symbol $M_e^G \equiv 4/5 M_e$ is sometimes used for this quantity. In this book, we use only the original Ferry definition given by Eq. (3.72). In reading scientific literature it is important to ascertain which definition is being used by the authors.

The degree of entanglement of a polymer is defined as the ratio of the molecular weight divided by the molecular weight between entanglements.

$$Z \equiv \frac{M}{M_e}$$

For a polymer to be considered highly entangled, Z should have a value of at least 30–40.

3.19 The Rouse-Bueche Model for Unentangled Melts

While we are interested here primarily in the behavior of high molecular weight, i.e., entangled, polymers, it will be useful for our discussion of a theory for entangled melts to summarize the molecular theory for unentangled melts. Rouse [43] developed a molecular theory for dilute polymer solutions in which a polymer molecule is modeled as a chain of N straight segments or *submolecules* that act as Hookean springs. Since the origin of the spring force is Brownian motion, the spring constant is proportional to the absolute temperature. These segments are connected by beads in which the mass of the molecule is assumed to be concentrated. The motion of the beads through the solvent gives rise to viscous resistance, which is described by a friction coefficient ζ. The combination of elastic spring and the viscous resistance gives rise to viscoelastic behavior. We have seen that for a Maxwell element or a Voigt element, each of which contains one spring constant and one viscous resistance constant, there is only one relaxation time. In the case of the segmented chain, however, there are many possible modes of deformation because of the flexibility of the chain, and this gives rise to multiple relaxation times.

Rouse's model for dilute solution behavior is not directly applicable to bulk polymers, and Bueche [44] suggested that for low molecular weight molten polymers in which there are no entanglements, the Rouse model could easily be modified to account for the fact that in a melt the polymer molecule is surrounded not by solvent but by other polymer molecules. The modified Rouse theory, which we call the Rouse-Bueche model, predicts that the relaxation modulus is given by a sum over terms involving the *Rouse relaxation times* τ_p as shown by Eq. (3.73).

$$G(t) = \frac{\rho RT}{M} \sum_{p=1}^{N} e^{t/\tau_p} \tag{3.73}$$

$$\tau_p = \frac{b^2 P^2 \zeta}{6\pi^2 p^2 kT} \quad p = 1, 2, 3, \ldots \tag{3.74}$$

where:

ζ translational friction coefficient per monomer unit

b a length characteristic of the chemical structure of the molecule, equal to $\left(\langle R^2 \rangle_0 / P\right)^{1/2}$

P degree of polymerization $= M/M_0$.

The *longest Rouse relaxation time* corresponds to $p = 1$ and is an important parameter in polymer rheology; for an unentangled polymer it is given by Eq. (3.75).

$$\tau_R = \frac{b^2 P^2 \zeta}{6\pi^2 kT} \tag{3.75}$$

The storage and loss moduli are given by Eqs. (3.54) and (3.55) with all the relaxation moduli G_i equal to $\rho RT/M$ and the relaxation times equal to the Rouse times of Eq. (3.74).

The low-frequency limiting value of the loss modulus is equal to $\omega\eta_0$, and this fact can be used to derive an expression for the zero-shear viscosity of an unentangled polymer.

$$\eta_0 = \frac{\zeta \rho b^2 M N_A}{36 M_0^2} \tag{3.76}$$

where:

N_A Avogadro's number

$M_o = M/P$ monomer molecular weight

ρ density.

At molecular weights well below those necessary for entanglement, the monomeric friction coefficient ζ is an increasing function of the molecular weight, but there is a significant range of molecular weights over which it is nearly constant. This implies that within this range the viscosity increases linearly with molecular weight, and there is substantial experimental verification of this prediction for unentangled polymers.

Equation (3.76) can be used to rewrite the longest Rouse time in terms of η_0 as shown by Eq. (3.77).

$$\tau_R = \frac{6\eta_0 M}{\pi^2 \rho RT} \tag{3.77}$$

In the terminal zone, i.e., at very long times or very low frequencies, behavior is governed by the longest relaxation time τ_R, and the several viscoelastic moduli simplify to the following forms.

$$G(t) = \frac{\rho RT}{M} e^{-t/\tau_R} \tag{3.78}$$

$$G'(\omega) = \frac{\rho RT}{M} \omega^2 \tau_R^2 \sum_{p=1}^{N} \frac{1}{p^4} = 1.08\omega^2 \tau_R^2 \frac{\rho RT}{M} \tag{3.79}$$

$$G''(\omega) = \frac{\rho RT}{M} \omega \tau_R \sum_{p=1}^{N} \frac{1}{p^2} = 1.645\omega \tau_R \frac{\rho RT}{M} \tag{3.80}$$

The Rouse-Bueche prediction of the steady-state compliance is:

$$J_s^0 = \frac{0.40M}{\rho RT} \tag{3.81}$$

Equation (3.81) indicates that the steady state compliance is proportional to the molecular weight for a linear, monodisperse polymer when there are no entanglements ($M < M'_C$). This is in sharp contrast to the behavior observed for highly-entangled ($M \gg M'_C$), linear, monodisperse melts, where J_s^0 is found to be independent of molecular weight. Also, this property is strongly influenced by the molecular weight distribution, as is shown in Chap. 7.

At higher frequencies (shorter times) in the *transition zone*, it has been shown [1, p. 189] that if the 2 or 3 longest relaxation times are ignored a relaxation modulus that approximates the Rouse prediction over a limited portion of the zone is:

$$G(t) = Ct^{-1/2} \tag{3.82}$$

where

$$C \equiv \sqrt{\frac{3\rho RT\eta_0}{2\pi M}}$$

Since η_0 is proportional to M, C is independent of the molecular weight, and Eq. (3.82) implies that:

$$G'(\omega) = G''(\omega) = \sqrt{\pi/2}\, C\omega^{1/2} \tag{3.83}$$

These approximate forms are only valid over a limited portion of the transition zone, because at very short times or high frequencies, the segmented chain model of a polymer molecule is no longer valid.

Figure 3.21 is a log–log plot of dimensionless relaxation modulus versus dimensionless time, as predicted by the Rouse-Bueche model. We note that at short times, the slope on this log–log plot has a slope of $-1/2$, in accord with Eq. (3.82), and at long times the exponential behavior given by Eq. (3.78) is obeyed. Figure 3.22 is a log–log plot of dimensionless storage and loss moduli versus dimensionless frequency. At the highest frequencies shown, the behavior is as predicted by Eq. (3.83), and the terminal slopes are in accord with Eqs. (3.79) and (3.80).

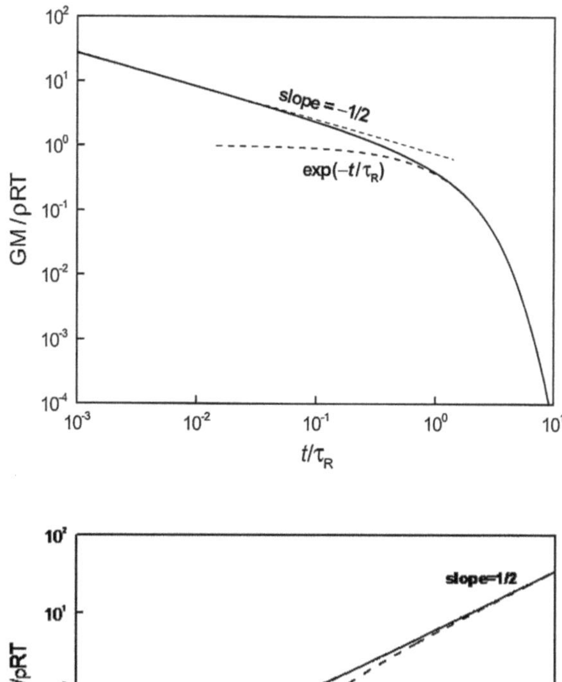

Fig. 3.21 Double logarithmic plot of dimensionless relaxation modulus versus dimensionless time for a linear polymer described by the Rouse-Bueche model, Eqs. (3.73) and (3.74). At the shortest times, the slope of −½ is in accord with Eq. (3.82), while the behavior at long times is shown by Eq. (3.78)

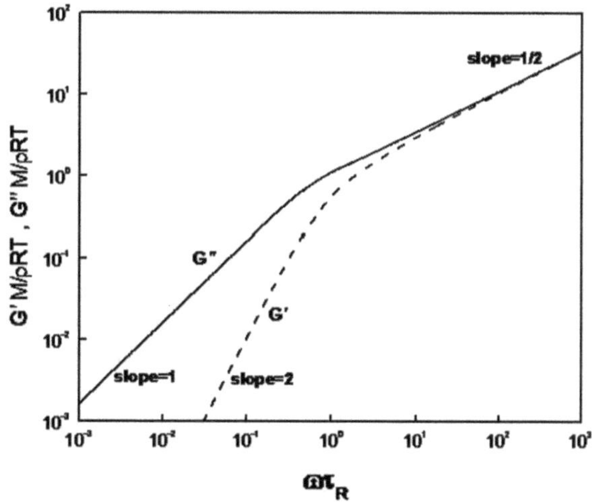

Fig. 3.22 Curves of dimensionless storage and loss moduli as predicted by the Rouse-Bueche model. At short times the slopes are in accord with Eqs. (3.79) and (3.80), while at long times the slopes of ½ are in accord with Eq. (3.83)

The Rouse-Bueche model can be extended to polydisperse systems as long as no species have molecular weights high enough to participate in entanglements [45, 46]. One surprising result is that the steady state compliance of a blend of two molecular weights can greatly exceed that of either component, especially when one component has a much higher molecular weight than the other and this component has the lower concentration in the blend [47].

3.20 Tube Models for Entangled Melts

We provide here only a brief introduction to this subject. A much more detailed treatment can be found in the book by Dealy and Larson [48]. While the Bueche-Rouse model does a good job of describing the behavior of polymers of low molecular weight, we have seen that as molecular weight increases, there are dramatic departures from this behavior. One example is the very large change in the dependence of viscosity on molecular weight when M approaches M_C. Whereas the Rouse-Bueche theory predicts that η_0 is proportional to M (Eq. 3.76), in the neighborhood of a molecular weight M_C the viscosity starts to increase much more rapidly, becoming proportional to about the 3.5 power of the molecular weight. Another example is the appearance of a plateau in the relaxation modulus of monodisperse samples. When these phenomena were first encountered, they were attributed to *entanglements*, and while it is now known that they are due simply to the *uncrossability* of long chains over each other, the term entanglements continues to be used.

The detailed modeling of the behavior of a very large number of very long molecules is not possible at present, but progress has been made using a *mean-field* approach in which attention is focused on a single *test molecule*, and its interactions with the surrounding molecules are averaged. The inspiration for this type of model came from Pierre-Gilles deGennes, a Nobel Prize winning French polymer physicist, who noted that in trying to relax the stress imposed on a long chain by strain, the displacements available to chain segments are initially restricted to very small motions, and that larger motions required for full relaxation could only occur by as a result of the segments moving along the length of the backbone. This motion was called *reptation* by deGennes [49], after the Latin *reptare,* to creep, from which the word reptile also derives.

Doi and Edwards used the reptation concept to develop a model for the viscoelasticity of molten, high molecular weight polymers [50, 51]. Considering a Rouse segmented chain they assumed that the effect of the restrictions on lateral motion imposed by the neighboring molecules sketched in Fig. 3.23, was equivalent to placing the molecule in a tube, as shown in Fig. 3.24. This tube has a diameter d and a length L. Because the model does not refer to specific points of entanglement, the molecular weight between entanglements M_e does not appear as a parameter. However, there is some basis for associating the number of entanglements M/M_e with the ratio L/d.

Fig. 3.23 Sketch of a typical polymer chain showing the segments of the surrounding chains that restrict its motion in the direction normal to its length

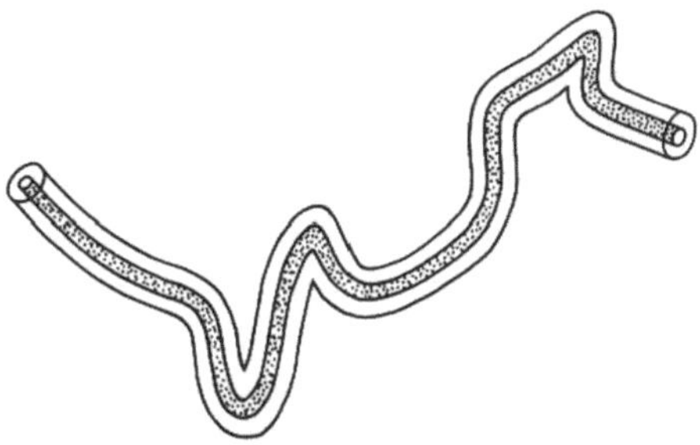

Fig. 3.24 Sketch showing the hypothetical tube pictured by Doi and Edwards to represent the restrictions of the surrounding chains on the one whose behavior is being modeled

At very short times, the only mechanism available is the redistribution of extensions among the segments *between* points where constraints are present. This rapid relaxation process has a characteristic time τ_e called the *equilibration time*. Once this process is completed further relaxation is strongly suppressed for a period of time, as the only mechanism available for the molecule to further relieve stress is *disengagement*, i.e., reptation out of its tube. This is a relatively slow process with a characteristic time τ_d, and for a long chain, $\tau_d \gg \tau_e$. Another relaxation process that can, in principle, occur in entangled melts is relaxation of the contour length, i.e., the retraction of the molecule within its tube. However, this only comes into play when a deformation is large and fast and is taken up in Chap. 4.

In this way the Doi-Edwards tube model accounts for the two distinct groups of relaxation times that are observed for high molecular weight, monodisperse, linear polymers. For times $t < \tau_e$ this theory agrees with the Rouse-Bueche theory.

In the plateau and terminal zones the model predicts:

$$G(t) = G_N^0 \frac{8}{\pi^2} \sum_{p=1\ odd}^{N} \frac{1}{p^2} e^{-p^2 t/\tau_d} \tag{3.84}$$

Because of the $1/p^2$ factor, this result is dominated by the first term with $p = 1$. This spectrum is much narrower than observed spectra and results from the fact that the original Doi-Edwards model includes only one relaxation mechanism in the plateau and terminal zones. As a result the predicted relaxation modulus at short times and the storage and loss moduli at high frequencies fall below experimental data, especially in the case of $G(t)$. We will see how the model has been modified to account for this below.

The predictions for the zero-shear viscosity and steady-state compliance are as follows.

$$\eta_0 = \frac{a^2 \varsigma M^3 G_N^0}{12 M_e M_0^2 kT} \tag{3.85}$$

$$J_S^0 = \frac{6}{5 G_N^0} \tag{3.86}$$

The strong effect of M on the viscosity is qualitatively correct, but the value of 3 for the exponent is less than the observed value of about 3.5. The steady-state compliance is independent of molecular weight, which is correct for high molecular weight melts and is in contrast with the prediction of the Rouse-Bueche model for unentangled melts that J_s^0 is proportional to molecular weight.

A central hypothesis of the Doi-Edwards model is that the tube retains its identity throughout the disengagement time for an individual molecule. For a highly-entangled monodisperse polymer, estimates of the time required for the tube to lose its identity through Brownian motion of the molecules comprising the tube is much larger than τ_d, and the hypothesis is valid. However, in a polydisperse system some of the molecules making up the tube are shorter than the longest molecules present. In this case, *constraint release* provides an additional mechanism for relaxation and thus speeds up the process [52]. The presence of long-chain branches leads to major theoretical complications. Reptation is no longer possible, and relaxation occurs primarily through the mechanism of *contour length fluctuations* [53, 54].

3.21 Molecular Weights for the Onset of Entanglement Effects

One manifestation of entanglements is a rather sharp change in the dependence of viscosity on molecular weight. Whereas the modified Rouse theory predicts that η_0 is proportional to M (Eq. 3.76), in the neighborhood of a certain molecular weight M_C, the viscosity starts to increase much more rapidly with molecular weight, often approaching a proportionality to about the 3.5 power of the molecular weight, as predicted by a tube model that accounts for contour length relaxations. The characteristic molecular weight M_C depends on molecular structure and is in the range of two to three times M_e. Entanglements also have a dramatic effect on compliance, which we can see by comparing Eqs. (3.81) and (3.86). Whereas the Rouse-Bueche model for unentangled chains predicts that J_s^0 is *proportional* to

M'_C (Eq. 3.81), J_s^0 of entangled, monodisperse, linear polymers is *independent* of M, as predicted by tube models (Eq. 3.86). The molecular weight at which this change in behavior is observed, M'_C is four to six times M_e.

References

1. Ferry J (1980) Viscoelastic properties of polymers, 3rd edn. Wiley, New York
2. Kraft M, Meissner J, Kaschta J (1995) Linear viscoelastic characterization of polymer melts with long relaxation times. Macromol 32:751–757
3. Cox WP, Merz EH (1958) Correlation of dynamic and steady flow viscosities. J Polym Sci 28:619–621
4. Ferri D, Lomellini P (1999) Melt rheology of randomly branched polystyrenes. J Rheol 43:1355–1372
5. Utracki LA, Gendron R (1984) Pressure oscillation during extrusion of polyethylenes. J Rheol 28:601–623
6. Venkatraman S, Okano M, Nixon AA (1990) Comparison of torsional and capillary rheometry for polymer melts: the Cox-Merz rule revisited. Polym Eng Sci 30:308–313
7. Honerkamp J, Weese J (1989) Determination of the relaxation spectrum by a regularization technique. Macromol 22:4372–4377
8. Honerkamp J, Weese J (1993) A nonlinear regularization method for the calculation of relaxation spectra. Rheol Acta 32:65–73
9. http://www.nlreg.com. Accessed 30 May 2012
10. Orbey N, Dealy JM (1991) Determination of the relaxation spectrum from oscillatory shear data. Journ Rheol 35:1035–1050
11. He C, Wood-Adams P, Dealy JM (2004) Broad frequency characterization of molten polymers. J Rheol 48:711–724
12. Rubinstein M, Colby RH (2003) Polymer physics. Oxford University Press, Oxford
13. Wood-Adams P, Costeux S (2001) Thermorheological behavior of polyethylene: effects of microstructure and long chain branching. Macromol 34:6281–6290
14. Li SW, Park HE, Dealy JM (2011) Evaluation of molecular linear viscoelastic models for polydisperse H-polybutadienes. Rheol 55:1341–1373
15. Graessley WW, Roovers J (1979) Melt rheology of four-arm and six-arm star polystyrenes. Macromol 12:959–965
16. Roovers J, Graessley WW (1981) Melt rheology of some model comb polystyrenes. Macromol 14:766–773
17. Stadler FJ, Kaschta J, Münstedt H (2008) Thermorheological behavior of various long-chain branched polyethylenes. Macromol 41:1328–1333
18. Honerkamp J, Weese J (1993) A note on estimating mastercurves. Rheol Acta 32:57–64
19. IRIS software is described at http://rheology.tripod.com
20. Keßner U, Kaschta J, Münstedt H (2009) Determination of method-invariant activation energies of long-chain branched low-density polyethylene. J Rheol 53:1001–1016
21. Ngai KL, Plazek DJ (2007) Temperature dependencies of the viscoelastic response of polymer systems. In: Mark JE (ed) Physical properties of polymer handbook. Springer, New York
22. Cole KS, Cole RH (1941) Dispersion and absorption in dielectrics I. Alternating current characteristics. J Chem Phys 9:341–353
23. Marin G, Graessley WW (1977) Viscoelastic properties of high molecular weight polymers in the molten state. 1. Study of narrow molecular weight distribution samples. Rheol Acta 16:527–533

24. Montfort JP, Marin G, Arman J, Monge P (1978) Blending law for binary blends of fractions of linear polystyrene. Polymer 19:277–284
25. Marin G (1998) Oscillatory rheometry. In: Collyer AA, Clegg DW (eds) Rheological measurements, 2nd edn. Chapman and Hall, London, pp 3–45
26. Marin G, Labaig JJ, Monge P (1975) Dynamic viscoelasticity of entangled polymers. Polymer 16:223–226
27. Labaig JJ, Monge P, Bednarick J (1973) Steady flow and dynamic viscoelastic properties of branched polyethylene. Polymer 14:384–386
28. Garcia-Franco CA, Mead DW (1999) Rheological and molecular characterization of linear backbone flexible polymers with the Cole–Cole model relaxation spectrum. Rheol Acta 38:34–47
29. Utracki LA (1989) Polymer alloys and blends. Hanser, Munich, p 189
30. Utracki LA, Schlund B (1987) Linear low density polyethylenes and their blends: part 2. Shear flow of LLDPE's. Polym Eng Sci 17:367–379
31. van Gurp M, Palmen J (1998) Time-temperature superposition for polymeric blends. Rheol Bull 67(1):5–8
32. Trinkle S, Friedrich C (2001) Van Gurp-Palmen plot: a way to characterize polydispersity of linear polymers. Rheol Acta 40:322–328
33. Walter P, Trinkle S, Mülhaupt R (2001) Influence of zirconocene structure and propene content on melt rheology of polyethene and ethene/propene copolymers. Polym Bull 46:205–213
34. Trinkle S, Walter P, Friedrich C (2002) Van Gurp-Palmen plot II—classification of long chain branched polymers by their topology. Rheol Acta 41:103–113
35. Lohse DJ, Milner ST, Fetters LJ, Xenidou M, Hadjichristidis N, Mendelson RA, Garcia-Franco CA, Lyon MK (2002) Well-defined, model long chain branched polyethylene. 2 melt rheological behavior. Macromol 35:3066–3075
36. Wang S, Wang SQ, Halasa A, Hsu WL (2003) Relaxation dynamics in mixtures of long and short chains: tube dilation and impeded curvilinear diffusion. Macromol 36:5355–5371
37. Liu C, He J, Ruymbeke E, Keunings R, Bailly C (2006) Evaluation of different methods for the determination of the plateau modulus and the entanglement molecular weight. Polym 47:4461–4479
38. Wu S (1987) Entanglements between dissimilar chains in compatible blends: poly(methyl methracrylate) and poly (vinylidine flouride). J Polym Sci Polym Phys 25:527–566
39. Wu S (1989) Chain structure and entanglements. J Polym Sci Polym Phys 27:723–741
40. Raju VR, Smith GC, Marin G, Knox JR, Graessley WW (1979) Properties of amorphous and crystalline hydrocarbon polymers. I. Melt rheology of fractions of linear polyethylene. J Polym Sci Polym 17:1183–1195
41. Fetters LJ, Lohse DJ, Richter D, Witten TA, Zirkel A (1994) Connection between polymer molecular weight, density, chain dimensions, and melt viscoelastic properties. Macromol 27:4639–4647
42. Fetters LJ, Lohse DJ, Colby RH (2007) Chain dimensions and entanglement spacings. In: Mark JE (ed) Physical properties of polymer handbook. Springer, New York
43. Rouse PE (1953) A theory of the linear viscoelastic properties of dilute solutions of coiling olymers. J Chem Phys 21:1272–1280
44. Bueche F (1952) Viscosity self-diffusion and allied effect in solid polymers. J Chem Phys 20:1959–1964
45. Berry GC, Fox TG (1967) The viscosity of polymers and their concentrated solutions. Adv Polym Sci 5:261–357
46. Ninomiya K, Ferry JD, Oyanagi Y (1963) Viscoelastic properties of polyvinyl acetates. ii. creep studies of blends. J Phys Chem 67:2297–2308
47. Leaderman H, Smith RG, Williams LC (1959) Rheology of polyisobutylene.3. elastic recovery, non-newtonian flow, and molecular weight distribution. J Polym Sci 36:233–257
48. Dealy JM, Larson RG (2006) Structure and rheology of molten polymers. Hanser-Gardner Publications, Cincinnati (Second, electronic, edition to appear in 2012)

49. de Gennes PG (1971) Reptation of a polymer chain in the presence of fixed obstacles. J Chem Phys 55:572–580
50. Doi M, Edwards SF (1978) Dynamics of concentrated polymer systems. 1. Brownian-motion in equilibrium state. J Chem Soc Faraday Trans II 74:1789–1801; 2. Molecular-motion under flow. 74:1802–1818 (1978); (1979); 4. Rheological properties. 75:38–54
51. Doi M, Edwards SF (1986) The theory of polymer dynamics. Oxford University Press, Oxford
52. Doi M, Graessley WW, Helfand E, Pearson DS (1987) Dynamics of polymers in polydisperse melts. Macromol 20:1900–1906
53. Doi M, Kuzuu NY (1980) Rheology of star polymers in concentrated-solutions and melts. J Polym Sci Polym Lett 18:775–780
54. Pearson DS, Helfand E (1984) Viscoelastic properties of star-shaped polymers. Macromol 17:888–895

Chapter 4
Nonlinear Viscoelasticity: Phenomena

Abstract While linear viscoelastic properties are very helpful in establishing the molecular structure of polymers, the deformations that occur in processing machinery are always large and rapid, and under these conditions melt behavior is decidedly nonlinear. One manifestation of this is that the response to a shearing deformation no longer provides a complete description of rheological behavior. The molecular processes that give rise to nonlinear viscoelasticity are described as well as several large and rapid deformations that used to obtain information about nonlinear behavior. These include large-amplitude oscillatory shear and extensional flow.

4.1 Introduction

The response of a polymer to large, rapid deformations is nonlinear, i.e., the stress depends on the magnitude, the rate and the kinematics of the deformation. The Boltzmann superposition principle no longer applies, and nonlinear behavior cannot be predicted from linear properties. There exists no general model, i.e., no universal *constitutive equation* or *rheological equation of state* capable of describing nonlinear behavior. These are discussed in Chap. 5, but since most readers will be interested more in relating rheological behavior to molecular structure and processability, rather than the computational simulation of complex flows, we will begin this chapter with an interpretation of nonlinear behavior in terms of a model based on molecular phenomena. The role of rheology in melt forming operations is described in Chap. 8.

The nonlinear behavior of linear polymers is, in general, not as sensitive to MWD as linear behavior, but both the linear and nonlinear properties are strongly affected by long-chain branching structure. Using only data from shear-flow tests, it is often difficult to distinguish between the effects of molecular weight distribution and long-chain branching, but extensional flow behavior is very sensitive to branching when there are chain segments having branch points at both ends.

J. M. Dealy and J. Wang, *Melt Rheology and its Applications in the Plastics Industry*, Engineering Materials and Processes, DOI: 10.1007/978-94-007-6395-1_4,

4.2 Nonlinear Phenomena from a Tube Model Point of View

In Chap. 3 we saw that tube models are based on a representation of the constraints imposed on a highly entangled polymer molecule (*test chain*) in which the surrounding chains are imagined to constitute a *tube* [1]. We begin by reviewing what was said in Chap. 3 about how tube models describe small or slow deformation. In response to a sudden deformation, as in a stress relaxation test, the tube is deformed, and the distribution of orientations of test chain segments is shifted from its equilibrium distribution. The relaxation of the test chain back to its undeformed, equilibrium configuration is constrained by its confinement within the tube. When the imposed deformation is very small the first relaxation process is equilibration within the tube, as mentioned in Chap. 3. Equilibration involves the redistribution of stress among chain segments. Further relaxation occurs primarily as a result of the test chain escaping the constraints of the tube, and this requires it to slither along or *reptate* out of its tube. This is a much slower process and is the reason that there is a plateau in the relaxation modulus for polymers with high molecular weights and narrow molecular weight distributions. In other words, in the plateau region of relaxation times there is limited possibility for relaxation. These relaxation mechanisms show up in the linear relaxation spectrum $H(\tau)$ in the form of peaks, as was shown in Fig. 3.7. If the molecular weight is not narrow, the shorter molecules making up the tube will relax fast enough to cause the expansion of the tube. This process, called *constraint release* accelerates the relaxation of a long molecule within the expanding tube. This results in significant relaxation in what would be the plateau zone for a monodisperse sample of the same polymer.

If the deformation is not very small or slow, the orientation of the chain segments are sufficiently large to cause a nonlinear response. As shown in Chap. 5, this effect can be accounted for in rheological models by replacing the infinitesimal strain tensor with one able to describe large deformations; no new relaxation mechanism needs to be invoked. Moderately nonlinear effects related to orientation can be modeled in this manner.

In continuous shear flow at significant rates, another relaxation mechanism has been proposed to account for the monotonic increase of shear stress with shear rate at a steady shear rate, and this is *convective constraint release*, which results from the loss of entanglements of a given chain with the chains on neighboring streamlines due to the fact that they are moving at different velocities. In shear flows at very high strain rates and in extensional flows at lower strain rates, another phenomenon occurs, and that is chain stretch. In shear chain stretch is rarely observed except in stress relaxation at large strains, but stretch is much more easily generated in extensional flow, especially when there is long-chain branching.

4.3 Nonlinear Stress Relaxation

4.3.1 Chain Retraction and the Damping Function

In an ideal step strain test the shear rate is infinite. Of course this is not physically possible, but the deformation is still so fast that there is no time for relaxation, and there is no continuous flow, so convective constrain release does not come into play. When the strain is large a chain is stretched beyond its equilibrium length, but as soon as the deformation stops, there is a very fast *retraction within the tube*, sometimes called *contour length relaxation*. Figure 4.1 [2] illustrates this phenomenon schematically. The chain segment at left, which has an initial length R_1, is initially stretched but then very rapidly retracts to its original length. Following this fast process relaxation continues, and in a monodisperse polymer this is primarily by reptation. The result is a relaxation modulus curve that has an initial very rapid decrease, due to retraction, followed by a curve that has the same shape as that for linear behavior.

These features can be seen in Fig. 4.2, which shows the relaxation modulus following strains of several sizes for a monodisperse polystyrene solution [4]. Except at the shortest times and smallest strains, the modulus curves drop to successively lower levels as the strain is increased and appear to be superposable by vertical shifting. Figure 4.3 shows the result of such shifting, and the superposition is excellent at times greater than τ_k, which is very small. This is essentially the time required for retraction.

Superposability implies that the nonlinear relaxation modulus can be separated into time-dependent and strain-dependent factors, as shown by Eq. (4.1).

$$G(t, \gamma) = G(t)h(\gamma) \tag{4.1}$$

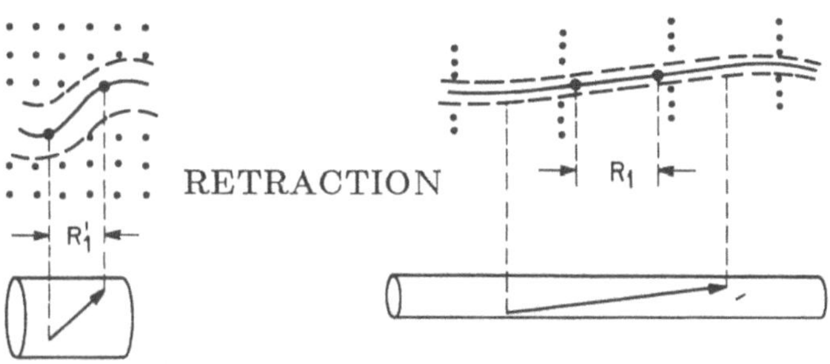

Fig. 4.1 Contour length relaxation. In response to a sudden strain, the chain and the tube formed by the surrounding molecules initially deform affinely, i.e. in the same way as the bulk, and the chain then retracts very rapidly to its equilibrium length within the tube, but the chain segments between entanglements keep the orientation of the deformed tube. From Larson [2]

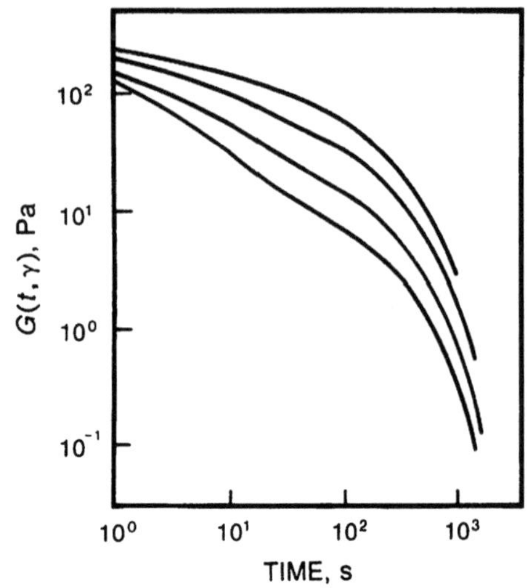

Fig. 4.2 Relaxation moduli at several strains. For the smallest strain (*top curve*) the response is linear, but as the strain is increased, it falls further and further below the linear data. After a rapid initial relaxation during time τ_k, further relaxation proceeds as for the linear case, so that the remainder of the high-stress curves are parallel to that for the smallest strain. Data of Osaki et al. [3]

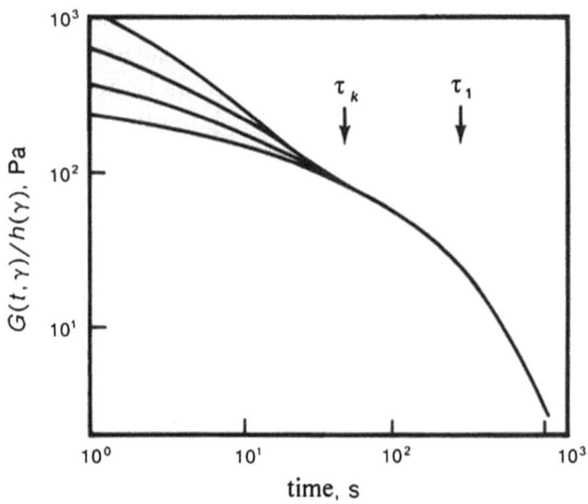

Fig. 4.3 The data of Fig. 4.5 shifted vertically by use of the damping function so that data at times longer than τ_k collapse onto a single curve. τ_1 is the longest relaxation time, where the terminal zone begins. Data of Osaki et al. [3]

This type of stress relaxation is said to exhibit *time–strain separability*, and the strain-dependent factor $h(\gamma)$ is called the *damping function*. This can be thought of as is the fraction of the initial stress not relaxed by retraction.

The original Doi-Edwards tube model yields a fairly accurate prediction of the damping function for monodisperse, entangled polymers. Although this prediction cannot be expressed in an analytical form, Larson [4, p. 143] showed that it is very similar to Eq. (4.2) with $a = 0.2$.

$$h(\gamma) = \frac{1}{1 + a\gamma^2} \tag{4.2}$$

The Doi–Edwards prediction of $h(\gamma)$ is independent of molecular weight for linear, entangled, monodisperse polymers with M between 40 and 50 M_e Polydispersity decreases damping, i.e., the damping function has a larger value for a broad-distribution polymer [5, 6], but time–strain separability fails when polydispersity becomes large. The damping function is also affected by branching [5, 6–10], but the dependence is not strong.

There have been reports of materials that exhibit very severe damping compared to the DE prediction, for example in melts with $M/M_e > 50$ [11]. However, Sanchez-Reyes and Archer [12] showed that this anomalous relaxation behavior was the result of slip between bulk polymer and molecules strongly adsorbed at the wall.

4.3.2 Normal Stress Relaxation

In defining material functions to describe stresses in simple-shear deformations, we use the frame of reference that was illustrated in Fig. 2.1. While tensor notation is introduced in Chap. 5, we will need here only the components shown if Fig. 4.4. The shear stress is component σ_{21} (shown in Chap. 5 to be equal to σ_{12}), and the three normal stresses are: σ_{11}, in the direction of flow (x_1), σ_{22} in the direction of the gradient (x_2), and σ_{33}, in the *neutral* (x_3) direction. As this is by definition a two-dimensional flow, there is no velocity and no velocity gradient in the x_3 direction. In discussing shear flow behavior, we refer to the shear stress as σ without subscripts. As was pointed out in Chap. 2, in an incompressible material, normal stresses are themselves of no rheological significance, because as long as they are equal in all directions they cause no deformation. However, differences between normal stress components are significant, because they do cause deformation. As noted in Chap. 2 for simple shear, the two rheologically significant differences are the first and second normal stress differences N_1 and N_2.

$$N_1 \equiv \sigma_{11} - \sigma_{22} \tag{2.4}$$

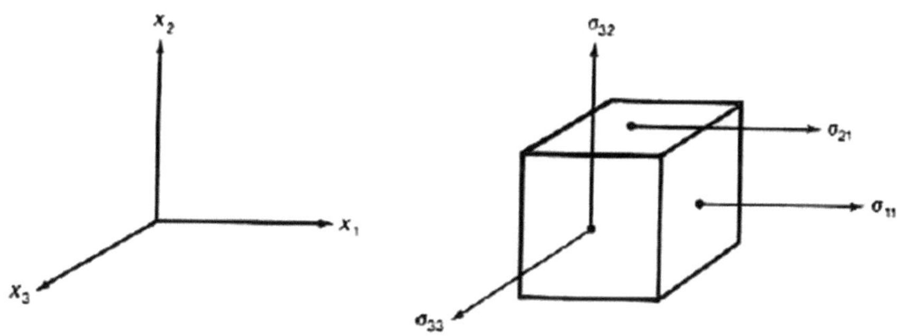

Fig. 4.4 Stress components referred to in this chapter. When the two subscripts are equal, we have a normal stress, and when they different it is a shear stress. The first index identifies the face on which the stress acts, and the second indicates the direction of the stress

$$N_2 \equiv \sigma_{22} - \sigma_{33} \tag{2.5}$$

In Chap. 5 it is shown that a very simple model of nonlinear viscoelasticity, the rubberlike liquid model, predicts that stress relaxation after a step shear strain is given by [13]:

$$N_1(t) = \gamma^2 G(t) = \gamma\sigma(t) \tag{4.3}$$

In fact it has been found that at sufficiently small strains, the *stress ratio* (N_1/σ) does become equal to the strain as indicated by this relationship, which is called the Lodge–Meissner rule [14]. It is similar to the behavior of an isotropic, perfectly elastic solid [15, p. 78], which has a constant modulus of elasticity:

$$\sigma = G\gamma \qquad \text{(linear elasticity)} \tag{4.4}$$

$$N_1 = G\gamma^2 \qquad \text{(linear elasticity)} \tag{4.5}$$

Thus, the stress ratio (N_1/σ) is also equal to the strain for a perfectly elastic solid.

The Doi-Edwards theory [1, p. 253] predicts that the stress ratio is equal to the strain for any strain.

$$\frac{N_1(t,\gamma)}{\sigma(t,\gamma)} = \gamma \tag{4.6}$$

This implies that the Lodge-Meissner rule (Eq. 4.4) is valid at large strains. If time-strain superposability is valid, Eq. (4.6) becomes:

$$N_1(t,\gamma) = \gamma^2 h(\gamma) G(t) \tag{4.7}$$

This relationship has been found to describe experimental data for LDPE for γ up to 30 [16], but the precise measurement of transient normal stress differences is subject to large errors due to instrument compliance and temperature variation.

It was pointed out above that wall slip can cause a larger error in the determination of the strain in step-strain experiments, and the true strain may be much smaller than the nominal strain inferred from the displacement of a rheometer surface. The observation that N_1/σ is independent of time does not, by itself, imply that there is no slip unless this ratio is also equal to the nominal strain applied. And when the Lodge-Meissner rule is not obeyed, it is often taken as evidence that slip occurred, and the stress ratio N_1/σ is used in place of the nominal strain as the independent variable in reporting shear stress and normal stress difference data [17].

The second normal stress difference has been found to be negative with a magnitude less than that of N_1. A useful material function is the *normal stress ratio* $\Psi(t,\gamma)$ defined as follows:

$$(t,\gamma) \equiv \frac{-N_2(t,\gamma)}{N_1(t,\gamma)} = \frac{-{}_2(t,\gamma)}{{}_1(t,\gamma)} \tag{4.8}$$

For strains less than one, $\Psi(\gamma)$ approaches a limiting, zero-strain value $\Psi(0)$, often around 0.25.

4.4 Dimensionless Groups Used to Plot Rheological Data

4.4.1 The Deborah Number

In plotting the nonlinear response to transient strains other than step strain, and for comparing data with theoretical predictions, it is sometimes useful to show results in terms of a dimensionless ratio called the *Deborah number* (De). This is a measure of the degree to which elastic behavior is expected in a flow that is unsteady from the point of view of a material element. In other words, it reflects the rate at which elastic energy is either stored or released during an experiment. This dimensionless group is the ratio of a time arising from the fluid's viscoelasticity, i.e., a relaxation time that we call here τ_r, to a time that is a measure of the duration of the deformation. In simple shear, at steady state when all stresses (and the stored elastic energy) are constant with time, the duration of the flow is unlimited, and De is zero. More generally, De is zero in all flows with constant stretch history (steady from the point of view of a material element) when the stresses are steady. Thus, it is only in transient flows, i.e., when the deformation is unsteady from the point of view of a fluid element, that the Deborah number non-zero.

It is difficult to write a concise definition of a time constant that governs the rate at which stored elastic energy changes in a given deformation without reference to a specific flow, and we therefore give a definition of the Deborah number in general terms

$$\mathrm{De} \equiv \frac{\tau_r}{\text{char. time of transient deformation}} \tag{4.9}$$

In an oscillatory flow, the characteristic time of the deformation is the reciprocal of the frequency. In start-up flow, De is time-dependent; at early times the stress changes rapidly, and elasticity plays a key role, but at long times, as the strain-rate approaches a steady value, De approaches zero. Thus, De $= \tau_r/t$. This implies that in an ideal start-up flow De is initially infinity. This reflects the fact that elasticity arising from intra-molecular motions cannot operate instantaneously, and the only mechanism available arises from a glassy mechanism.

4.4.2 The Weissenberg Number

The Weissenberg number (Wi) is a measure of the degree of nonlinearity or orientation that is expected as the result of flow. This is defined as the product of a time scale governing the onset of rheological nonlinearity, let's call it λ, and the rate of strain of the deformation. We saw in Chap. 2 that a fluid with a shear-rate dependent viscosity must have at least one material constant with units of time. For example, one might select the reciprocal of the shear rate at which the viscosity falls to ninety percent of its zero-shear value. Obviously, the degree to which a melt deviates from Newtonian behavior depends on rate of the deformation. Thus the Weissenberg number in simple shear flow is defined as:

$$\mathrm{Wi} \equiv \dot{\gamma}\lambda \tag{4.10}$$

For single-phase, low-molecular-weight fluids, the time constant of the material is extremely short, so that the Weissenberg number is also extremely small. But for molten, high-molecular weight polymers, λ can be quite large. For molten polymers, the Weissenberg number also indicates the degree of anisotropy generated by the deformation. And in the case of steady simple shear, the normal stress differences are manifestations of anisotropy and thus of nonlinear viscoelasticity. Therefore, the Weissenberg number also governs the degree to which the normal stress differences differ from zero. The Weissenberg number is easily defined for a flow with constant stretch history. For example, in steady uniaxial extension at a Hencky strain rate $\dot{\varepsilon}$, $\mathrm{Wi} = \dot{\varepsilon}\lambda$.

For deformations in which linear viscoelastic behavior is exhibited, Wi is zero. However, there are flows of practical importance in which both Wi and De are nonzero and are sometimes even directly related to each other. This causes confusion, as authors often use the two groups interchangeably. This situation arises, for example, in the flow from a reservoir into a much smaller channel, either a slit or capillary. A Weissenberg number can be defined for this flow as the product of the characteristic time of the fluid and the shear rate at the wall of the flow channel. However, entrance flow is clearly not a flow with constant stretch history, and the Deborah number is thus non-zero as well and depends on the rate of convergence of the flow.

The utility of these dimensionless parameters is limited by the difficulty of defining a meaningful characteristic time of the melt. For high molecular weight, monodisperse samples, possible choices are the longest Rouse time and the equilibration and disengagement times of the tube model. For example, the original tube model predicts that $\Psi(\dot{\gamma})$ is a universal function of $(\dot{\gamma}\tau_d)$ for all entangled polymers. But for polydisperse materials, these well-defined times are not available.

4.5 The Viscosity in Terms of the Tube Model

The original Doi-Edwards model predicted that the shear stress in steady shear increases from zero and goes through a maximum and a minimum before increasing again. This type of behavior has never been observed, and this remained a basic deficiency of tube models until Ianniruberto and Marrucci [18] introduced the concept of *convective constraint release* (CCR). In steady shear flow, molecules on neighboring streamlines move at different speeds, and this carries away entanglements at a rate comparable to the reciprocal of the shear rate.

If the shear rate is increased to a level near the reciprocal of the Rouse time of the chain, chain stretch should become active, as the rate would be too fast for CCR to maintain equilibrium. But this effect has not been observed in melts because of the very large value of M/M_e required for $1/\tau_R$ to be within the range of shear rates that are experimentally accessible.

4.6 Transient Shear Tests at Finite Rates

4.6.1 Stress Growth and Relaxation in Steady Shear

In *start-up of steady simple shear*, the measured stresses are divided by the imposed shear rate or its square to obtain the *shear stress growth coefficient* and the *first normal stress growth coefficient*, which are defined as follows:

$$\eta^+(t,\dot\gamma) \equiv \sigma(t,\dot\gamma)/\dot\gamma \qquad (4.11)$$

$$\Psi_1^+(t,\dot\gamma) \equiv N_1(t,\dot\gamma)/\dot\gamma^2 \qquad (4.12)$$

At short times, when the deformation is small, we expect the shear stress growth coefficient to follow the prediction of the Boltzmann superposition principle. And if the shear rate is sufficiently low, the entire transient will be governed by this principle, and $\eta^+(t,\dot\gamma)$ will be the same as $\eta^+(t)$, which was given as Eq. (3.10) and is repeated here as (4.13).

$$\eta^+(t) \equiv \frac{\sigma}{\dot\gamma} = \int_0^t G(s)ds \qquad (4.13)$$

The shear stress growth coefficient for an entangled polystyrene solution is shown in Fig. 4.5 at several strain rates [19]. At the lowest strain rate, the behavior is linear, corresponding to Eq. (4.13), and there is no overshoot. The strains γ_S and γ_N at which the maxima occur in the shear stress and first normal stress difference are sometimes used as empirical measures of nonlinearity. After a sufficient length of time at constant shear rate, all the stresses become independent of time, and three rheologically meaningful material functions of shear rate can be determined. These *viscometric functions* are the viscosity and the first and second normal stress differences, which were first presented in Chap. 2. Stress growth data have rarely been used to characterize molten polymers [20].

There is also a *shear stress decay coefficient* $\eta^-(t,\dot\gamma)$ that describes stress relaxation following the cessation of steady simple shear at a time $t = 0$, but it has not been found useful for polymer characterization. An interesting variant of start-up and cessation of steady shear is *interrupted shear*, which is used to monitor

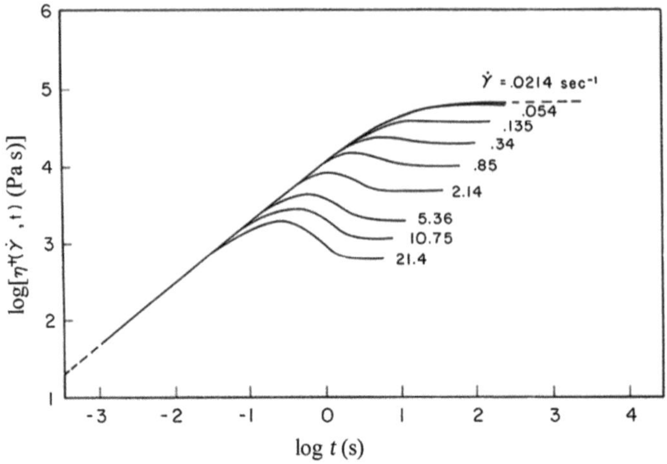

Fig. 4.5 Shear stress start-up function for an entangled polystyrene solution at a number of shear rates. The data at the lowest shear rate are in accord with Eq. (4.13) for linear behavior, and as the shear rate increases, the data start out along the linear curve, exhibit an overshoot, and then descend to a steady value, which is the viscosity at that shear rate

relaxation processes that occur at very long times. The time at which data can no longer be obtained in a relaxation experiment is governed by the sensitivity of the stress or strain transducer used to monitor the process. In order to obtain information about relaxation that occurs after the stress signal has fallen into the noise level, a series of start-up and cessation tests are performed with increasing *rest time*, i.e., the time between one cessation and the next start-up of flow, t_r. The start-up flow is allowed to continue until the steady-state stress σ_{ss} is reached. In the nonlinear regime, there will be a peak stress $\sigma_m(t_r)$ that is a function of the rest time t_r. This stress will have its maximum value $\sigma_m(\infty)$ during the first test, which takes place when the sample has not been previously sheared, and unless the rest time is long, the maximum stress in the second test, started at a time t_r after the first ended, will be smaller. In the limit of zero rest time, there will be no overshoot, so that the stress will return immediately to its steady-state value, i.e., $\sigma_{ss} = \sigma_m(0)$. Only after a very long rest time is the initial maximum stress recovered. There have been several reports of the use of this technique to characterize polymeric liquids [21–24].

4.6.2 Nonlinear Creep

If a creep experiment is carried out at a constant stress sufficiently large that the sample is strained into the nonlinear regime before achieving a steady strain rate, the compliance function will be a function of stress as well as time. Data from this type of experiment have been reported by Agarwal and Plazek [25], but creep has not been widely used for the study of nonlinear viscoelasticity.

4.6.3 Large-Amplitude Oscillatory Shear

In large amplitude oscillatory (LAOS) the shear strain is sinusoidal, as in the measurement of the storage and loss moduli, but the strain amplitude is sufficiently large that the response is not governed by the Boltzmann superposition principle. As a result the stress is not sinusoidal and cannot be interpreted in terms of one in-phase and one out-of-phase component. Large-amplitude oscillatory shear has the interesting feature that the Weissenberg and Deborah numbers can be varied independently. The Weissenberg number is $\dot{\gamma}_o\tau$, where τ is a time describing the first onset of nonlinearity as the frequency increases from zero, and the Deborah number is $\omega\tau$, where τ is a relaxation time. Hyun et al. [26] have provided a thorough review of the use of this strain history, and Giacomin et al. [27] have summarized published analytical predictions for a number of constitutive equations for LAOS. In using SAOS for linear viscoelastic measurements, one can oscillate either the strain or the stress and obtain the storage and loss moduli. However, because LAOS operates in the nonlinear regime, the two types of deformation do not yield the same information and cannot be converted one to the other.

If the material is isotropic in its rest state and has no yield stress, continuum mechanics considerations imply that the stress can be represented as a Fourier series with only odd terms.

$$\sigma(t) = \sum_{\mathrm{n=1,odd}} \sigma_{\mathrm{n}} \sin(n\omega\, t + \delta_{\mathrm{n}}) \tag{4.14}$$

Krieger and Niu [28] developed a method for signal averaging to determine the Fourier components by high-sensitivity Fourier transform, and Wilhelm et al. [29] have made extensive use of this technique to characterize molten polymers.

An interesting way to display LAOS data is as a plot of stress versus strain or strain rate. These are closed loops with skew symmetry. For rubbers in the linear regime, stress versus strain is preferred and the loops are ellipses with axes oriented in the x and y directions. For melts in the linear regime, the loops are ellipses with principal axes at an angle with the x and y directions; this angle corresponds to the loss angle. For nonlinear behavior, the loops are no longer ellipses but retain symmetry properties. Figure 4.6 shows such loops for a polyethylene (top) and a polystyrene (bottom) [30]. The strain rate amplitude for the upper left plot was within the linear range, and the loop is an ellipse. For the PS, the amplitude for the left hand plot was somewhat above the limit of linear viscoelasticity, and the loop is pointed at the ends. The first normal stress difference is always positive and its loop is symmetric about the vertical axis. The same is true for $N_3 \equiv N_1 + N_3$, and Fig. 4.7 shows an N_3 loop for an entangled polystyrene solution [31].

Each component of the series given by (4.14) can be rewritten as the sum of an in-phase and out-of-phase component as shown by Eq. (4.15).

$$\sigma(t) = \gamma_0 \sum_{n,\mathrm{odd}} \left[G'_n(\omega, \gamma_0)\sin(n\omega t) + G''_n(\omega, \gamma_0)\cos(n\omega t)\right] \tag{4.15}$$

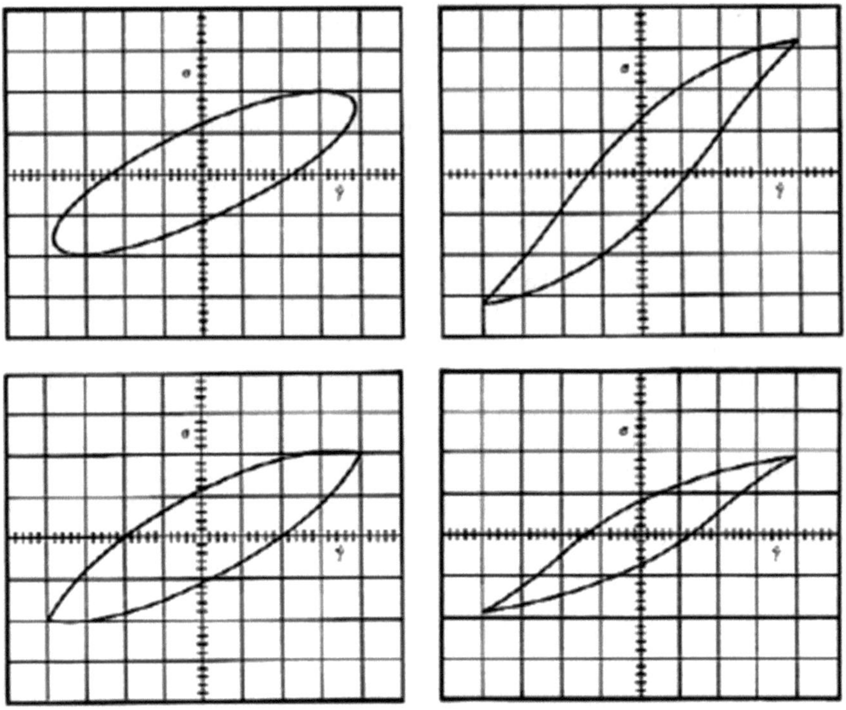

Fig. 4.6 Shear stress versus shear rate loops for a polyethylene (*top*) and a polystyrene (*bottom*). The strain-rate amplitude for the upper left is within the linear regime, and the loop is an ellipse. For the polystyrene, both loops are in the nonlinear regime, but the strain-rate amplitude is larger for the loop on the right. From Tee and Dealy [30]

Fig. 4.7 Third normal stress difference ($N_1 + N_2$) versus shear rate loop for an entangled polystyrene solution. Like N_1 this quantity is always positive, because N_2 is negative but smaller in magnitude. From Reimers and Dealy [31]

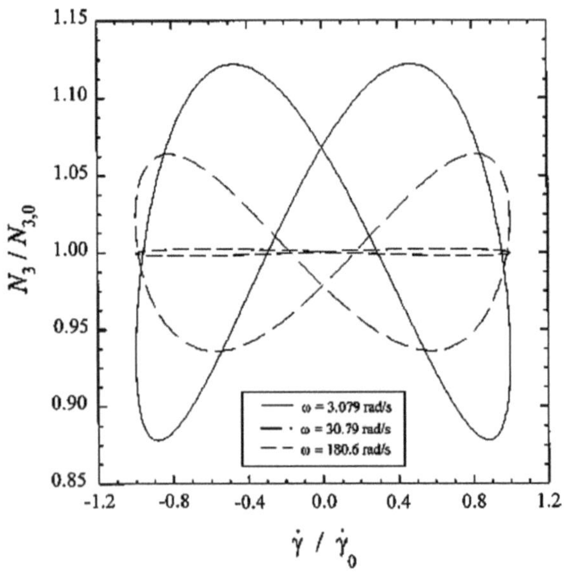

However, the terms no longer have the simple mechanical significance that they had for linear viscoelasticity. For example, all the energy dissipation is associated with the principal harmonic ($n = 1$).

The following power series has also been used to represent LAOS data:

$$\sigma(t) = \sum_{p,\text{odd}} \sum_{q.\text{odd}}^{p} \gamma_0^{\text{q}} \left[G_{\text{pq}} \sin q\omega t + G_{\text{pq}} \cos q\omega t \right] \tag{4.16}$$

For example, Pearson and Rochefort [32] developed series expansions for the fundamental frequency components of the torque for parallel disk fixtures in terms of the parameters of (4.16) and also expansions of the G_{pq} coefficients for the Doi-Edwards model. The terms of these series expansions are not useful for relating LAOS data to molecular structure.

Yet another approach involves the use of Chebyshev polynomials [33]. In this representation, time no longer appears as the independent variable and is replaced by the strain and strain rate.

It has been reported that LAOS data provide information about structural features of a polymer that cannot be inferred from the storage and loss moduli of linear viscoelasticity [26]. For example it has been suggested that long-chain branching can be inferred from the higher harmonics, although there is at present no theoretical basis for such a correlation, and polydispersity may blur the relationship. To facilitate data analysis the relative values of the stress amplitudes σ_{n} in Eq. (4.15) are described using the terminology $I_{\text{m/n}}$, defined as $\sigma_{\text{m}}/\sigma_{\text{n}}$. A quantity that has been used to correlate LAOS data with structure is the "nonlinearity parameter" Q, defined as follows:

$$Q(\omega, \gamma_0) \equiv \frac{I_{3/2}}{\gamma_0^2} \tag{4.17}$$

The present authors believe that the most useful rheological source of structural information is small-amplitude oscillatory shear data, especially at low frequencies, supplemented by uniaxial extensional data for long-chain branched polymers.

4.7 Extensional Flow Behavior: Introduction

Most laboratory tests of melt flow behavior involve shearing, and we will see in Chap. 7 that linear viscoelastic behavior is a rich source of information about molecular structure. However, no matter how many linear or nonlinear material functions we determine in shear, such information cannot be used to predict the response of a melt to extensional deformations, unless the strain rate is extremely low. Nonlinear behavior in extension provides information about long-chain branching that is not revealed by shear data. In addition, extensional flows play an important role in several melt forming processes.

As in the case of shear flow, the preferred test modes for the study of stretching deformations are those that generate a homogeneous strain. The shear free flow that has been most used in experimental rheology is uniaxial extension, also called simple extension, which is an axisymmetric flow with stretching in the direction of the axis of symmetry. The effect of uniaxial extension on the shape of an initially cubic fluid element is sketched in Fig. 4.8. Another axisymmetric extensional flow is biaxial extension, which can also be thought of as axial compression. In this deformation there is compression along the axis of symmetry and stretching in the radial direction, as shown in Fig. 4.9. Among non-symmetric stretching flows, the simplest is planar extension, in which the velocity in one direction is zero, as shown in Fig. 4.10. This type of flow can be approximated, for example, by stretching a thin-walled tube in its axial direction while providing sufficient internal pressure to keep the diameter constant.

In uniaxial extension at significant rates molecules tend to be oriented in the axial direction, while in biaxial extension the tendency is to align molecules in the radial direction, and this implies a symmetrical distribution of alignments in planes perpendicular to the axis of symmetry. In other words, within one of these planes, there is no overall preferred orientation in any one direction. Planar extension is intermediate between these two extremes, with a tendency to orient molecules

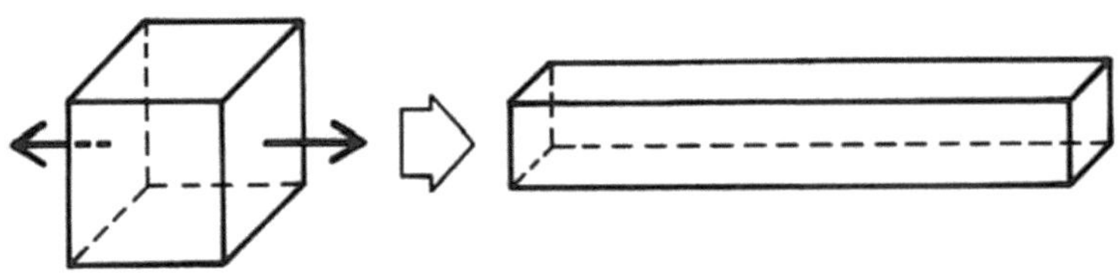

Fig. 4.8 Uniaxial (simple) extension acting on a cubical fluid element

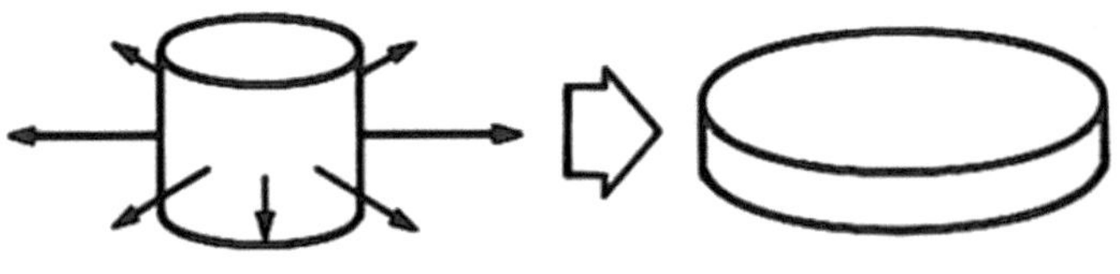

Fig. 4.9 Biaxial (equibiaxial) extension acting on a cylindrical fluid element

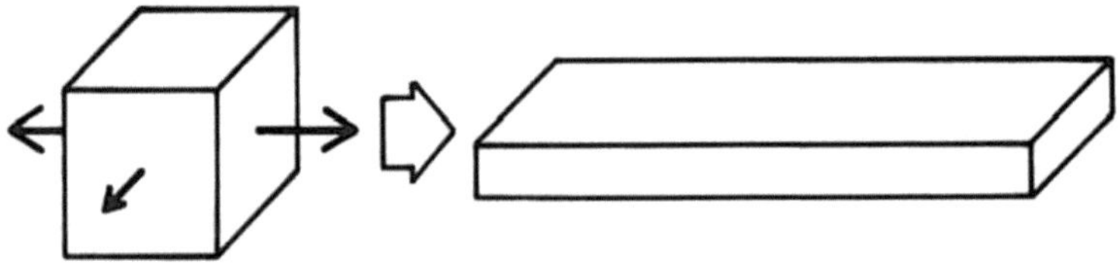

Fig. 4.10 Planar extension acting on a cubical fluid element. The tension in the lateral direction is just sufficient to keep the sample from shrinking in that direction

mainly parallel to the direction of stress and to a lesser degree in the transverse direction. Uniaxial extension is the deformation for which strain hardening is most apparent and is the one most sensitive to long-chain branching.

Nearly all melt processing operations involve some kind of extensional flow, as there is always stretching along streamlines at one or more stages of the process. For example, whenever there is a converging flow, as in a die, there is acceleration, and this implies the presence of uniaxial or planar extension, depending on the flow geometry. In a diverging channel, there will be extension in directions normal to the main flow direction implying biaxial extension. Converging and diverging die flow also have important shearing components, and extensional flow properties are not as important. It is in extrudate flows with tensile stresses that these properties play an important role. These include film blowing, sheet extrusion and fiber spinning.

Uniaxial extension is an axisymmetric deformation in which a tensile stress is applied in one direction, which we will call it the z-direction, while the free surfaces of the sample are under a uniform normal stress, the ambient pressure. The quantity measured is the net tensile stress σ_E defined as $(\sigma_{zz} - \sigma_{rr})$, which is the applied axial stress minus that acting on the free surfaces, which is often negligible.

Step strain in extension is not practical for melts, and the experiment usually carried out is start-up of steady simple extension at constant *Hencky strain rate* $\dot{\varepsilon}$. The Hencky strain rate can be defined in terms of the length L of the sample as shown by Eq. (4.18).

$$\dot{\varepsilon} = \mathrm{d}\ln L/\mathrm{d}t \tag{4.18}$$

This strain rate is a measure of the speed with which material elements on a streamline are separated from each other. Note that both the velocity of the end of the sample and the sample length increase exponentially with time at constant Hencky strain rate.

$$L(t) = L(0)\exp(\dot{\varepsilon}t) \tag{4.19}$$

But at a fixed distance z from the $z = 0$ plane where $v_z = 0$, the velocity is constant and proportional to the Hencky strain rate:

$$v_z = \dot{\varepsilon}z \tag{4.20}$$

The material function usually reported is the *tensile stress growth coefficient* defined by Eq. (4.21).

$$\eta_E^+(t, \dot{\varepsilon}) \equiv \sigma_E(t, \dot{\varepsilon})/\dot{\varepsilon} \tag{4.21}$$

The Boltzmann superposition principle can be used to derive the linear response, which is exhibited at sufficiently small strains or strain rates:

$$\lim_{\dot{\varepsilon}\to 0} \eta_E^+(t, \dot{\varepsilon}) = \eta_E^+(t) = 3\int_0^t G(s)ds = 3\eta^+(t) \tag{4.22}$$

The linear relaxation modulus *G*(*t*) can be inferred from oscillatory shear data, and it is useful to compare extensional flow data with the behavior predicted by Eq. (4.22) as is explained below.

If steady extensional flow is achieved, the *extensional viscosity* can be determined:

$$\lim_{t\to\infty} \eta_E^+(t,\dot{\varepsilon}) \equiv \eta_E(\dot{\varepsilon}) \tag{4.23}$$

and in the limit of vanishing strain rate, the extensional viscosity becomes simply three times the zero-shear viscosity:

$$\lim_{\dot{\varepsilon}\to 0} \eta_E(\dot{\varepsilon}) = 3\int_0^\infty G(s)\,ds = 3\eta_0 \tag{4.24}$$

This limiting case describes the behavior of a Newtonian fluid, $\eta_\mathrm{E} = 3\eta_0$. The extensional flow behavior of Newtonian fluids was first noted by Trouton, and the quantity $\eta_E(\dot{\varepsilon})/3\eta_0$ is sometimes referred to as the *Trouton ratio* and used to normalize extensional viscosity data. However, this term is also used for several other ratios, including $\eta(\dot{\varepsilon})/3\eta_0$; $\eta(\dot{\varepsilon})/\eta(\dot{\gamma}=\dot{\varepsilon})$ and $\eta(\dot{\varepsilon})/\eta(\dot{\gamma}=\sqrt{3}\dot{\varepsilon})$, the last of which compares the extensional and shear viscosities at equal values of the second invariant of the rate of strain tensor. Such normalizations do not reveal anything of basic interest.

There are three possible types of behavior for the extensional viscosity function. It can exhibit the linear behavior shown by Eq. (4.24); it can exhibit *extension thickening* (η_E increasing with $\dot{\varepsilon}$); or it can exhibit *extension thinning* (η_E decreasing with $\dot{\varepsilon}$). It is important not to confuse this classification scheme with that used to describe the transient response to the start-up of steady simple extension.

Because of the strong tendency of melts to neck and rupture at high stretch rates, one does not usually observe a true plateau in $\eta_E^+(t,\dot{\varepsilon})$ that can be used to determine $\eta_E(\dot{\varepsilon})$, and instead there is a maximum where necking begins. For many years it was thought that this maximum could be taken to be $\eta_E(\dot{\varepsilon})$. However, more recent observations [34–36] suggest that for highly-branched melts, particularly LDPE, the maximum is not a steady state but is followed by a true plateau at a lower stress. As a result, previously reported extensional viscosity data for LDPE should be viewed with caution. Figure 4.11 shows data obtained by Huang et al. [34] for an LDPE using a filament stretching rheometer (FSR). There is very strong strain hardening with a maximum followed at some strain rates by an apparent approach to a steady value. Since the FSR is not suitable for routine use, for industrial applications one should rely on the transient function $\eta_E^+(t,\dot{\varepsilon})$, which can be determined using the easy-to-use devices described in Chap. 6.

In reporting results of extensional flow tests, it is common practice to compare the nonlinear material function $\eta_E^+(t,\dot{\varepsilon})$ with the linear response given by Eq. (4.22). If both data sets are accurate, the nonlinear response should agree with the linear one at short times and low strain rates. Furthermore, the way in which the

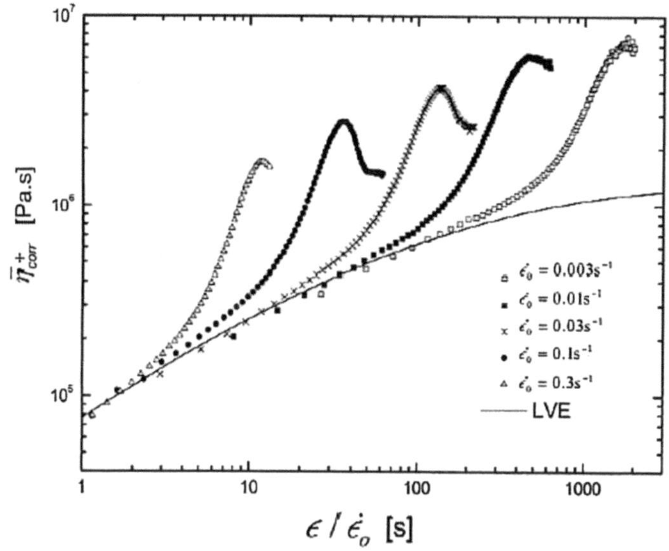

Fig. 4.11 Tensile stress growth coefficient for a LDPE. After a maximum the data fall to a plateau value equal to the extensional viscosity. From Huang et al. [34]

nonlinear response departs from the linear one can be used to classify extensional flow behavior. If the nonlinear data rise above the linear curve at some point, the melt is said to be *strain hardening*, and if they fall below, it is said to be *strain softening*. Figure 4.12 shows the data of Münstedt and Laun [37] for a low-density polyethylene made in a tubular reactor, and we see that this material exhibits strong

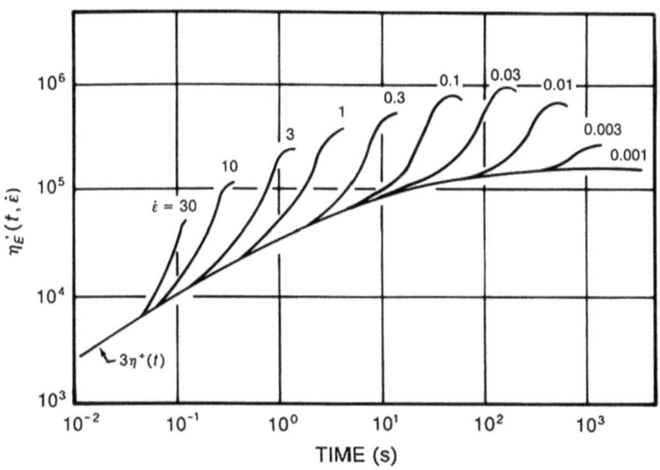

Fig. 4.12 Tensile stress growth coefficient for a LDPE at many strain rates. At the lowest strain rate the data follow the curve for linear behavior, which is given by Eq. (4.22). From Münstedt and Laun [37]

strain hardening. Linear polymers containing no very high-molecular-weight components exhibit strain softening. This classification system has inspired the use of a *reduced* stress growth coefficient [38] $\eta_{ER}^{+}(t, \dot{\varepsilon})$, or *degree of strain hardening* χ [39], defined as follows:

$$\eta_{ER}^{+}(t, \dot{\varepsilon}) \equiv \frac{\eta_{E}^{+}(t, \dot{\varepsilon})}{3\eta^{+}(t)} \tag{4.25}$$

When there is strain hardening, the value of this function is greater than one and vice versa. The use of this function to compare data for several polymers shows the comparative strain hardening normalized with respect to the effect of the linear relaxation spectrum. Wood-Adams et al. [38] used this function to compare data for a series of metallocene polyethylene homopolymers having very low levels of long-chain branching and found that once the effect of the linear spectrum was eliminated in this way, the differences between the samples became insignificant except at the longest times.

When comparing extensional flow data with the LVE prediction, care must be taken in the calculation of the linear prediction. If the data used to establish the relaxation spectrum do not include very short-time (high-frequency) data, the initial portion of the curve will not be correct. It may thus be better to use data from start-up of steady simple shear to measure $\eta^{+}(t)$ directly rather than inferring it from complex modulus data.

4.8 Extensional Flow Characterization of Melts

4.8.1 Linear Polymers

Tube models predict that chain stretch should occur in an entangled, linear, monodisperse melt when the rate of deformation is greater than $1/\tau_R$ and that τ_R depends on the ratio (M_C/M). For polyolefins, monodisperse samples cannot be made, and in addition they have small values of M_C, so an extremely high value of M would be required to generate significant chain stretch at a practically achievable strain rate. For polystyrene, on the other hand, M_C and τ_R are much larger, and it is possible to generate chain stretch and thus strain hardening in a practical experiment. Studies of linear, high-MW polystyrene melts [40] using a filament-stretching rheometer have, indeed, revealed marked strain hardening at high strain rates. While strain hardening is not expected to be observed in polydisperse polymers, there have been reports that a small amount of relatively high molecular weight material in polystyrenes [41–43] and polyolefins [44, 45] can lead to strain hardening. This phenomenon is discussed in more detail in Chap. 7.

4.8.2 Long-Chain Branched Polymers with Known Structures

Star polymers do not exhibit strain hardening [46], and the presence of inner chain segments, i.e., "backbones" with branch points at both ends, are believed to be essential to produce this effect. Data on a series of long-chain branched metallocene polyethylenes supporting this concept are presented in Chap. 7.

4.8.3 Randomly Branched Polymers and LDPE

Linear polymers that are strain-softening can be made strongly strain-hardening by the introduction of long-chain branches. This has been demonstrated by treating linear polymers by electron radiation [47] or crosslinking agents [48, 49]. Hingmann and Marczinke [48] estimated that fewer than three branches per molecule are required in the polypropylene they studied to produce pronounced strain hardening.

Gabriel and Münstedt [50] compared polyolefins having various branching structures and concluded that both the number of branches per molecule and the number of entanglements per branch play crucial roles in strain hardening. They believe that if a polyethylene exhibits strain hardening but has a zero-shear viscosity dependence on M_w that is typical of linear polymers (Eq. 2.19), it is very likely a linear polymer containing some very high-molecular-weight material.

The structure and rheology of low-density polyethylene (LDPE) is of practical interest because of its importance as a commercial polymer, especially in film-blowing and extrusion-coating. It is the most strain hardening in extension of all commercial polymers, and this endows it with superior melt strength. This can be seen in Fig. 4.12 which shows data for an LDPE made in a tubular reactor [37]. The difficulties that arise in measuring the extensional viscosity defined by Eq. (4.23) were explained in Sect. 4.7, while the relationship between the structure and rheology of LDPE is discussed in detail in Chap. 7.

4.9 Shear Modification

It was observed many years ago that after high-rate shearing of LDPE its rheological behavior is altered for a considerable time thereafter [51], a phenomenon called *shear modification*. For example if the melt is extruded and immediately converted to pellets, when these pellets are remelted the properties of the melt are different from those of the original polymer. But if the re-melted polymer stands in its molten state for a sufficiently long time, or if it is dissolved in a solvent and re-precipitated, it regains its original properties. Shearing in a mixer for a period of one hour was observed to reduce the viscosity, first normal stress difference, and extrudate swell by about thirty percent [52]. The effect of shearing on extensional flow behavior is particularly strong. Leblans and Bastiaansen [53] found that five passes through a twin-screw extruder substantially decreased the tendency for strain hardening and reduced the steady-state extensional viscosity by a factor of three. It has been hypothesized that this phenomenon results from the alignment of the long branches along the backbone. In terms of the tube picture, the branches are drawn into the tube of the backbone during deformation, and a long time is required for the complex molecule to recover its equilibrium configuration. Leblans and Bastiaansen [53] found that the Doi-Edwards strain measure provided a good description of the extensional behavior of sheared LDPE, making its response similar to that of linear HDPE. This supports the idea that the long branches are drawn into the tube with the backbone by pre-shearing. Bourrigaud et al. [54] modified the tube model for a pom-pom polymer to simulate this phenomenon. Their results indicated that strain hardening can be eliminated by sufficient shear modification and that extensional flow itself should have a much stronger modifying effect than shear.

Yamaguchi and Gogos [55] observed a strong effect of preshearing on melt strength. Yamaguchi et al. [56] compared several shearing devices and reported that the continuous shearing that occurs in a twin-screw extruder with conveying screws or in an internal batch mixer is much more effective for shear modification than that in a two-roll mill or an extruder equipped with kneading blocks. Yamaguchi and Takahashi [57] found that the melt strength of autoclave LDPE is significantly more sensitive to shear history than that produced in a tubular reactor.

The practical application of shear modification in a plastics forming operation, for example to improve the surface of blown film, is not practical because of the substantial time and energy required to perform the shear modification.

4.10 Time-Temperature Superposition of Nonlinear Properties

Graessley [58, p. 239] provides a discussion of the application of time-temperature superposition to nonlinear properties, and we give here only a brief summary. The viscosity function is often found to obey time-temperature superposition. If the vertical shift factor is neglected, the zero-shear viscosity can be used as the horizontal shift factor to obtain a master curve by plotting

$$\frac{\eta(T,\dot{\gamma})}{\eta_o(T)} \quad \text{versus} \quad \eta_0(T)\dot{\gamma} \tag{4.27}$$

The other viscometric functions can also be treated in this manner. But viscometric functions are not functions of time; what about time-dependent properties? Constitutive equations for melts almost universally build on the same spectra

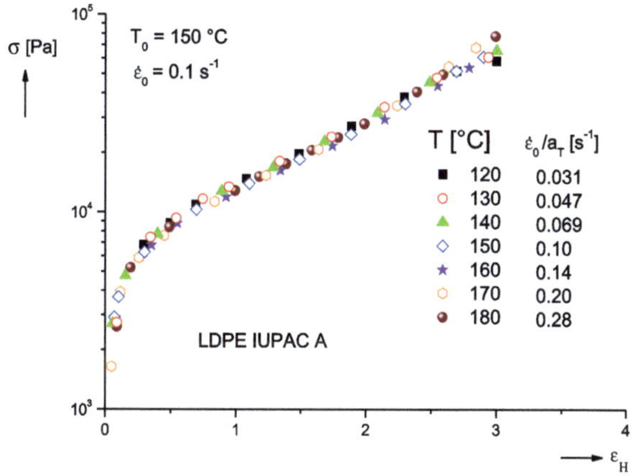

Fig. 4.13 Time-temperature superposition applied to data for tensile stress growth coefficient. From Münstedt and Laun [37]

of relaxation times that are used to describe linear viscoelasticity. This implies that if time-temperature superposition brings linear data together, it should also work for nonlinear properties using the same shift factors. The fundamental assumption is that the relaxation times are independent of the state of stress. For example, a temperature-invariant representation of the shear stress-growth coefficient is based on the following relationship:

$$\frac{b_T}{a_T}\eta^+(t/a_T, \dot{\gamma}a_T, T) = \eta^+(t, \dot{\gamma}, T_0) \tag{4.28}$$

Münstedt and Laun [37] applied time-temperature superposition to extensional stress-growth data for an LDPE, and their excellent result is shown in Fig. 4.13.

References

1. Doi M, Edwards SF (1986) The theory of polymer dynamics. Oxford University Press, New York
2. Larson RG (1984) A constitutive equation for polymer melts based on partially extending strand convection. J Rheol 28:545–571
3. Osaki K, Nishizawak K, Kurata M (1982) Material time constant characterization of nonlinear viscoelasticity of entangled polymeric solutions. Macromol 15:1068–1071
4. Larson RG (1988) Constitutive equations for polymer melts and solutions. Butterworth-Heinemann, Boston
5. Wagner MH, Rubio P, Bastian H (2001) The molecular stress function model for polydisperse polymers melts with dissipative convective constraint release. J Rheol 45:1387–1412
6. Urakawa O, Takahashi M, Masuda T, Ebrahimi NG (1995) Damping functions and chain relaxation in uniaxial and biaxial extensions: comparison with the Doi-Edwards theory. Macromol 28:7196–7201
7. Yamaguchi M, Takashi M (2001) Rheological properties of low-density polyethylenes produced by tubular and vessel processes. Polymer 42:8663–8670

8. Islam MT, Archer LA, Varshney SK (2001) Linear rheology of entangled six-arm and eight-arm polybutadienes. Macromol 34:6438–6449
9. McLeish TCB et al (1999) Dynamics of entangled H-polymers: theory, rheology, and neutron scattering. Macromol 32:6734–6758
10. Kasehagen LJ, Macosko CW (1998) Nonlinear shear and extensional rheology of long-chain randomly branched polybutadiene. J Rheol 42:1303–1327
11. Morrison FA, Larson RG (1992) A study of shear-stress relaxation anomalies in binary mixtures of monodisperse polystyrenes. J Polym Sci Phys Ed 30:943–950
12. Sanchez-Reyes J, Archer LA (2002) Step shear dynamics of entangled polymer liquids. Macromol 35:5194–5202
13. Mills NJ (1969) The rheological properties of PDMS. Eur Poly J 5:675–695
14. Lodge AS, Meissner J (1972) On the use of instantaneous strains to test the Gaussian network hypothesis. Rheol Acta 11:351–352
15. Lodge AS (1978) Elastic liquids. Academic Press, New York, chap 6
16. Laun HM (1978) Description of the non-linear shear behavior of a low-density polyethylene melt. Rheol Acta 17:1–15
17. Brown EF, Burghardt WR, Kahvand H, Venerus DC (1995) Comparison of optical and mechanical measurements of second normal stress difference relaxation following step strain. Rheol Acta 34:221–234
18. Ianniruberto G, Marrucci G (2002) A multi-mode CCR model for entangled polymers with chain stretch. J Non-Newt Fl Mech 102:383–395
19. Menezes EV, Graessley WW (1980) Study of the nonlinear response of a polymer solution to various uniaxial shear flow histories. Rheol Acta 19:38–50
20. Schweizer T, van Meerveld J, Öttinger HC (2004) Nonlinear shear rheology of polystyrene melt with narrow molecular weight distribution–Experiment and theory. J Rheol 48: 1345–1363
21. Stratton RA, Butcher AF (1973) Stess relaxation upon cessation of steady flow and the overshoot effect of polymer solutions. J Polym Sci Polym Phys Ed 11:1747–1758
22. Tsang WKW, Dealy JM (1981) The use of large transient deformations to evaluate rheological models for molten polymers. J Non-Newt Fl Mech 9:203–222
23. Seay CW, Baird DG (2004) The use of interrupted stress growth in assessing branching structure in metallocene catalyzed polyethylene. In Proceedings of the 14th International Congress Rheology paper MS10, The Korean Society of Rheology, Seoul
24. Robertson CG, Warren S, Plazek DJ, Roland CM (2004) Re-entanglement kinetics in sheared polybutadiene solutions. Macromol 37:10018–10022
25. Agarwal PK, Plazek D (1977) Shear creep recovery behavior of IUPAC low-density polyethylenes. J Appl Polym Sci 21:3251–3260
26. Hyun K, Wilhelm M, Klein CO, Cho KS, Nam JG, Ahn KH, Lee SJ, Ewoldt RH, McKinley GH (2011) A review of nonlinear oscillatory shear tests: Analysis and application of large amplitude oscillatory shear (LAOS). Prog Polym Sci 36:1697–1753
27. Giacomin AJ, Bird RB, Johnson LM, Mix AW (2011) Large-amplitude oscillatory shear from the corotational Maxwell model. J Non-Newt Fl Mech 166:1081–1099
28. Krieger IM, Niu TF (1973) A rheometer for oscillatory studies of nonlinear fluids. Rheol Acta 12:567–571
29. Wilhelm M, Reinheimer P, Ortseifer M (1999) High sensitivity fourier-transform rheology. Rheol Acta 38:349–356
30. Tee TT, Dealy JM (1975) Nonlinear viscoelasticity of molten polymers. Trans Soc Rheol 19:595–615
31. Reimers M, Dealy JM (1998) Sliding plate rheometer studies of concentrated polystyrene solutions: nonlinear viscoelasticity and wall slip of two high molecular weight polymers in tricresyl phosphate. J Rheol 42:527–548
32. Pearson DS, Rochefort WE (1982) Behavior of concentrated polystyrene solutions in large-amplitude oscillating shear fields. J Polym Sci 20:83–98

33. Ewoldt RH, Hosoi AE, McKinley GH (2008) New measures for characterizing nonlinear viscoelasticity in large amplitude shear. J Rheol 52:1427–1428
34. Huang Q, Skov AL, Rasmussen HK, Hassager O, Harlen O, Hoyle DM, McLeish TC, Hassel D, Lord TD, Mackley MR (2011) Stress maximum and steady extensional flow of branched polymer melts. Society of Rheology 83rd Ann Mtg, Cleveland, Ohio, October 2011
35. Rasmussen HK, Nielsen JK, Bach A, Hassager O (2005) Viscosity overshoot in the start-up uniaxial elongation of low density polyethylene melts. J Rheol 49:369–381
36. Burghelea TI, Starý Z, Münstedt H (2011) On the 'viscosity overshoot' during uniaxial extension of a low density polyethylene. J Non-Newt Fl Mech 166:1198–1209
37. Münstedt H, Laun HM (1979) Elongational behavior of a low density polyethylene melt II. Transient behavior in constant stretching rate and tensile creep experiments. Rheol Acta 18:492–504
38. Wood-Adams PM, Dealy JM, deGroot AW, Redwine OD (2000) Effect of molecular structure on the linear viscoelastic behavior of polyethylene. Macromol 33:7489–7499
39. Vega JF, Fernández M, Santamaráa A, Muñoz-Escalona, Lafuente P (1999) Rheological criteria to characterize metallocene catalyzed polyethylenes. Macromol Chem Phys 200:2257–2268
40. Bach A, Almdal K, Rasmussen HK, Hassager O (2003) Elongational viscosity of narrow molar mass distribution polystyrene. Macromol 36:5174–5179
41. Münstedt H (1980) Dependence of the elongational behavior of polystyrene melts on molecular weight and molecular weight distribution. J Rheol 24:847–867
42. Frank A, Meissner J (1984) The influence of blending polystyrenes of narrow molecular weight distribution on melt creep flow and creep recovery in elongation. Rheol Acta 23:117–123
43. Minegishi A, Nishioka A, Takahashi T, Masubuchi Y, Takimoto J, Koyama K (2001) Uniaxial elongational viscosity of PS/a small amount of UHMW-PS blends. Rheol Acta 40:329–338
44. Linster JJ, Meissner J (1996) Melt elongation and structure of linear polyethylene (HDPE). Polym Bull 16:187–194
45. Sugimoto M, Masubuchi Y, Takimoto J, Koyama K (2001) Melt rheology of polypropylene containing small amounts of high-molecular weight chain. 2. Uniaxial and biaxial extensional flow. Macromol 34:6056–6063
46. Ye X, Sridhar T (2001) Shear and extensional properties of three-arm polystyrene solutions. Macromol 34:8270–8277
47. Kurzbeck C, Oster F, Münstedt H, Nguyen TQ, Gensler R (1999) Rheological properties of two polypropylenes with different molecular structure. J Rheol 43:359–374
48. Hingmann R, Marczinke BL (1994) Shear and elongational flow properties of polypropylene melts. J Rheol 38:573–587
49. Kasehagen LJ, Macosko CW (1998) Non-linear shear and extensional rheology of long-chain randomly branched polybutadiene. J Rheol 42:1303–1327
50. Gabriel C, Munstedt H (2003) Strain hardening of various polyolefins in uniaxial elongational flow. J Rheol 47:619–630
51. Hanson DE (1969) Shear modification of polyethylene. Polym Eng Sci 9:405–414
52. Rokudai M, Mihara S, Fujiki TJ (1979) Influence of shearing history on the rheological properties and processability of branched polymers. J Appl Polym Sci 32:463–471
53. Leblans PJR, Bastiaansen C (1989) Shear modification of low-density polyethylene: its origin and its effect on the basic rheological functions of the melt. Macromol 22:3312–3317
54. Bourrigaud S, Marin G, Poitou A (2003) Shear modification of long-chain branched polymers: a theoretical approach using the pom-pom model. Macromol 36:1388–1394
55. Yamaguchi M, Gogos CG (2001) Quantitative relation between shear history and rheological properties of LDPE. Adv Polym Technol 20:261–269
56. Yamaguchi M, Todd D, Gogos CG (2003) Rheological properties of LDPE processed by conventional processing machines. Adv Polym Technol 22(3):179–187
57. Yamaguchi M, Takashi M (2001) Rheological properties of low-density polyethylenes produced by tubular and vessel processes. Polymer 42:8663–8670
58. Graessley WW (2008) Polymeric liquids and networks: dynamics and rheology. Garland Science, New York

Chapter 5
Nonlinear Viscoelasticity: Models

Abstract This chapter deals with the theory and modeling of nonlinear behavior. The study of its contents is not essential for an understanding of the other chapters, and it will be of interest primarily to those involved in the modeling of melt flow processes in which viscoelasticity plays an important role, such as blow molding and film blowing. Tensor representations of stress and strain are introduced and used to present several models of nonlinear behavior.

5.1 Introduction

This chapter will be of limited interest to most engineers and scientists working in the plastics industry. It is certainly true that the deformations that occur in melt processing operations are never very small or very slow, and the rheological behavior is thus nonlinear. Internal flows dominated by shear deformation are governed by the viscometric functions, which were discussed in Chap. 2. Flows of this type occur, for example, in extruders, channels, and the filling of a mold. Thus, while such deformations are in the nonlinear regime, elasticity does not play a central role. On the other hand, elasticity plays an important role in processes involving free surfaces such as film blowing, profile extrusion, and blow molding. At the present time, the development of models for such processes is limited by the availability of constitutive equations that are convenient for computation but also make reliable predictions for large, rapid deformations. In this chapter we present the basic concepts involved in the formulation of models of nonlinear behavior. A much more detailed treatment of this subject can be found in books by Larson [1] and by Larson and Dealy [2].

The Boltzmann principle indicates that knowledge of the linear relaxation modulus $G(t)$ is sufficient to calculate the stress for any very slow or very small shearing deformation. And in this chapter we will develop a tensorial form of this principle that can be used to determine the stress resulting from non-shear

J. M. Dealy and J. Wang, *Melt Rheology and its Applications in the Plastics Industry*, Engineering Materials and Processes, DOI: 10.1007/978-94-007-6395-1_5,

deformations that are very small or very slow. Then we explore methods for describing the stress arising from deformations that are large and rapid, i.e., for describing nonlinear viscoelastic behavior. A model that relates the stress tensor to the deformation tensor for any type of deformation is called a *constitutive equation* or *rheological equation of state*.

For many years, equations of state were empirical in nature, based on general continuum mechanics principles regarding acceptable ways of describing large, rapid deformations. These contain empirical constants or functions that can only be determined by fitting experimental data, and this greatly limits the range of application of the model, because laboratory techniques used to explore nonlinear behavior are quite limited in terms of kinematics and deformation rates.

More recently, mean field models, mostly based on a tube or slip-link picture, have shown promise in the modeling of nonlinear phenomena. However, in replacing the detailed modeling of individual molecular interactions by a much larger scale view of a molecule that interacts with its neighbors only at specific points along the chain, empiricism is still present but transferred to a mesoscopic scale. This has the advantage that the empiricism is based on a somewhat detailed picture of how molecules interact rather than on arbitrary functions. But to derive constitutive equations from mean field theories, additional assumptions are required, so that the final model is rather far removed from the molecular picture from which it arose.

The more complex the constitutive equation, the less useful it is for flow simulation. Simple integral and differential equations are required for modeling complex flows, and even if these start from molecular dynamics concepts, simplifying assumptions are required to arrive at such equations.

5.2 Tensor Notation

The description of large, rapid deformations of viscoelastic materials requires the use of tensors. For those readers who have had little if any experience in the use of tensor analysis, the very word tensor may suggest a mathematical system of impenetrable mystery. However, such readers should have no fear. While we do not offer here a complete course in tensor analysis, we do present in the next two brief sections everything you will need to know about tensors to understand the remainder of this chapter.

Certain physical quantities, such as force and velocity, are specified in terms of vectors, which have both a magnitude and a direction. For example, the velocity vector is specified by giving its components, v_1, v_2, and v_3 in the directions, x_1, x_2, and x_3. We will refer to a typical velocity component as v_i, where i can be 1, 2, or 3. While the magnitude is a physical attribute of a vector that does not depend on the choice of a particular coordinate system, the components of the vector do depend on the coordinate system selected to describe the flow. There is a simple rule that tells us how to use the components of a vector in one coordinate system to calculate the components of that vector in a second coordinate system that is rotated with respect to the first. If vectors are adequate to describe the velocity of a body and the force acting on it, why are they not sufficient for describing rheological phenomena? The answer is that rheology deals not with motion per se, but with deformation, and specifically with the relationship between the deformation of a fluid element and the surface forces exerted on this element by the surrounding fluid. Tensors are necessary to describe deformations and surface forces, and the tensors used to represent them are the *strain tensor* and the *stress tensor.*

Like a vector, a tensor can be represented in terms of its components, and the values of these components depend on the choice of the coordinate system. Furthermore, there is a rule for using the components of a tensor in one coordinate system to calculate the components of that tensor in another coordinate system, rotated with respect to the first. The existence of this rule shows that a tensor has a basic physical significance that transcends any particular choice of coordinate system. However, unlike a vector, the physical significance of a tensor cannot be described in terms of a directed line segment, i.e., in terms of a magnitude and a direction.

5.3 The Stress Tensor

The concepts of strain and stress were introduced in Chap. 1 in regard to specific components, the extensional stress for elongation and the shear stress for a shear. But stress and strain are not scalar variables like temperature and pressure that have single values at each point in space. Stress is directional like velocity, but it acts on three types of surface, one normal to each coordinate direction. Whereas a component of a vector is indicated by a single subscript, for example v_1, stress requires two indices, one indicating the direction of the stress and the other indicating the surface on which it acts. A typical stress component is written as σ_{ij}, where each index can take on values of 1, 2, or 3 corresponding to the three coordinate directions The second index indicates the direction in which the force acts, and the first index indicates the type of surface on which it acts. When the stress acts in a direction normal to a surface (i = j) it is a *normal stress*, and when it acts tangentially to a surface (i $\neq$ j) it is a *shear stress*. In general, then, nine quantities are required to describe the state of stress at a point in a fluid. To write the values of all the components, it is convenient to use matrix notation.

$$\sigma_{ij} = \begin{bmatrix} \sigma_{11} & \sigma_{12} & \sigma_{13} \\ \sigma_{21} & \sigma_{22} & \sigma_{23} \\ \sigma_{31} & \sigma_{32} & \sigma_{33} \end{bmatrix} \tag{5.1}$$

To understand the physical significance of the nine components of the stress tensor, consider the small cubical element of material shown in Fig. 5.1. The second subscript indicates the direction of the force and corresponds to the coordinate axis direction. For example, the stress component shown in Fig. 5.1

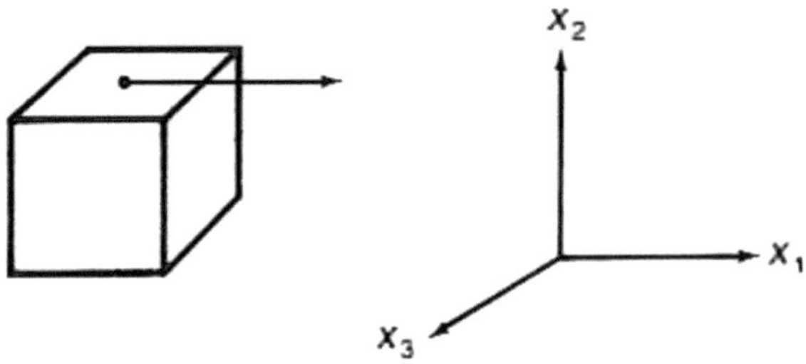

Fig. 5.1 Cubical fluid element showing stress component σ_{21}

acts in the x_1 direction, and the second subscript of this component is thus 1. The first subscript indicates the face on which the component acts, and, this is indicated by the coordinate direction normal to this face. Thus, the force shown acts on a face normal to the x_2 direction, and the first subscript of this component of the stress is thus 2. The stress component shown is thus σ_{21}. To complete our definition of the components of stress, we need a sign convention. In this book we use the standard convention of mechanics, although the reader should be aware that the opposite convention is used by some rheologists [3]. In this book, the stress is positive when it acts in the positive x_j direction on a face having the higher value of x_i, i.e., the face further from the origin in the x_i direction. For example, the stress component shown in Fig. 5.1 is positive if the force acts in the direction of the arrow. This is because it acts in the positive x_2 direction on a face having the higher value of x_1. Figure 5.2 shows several additional stress components.

Since the components of the stress tensor describe the state of stress at a point in the material, the cubical element shown in Fig. 5.1 must shrink to an infinitesimal size. Thus, the two force vectors shown in Fig. 5.3 act in opposite directions at the same point. From Newton's law of action and reaction, these two forces must be equal in magnitude. They are thus both manifestations of the same stress component, σ_{11}, and both have positive values if they act in the directions indicated. Using our sign convention, a tensile stress has a positive value.

The principle of conservation of angular momentum can be applied to the infinitesimal material element to show that the stress tensor has the following property:

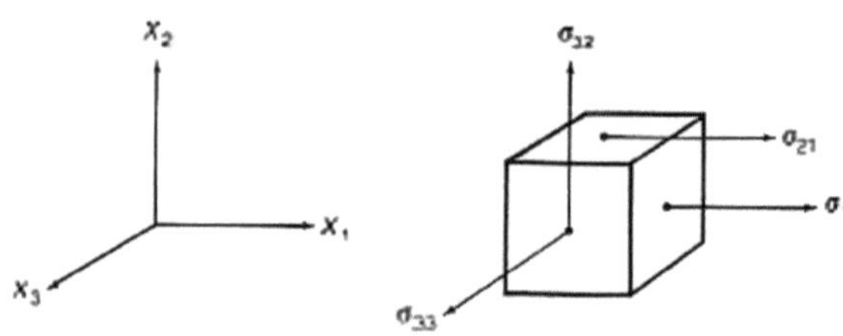

Fig. 5.2 Additional stress components. When two subscripts are equal, we have a normal stress, and when they different it is a shear stress. The *first index* identifies the face on which the stress acts, and the *second* indicates the direction of the stress

Fig. 5.3 The stresses acting at a point in opposite directions must be equal, from Newton's law of action and reaction. Thus, both the components shown are equal to σ_{11}

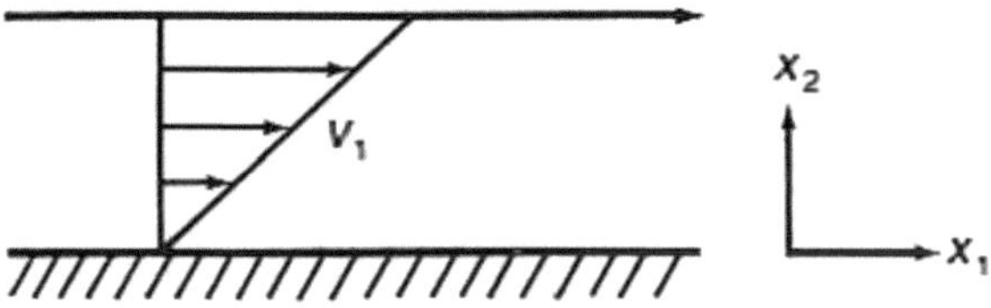

Fig. 5.4 Coordinate convention for simple shear. x_1 is the direction of motion, and x_2 is the direction of the velocity gradient

$$\sigma_{ij} = \sigma_{ji} \tag{5.2}$$

Thus, any two components that have the same subscripts or indexes, but in reverse order, have the same value and sign. A tensor that has this property is said to be *symmetric*. One result of this property is that a symmetric tensor has only six independent components rather than the nine that would be required to specify an asymmetric tensor. In other words, a second-order, symmetric tensor has only six independent components.

To make more concrete our discussion of stress, consider the shearing deformation shown in Fig. 5.4. There is a widely used convention in describing this flow; the direction of motion is x_1, while the velocity varies in the x_2 direction. To generate this deformation, a force is applied to the upper plate in the direction shown by the arrow. In the ideal case (no inertia or edge effects) this force generates a uniform deformation and stress in the sample. Since the force is in the x_1 direction and acts on a face perpendicular to the x_2 direction, the stress generated by the force F is σ_{21}. Obviously this is a shear stress, and due to the symmetry of the stress tensor it is equal to σ_{12}. We use the symbol σ, with no subscripts, to indicate this stress component in simple shear. Thus for simple shear,

$$\sigma_{21} = \sigma_{12} \equiv \sigma = F/A \tag{5.3}$$

where A is the area of the sample in contact with the plates.

The other shear stress components are zero:

$$\sigma_{13} = \sigma_{31} = \sigma_{23} = \sigma_{32} = 0. \tag{5.4}$$

We can now use matrix notation to specify the state of stress at a point in a material subjected to simple shear:

$$\sigma_{ij} = \begin{bmatrix} \sigma_{11} & \sigma_{12} & 0 \\ \sigma_{21} & \sigma_{22} & 0 \\ 0 & 0 & \sigma_{33} \end{bmatrix} \tag{5.5}$$

And we recall that $\sigma_{ij} = \sigma_{ji}$, so there are actually only four independent components. Don't worry about the three normal stress components now; we will deal with them a little further on.

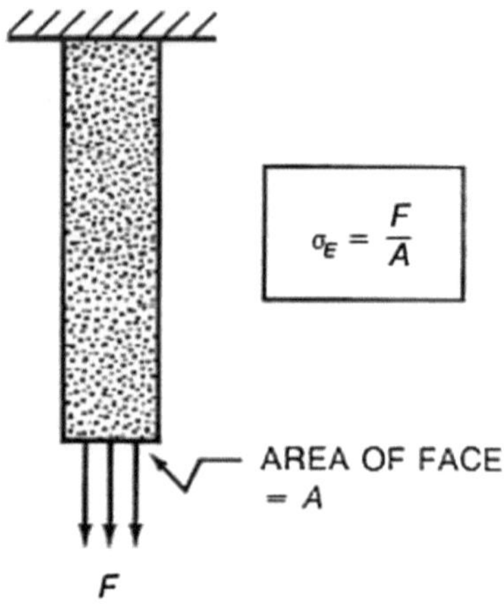

Fig. 5.5 Uniaxial (simple) extension

Another example of a deformation that is of central interest in rheology is simple or uniaxial extension. This is illustrated in Fig. 5.5. If we let x_1 be the direction of the applied force, the stress component resulting from this force will be σ_{11}, which is a normal stress. If it acts in the direction shown, it is positive, i.e., a tensile stress. There are no shear stress components in this case, and the components of the stress tensor are as shown below:

$$\sigma_{ij} = \begin{bmatrix} \sigma_{11} & 0 & 0 \\ 0 & \sigma_{22} & 0 \\ 0 & 0 & \sigma_{33} \end{bmatrix} \quad (5.6)$$

While all materials are compressible to some extent, in the case of molten plastics, quite high pressures are required to produce a significant change in the volume of a sample. Thus, for many practical purposes melts can be considered to be incompressible. Now consider what happens when we subject an incompressible material to a compressive stress that is equal in all directions, i.e., an *isotropic* stress that we call pressure P. But since pressure is positive when acting to compress a material, while a tensile stress is positive when it acts to stretch the material, we need a minus sign to relate the normal stress to the pressure. The components of the stress tensor in this situation are thus as shown by Eq. (5.7).

$$\sigma_{ij} = \begin{bmatrix} -P & 0 & 0 \\ 0 & -P & 0 \\ 0 & 0 & -P \end{bmatrix} \quad (5.7)$$

An incompressible fluid at rest will not be deformed by an isotropic stress, i.e., it will not change its size or shape. Thus, an isotropic stress field is of no rheological significance. Only when there are shear stresses acting, as in simple shear flow, or when the normal stress components are *different from each other*, will deformation occur in an incompressible material. Another way of saying this is that if a rheological measurement on an incompressible material is repeated at several ambient pressures, for example by placing the rheometer in a hyperbaric chamber, the measurements at various pressures will yield exactly the same values of rheological properties. This means that for an incompressible material a single normal component of stress has no rheological significance. In the case of simple shear the stress quantities of rheological significance as defined in Chap. 2 are the shear stress σ and the first and second normal stress differences N_1 and N_2.

$$\sigma = \sigma_{21} = \sigma_{12} \tag{5.8}$$

$$N_1 = \sigma_{11} - \sigma_{22} \tag{5.9}$$

$$N_2 = \sigma_{22} - \sigma_{33} \tag{5.10}$$

For Newtonian fluids the two normal stress differences are zero, but in entangled polymeric liquids they are nonzero. One manifestation of the first normal stress difference is observed when a polymeric fluid is sheared by placing it between two flat parallel disks and rotating one disk. The fluid exerts a normal force or thrust on the disks that tries to push them apart, while a Newtonian fluid exerts no such normal force. And the second normal stress difference contributes to coextrusion instabilities.

In the case of simple extension in the x_1 direction, there is only one significant stress variable, which is a normal stress difference (the shear stresses are zero), and there is axial symmetry, so $\sigma_{22} = \sigma_{33}$. This normal stress difference is called the *principal stretching stress*, σ_E, which is defined in Eq. (5.11).

$$\sigma_E = \sigma_{11} - \sigma_{22} = \sigma_{11} - \sigma_{33} \tag{5.11}$$

Because of the relatively high resistance to deformation of molten, high-molecular-weight polymers, the tensile stress σ_{11} is normally much larger than the ambient pressure, which is equal to $-\sigma_{22}$ and $-\sigma_{33}$. Thus, for most practical purposes, $\sigma_\mathrm{E} = \sigma_{11}$.

5.4 A Strain Tensor for Infinitesimal Deformations

We begin our discussion of strain tensors by developing one that can be used to write the Boltzmann superposition principle in a more general form that describes responses to deformations of arbitrary kinematics, i.e., shear, extension, or any combination of these.

The simplest description of deformation of a fluid element is one based on position vectors of fluid particles within this element. Figure 5.6 shows the position vector $x_\mathrm{A}(t_1)$, for the particle A at time t_1 as well as the position vector $x_\mathrm{A}(t_2)$, for the same fluid particle at another time t_2. The use of two times, or two states of the fluid element, is essential in discussing strain, as strain is a measure of the difference between two states. We will take t_1 to represent a time when the fluid is in a *reference configuration*, with which the configuration at any other time t_2 is to

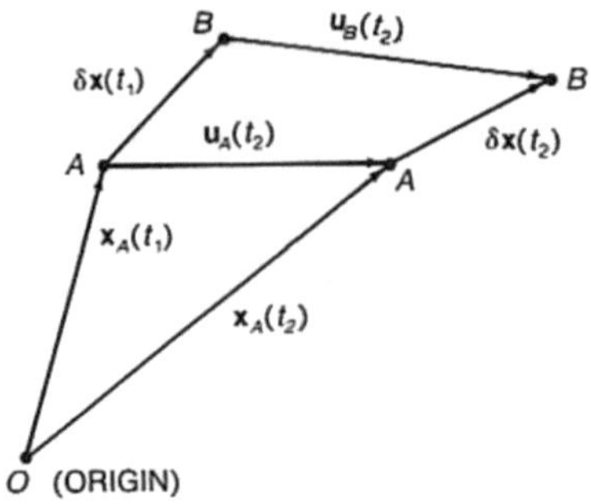

Fig. 5.6 Position vectors of particle A at times $t_1[\mathbf{x}_A(t_1)]$ and $t_2[\mathbf{x}_A(t_2)]$, and displacement vectors at time t_2 of neighboring particles A $[\mathbf{u}_A(t_2)]$ and B $[\mathbf{u}_B A(t_2)]$ relative to their positions at time t_1. The position vectors of particle B at t_1 and t_2 are not shown; they would be lines connecting the origin with the two points labeled B

be compared. Thus, we will say that a material element is in a strained state at time t_2, if it undergoes deformation in the time interval between t_1, and t_2.

To describe deformation it is necessary to look at two fluid particles and the change that occurs in their relative positions between times t_1 and t_2. In Fig. 5.6 two such particles are shown, with the position of the second particle with respect to the first given by the vector $\boldsymbol{\delta}\mathbf{x}(t_1)$ at time t_1 and by $\boldsymbol{\delta}\mathbf{x}(t_2)$ at time t_2. The displacement vector $\mathbf{u}(t_2)$ is the position of a particle at time t_2 relative to its position at time t_1. The relative variation of the $u(t_2)$ vectors for the two fluid particles obviously provides information about the deformation of the fluid element containing these particles. In the rigid displacement of the element shown in Fig. 5.7 there is no deformation, and the displacement vectors for the two particles are the same. This suggests that the **gradient** of **u** may be useful as a quantitative measure of strain. This gradient has nine components:

$$\frac{\partial u_i(t_2)}{\partial x_j(t_1)}$$

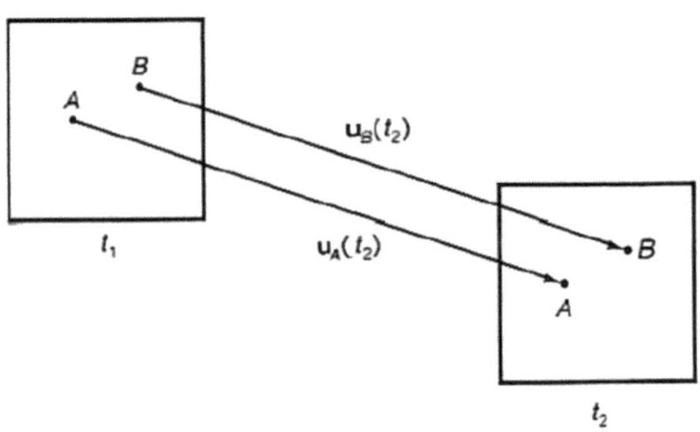

Fig. 5.7 Displacement vectors $\mathbf{u}_A(t_2)$ and $\mathbf{u}_B(t_2)$ of two particles located in a fluid element for rigid translation of the element. There is no deformation of the element in this process

These nine quantities are the components of a tensor, and they clearly contain all the information necessary to describe the relative displacements of the two neighboring particles. While this tensor is not suitable as a general measure of strain, if we limit our attention to flows in which this gradient is always very small, a useful measure of strain can be defined as follows:

$$\gamma_{ij}(t_1,t_2) \equiv \frac{\partial u_i(t_2)}{\partial x_j(t_1)} + \frac{\partial u_j(t_2)}{\partial x_i(t_1)} \tag{5.12}$$

where $\gamma_{ij}(t_1,t_2)$ is the infinitesimal strain tensor at rime t_2 relative to the configuration of the same fluid element at time t_1. Like the stress, this is a symmetric tensor, because if the order of the indexes i and j is reversed, the value of each component remains unchanged.

The *strain rate tensor*, also called the *rate of deformation tensor*, is defined by Eq. (5.13).

$$\dot{\gamma}_{ij} \equiv d\gamma_{ij}/dt \tag{5.13}$$

Since the components v_i of the velocity vector are given by:

$$v_i = du_i/dt \tag{5.14}$$

Equation (5.13) can be expanded as shown by Eq. (5.15).

$$\dot{\gamma}_{ij} = \frac{\partial v_j}{\partial x_i} + \frac{\partial v_i}{\partial x_j} \tag{5.15}$$

This definition of a rate of deformation is not limited to use in describing deformations that are infinitesimal in total magnitude, as its definition requires only a measure of strain valid during an infinitesimal interval of time.

As an example of the use of the infinitesimal strain tensor, we note that for simple shear of magnitude γ:

$$\gamma_{12} = \gamma_{21} = \frac{\partial u_1}{\partial x_2} + \frac{\partial u_2}{\partial x_1} \tag{5.16}$$

The variation with time of the components of the position vector for simple shear are:

$$x_1(t_2) = x_1(t_1) + x_2(t_1)[\gamma(t_2) - \gamma(t_1)] \tag{5.17a}$$

$$x_2(t_2) = x_2(t_1) \tag{5.17b}$$

$$x_3(t_2) = x_3(t_1) \tag{5.17c}$$

The only nonzero component of the displacement vector is u_1.

$$u_1(t_2) = x_1(t_2) - x_1(t_1) = x_2(t_1)[\gamma(t_2) - \gamma(t_1)]$$

and the components of the gradient of the displacement vector are:

$$\frac{\partial u_\mathrm{i}(t_2)}{\partial u_\mathrm{j}(t_1)} = \begin{bmatrix} 0 & [\gamma(t_2) - \gamma(t_1)] & 0 \\ 0 & 0 & 0 \\ 0 & 0 & 0 \end{bmatrix} \tag{5.18}$$

Thus, the only nonzero components of the infinitesimal strain tensor defined by Eq. (5.12) are:

$$\gamma_{12} = \gamma_{21} = \gamma(t_2) - \gamma(t_1) \equiv \gamma \tag{5.19}$$

and the only nonzero components of the rate of strain tensor are:

$$\dot{\gamma}_{12} = \dot{\gamma}_{21} \equiv \dot{\gamma} \tag{5.20}$$

Likewise, for simple extension, it can be seen that in the limit of infinitesimal deformation, the Hencky and linear strains become equal to each other, and that both become equal to $2(\partial u_1/\partial x_1)$. All of the components of the deformation gradient tensor for simple extension are shown below, where $\varepsilon \equiv \varepsilon(t_2) - \varepsilon(t_1)$.

$$\frac{\partial u_\mathrm{i}(t_2)}{\partial u_\mathrm{j}(t_1)} = \begin{bmatrix} \varepsilon & 0 & 0 \\ 0 & -\varepsilon/2 & 0 \\ 0 & 0 & -\varepsilon/2 \end{bmatrix} \tag{5.21}$$

Thus, the components of the infinitesimal strain tensor are:

$$\gamma_\mathrm{ij} = \begin{bmatrix} 2\varepsilon & 0 & 0 \\ 0 & -\varepsilon & 0 \\ 0 & 0 & -\varepsilon \end{bmatrix} \tag{5.22}$$

And the components of the strain rate tensor are:

$$\dot{\gamma}_\mathrm{ij} = \begin{bmatrix} 2\dot{\varepsilon} & 0 & 0 \\ 0 & -\dot{\varepsilon} & 0 \\ 0 & 0 & -\dot{\varepsilon} \end{bmatrix} \tag{5.23}$$

We are now in a position to write the constitutive equation for the Newtonian fluid, for which the strain tensor for small deformations is always applicable. Using the stress and strain *tensors* in place of the scalars sufficient for the special case of simple shear, we have:

$$\sigma_{ij} = \eta\dot{\gamma}_{ij} = \eta \begin{bmatrix} 0 & \dot{\gamma} & 0 \\ \dot{\gamma} & 0 & 0 \\ 0 & 0 & 0 \end{bmatrix}$$

This equation tells us how to calculate all the components of the stress in any flow of a Newtonian fluid. Applying this to uniaxial extensional flow:

$$\sigma_{ij} = \eta\dot{\gamma}_{ij} = \eta\begin{bmatrix} 2\dot{\varepsilon} & 0 & 0 \\ 0 & -\dot{\varepsilon} & 0 \\ 0 & 0 & -\dot{\varepsilon} \end{bmatrix} \tag{5.24}$$

And since the extensional viscosity is defined as the net stretching stress σ_E divided by the strain rate, it is given by:

$$\eta_E \equiv \frac{\sigma_\mathrm{E}}{\dot{\varepsilon}} = \frac{\sigma_{11} - \sigma_{22}}{\dot{\varepsilon}} = 3\eta \text{ (Newtonian fluid)} \tag{5.25}$$

The extensional viscosity of a Newtonian fluid is thus simply three times the (shear) viscosity.

5.5 The Boltzmann Superposition Principle in Tensor Form

The tensorial forms of the Boltzmann superposition principle are obtained by replacing the shear stress and shear rate in Eq. (3.9) by the stress tensor and the strain tensor developed above:

$$\sigma_{ij}(t) = \int_{-\infty}^{t} G(t - t')d\gamma_{ij}(t') \tag{5.26}$$

$$\sigma_{ij}(t) = \int_{-\infty}^{t} G(t - t')\dot{\gamma}_{ij}(t')dt' \tag{5.27}$$

Using Eq. (3.20) we can obtain the linear constitutive equation corresponding to a single Maxwell element.

$$\sigma_{\mathrm{ij}}(t) = \int_{-\infty}^{t} G_0\{\exp[-(t - t')/\tau]\}\dot{\gamma}_{\mathrm{ij}}(t')dt' \tag{5.28}$$

And the generalized (multimode)Maxwell model is:

$$\sigma_{\mathrm{ij}}(t) = \int_{-\infty}^{t} \sum_{k=1}^{N} G_\mathrm{k}\{\exp[-(t - t')/\tau_\mathrm{k}]\}\dot{\gamma}_{\mathrm{ij}}(t')dt' \tag{5.29}$$

where G_k and τ_k are the initial modulus and relaxation time corresponding to each Maxwell element.

There is no differential equation form of Eq. (5.27), but the special case of the Maxwell model can be written as follows:

$$\sigma_{\mathrm{ij}} + \tau \frac{\mathrm{d}\sigma_{\mathrm{ij}}}{\mathrm{d}t} = G\tau\dot{\gamma}_{\mathrm{ij}} \tag{5.30}$$

And if a discrete spectrum is used, as in the generalized Maxwell model (5.29), Eq. (5.30) is valid, with a separate differential equation for each relaxation time and the stress given by the sum of those given by these equations.

The tensorial form of the Boltzmann superposition principle can be used to show how extensional flow properties are related to the shear relaxation modulus. For an incompressible material undergoing start-up of steady uniaxial extension the components of the strain rate tensor are given by Eq. (5.23), which is shown again below.

$$\dot{\gamma}_{\mathrm{ij}} = \begin{bmatrix} 2\dot{\varepsilon} & 0 & 0 \\ 0 & -\dot{\varepsilon} & 0 \\ 0 & 0 & -\dot{\varepsilon} \end{bmatrix} \tag{5.23}$$

Thus, the net stretching stress is given by Eq. (5.31).

$$\sigma_{\mathrm{E}} \equiv \sigma_{11} - \sigma_{22} = \int_0^t G(t-t')[2\dot{\varepsilon} - (-\dot{\varepsilon})]\mathrm{d}t' = 3\dot{\varepsilon}\int_0^t G(t-t')\mathrm{d}t' = 3\dot{\varepsilon}\int_0^t G(s)\mathrm{d}s \tag{5.31}$$

And the tensile stress growth coefficient is:

$$\eta_E^+(t) \equiv \frac{\sigma_E(t)}{\dot{\varepsilon}} = 3\int_0^t G(s)\mathrm{d}s \tag{5.32}$$

But in Chap. 3 we showed that the shear stress growth coefficient is:

$$\eta^+(t) \equiv \frac{\sigma(t)}{\dot{\gamma}} = \int_0^t G(t-t')\mathrm{d}t' = \int_0^t G(s)\mathrm{d}s \tag{3.10}$$

Thus the function of time given by Eq. (3.32) is just three times the stress growth function for shear. This relationship is useful in the validation of extensional rheometer data.

$$\eta_E^+(t) = 3\eta^+(t) \tag{5.33}$$

5.6 Strain Tensors for Large, Rapid Deformations

The strain tensor defined by Eq. (5.12) is not suitable for the description of large deformations and must be replaced by a finite measure of strain. From a physical point of view, the complexity that arises in describing large, rapid deformation of viscoelastic materials arises from the fact that stress relaxation occurs in a material element, not at a stationary point in space. In other words, the stress in each fluid element depends on the strain history of that fluid element, so each fluid element has its own relaxation "clock". As a result, in order to meet the most basic requirements for a strain measure to describe nonlinear phenomena it must track fluid elements.

Some general principles of continuum mechanics can be used to identify likely candidates. The most important one is the *principle of material indifference*, which states that a material's rheological behavior reflects a basic physical property and therefore cannot depend on the frame of reference used to describe the behavior.

5.6.1 The Cauchy and Finger Tensors

We present in this section two strain measures that meet the requirements of material indifference and that have been found useful in describing nonlinear viscoelastic behavior. These are the Cauchy tensor $C_{ij}(t_1, t_2)$ and the Finger tensor: $B_{ij}(t_1, t_2)$. The time arguments have the following significance: t_1 is the time at which a material element is in its reference configuration, and t_2 is the time at which the strain is evaluated relative to the configuration of the element at time t_1. These tensors are developed from first principles in Appendix B for readers who are particularly interested, but to follow the contents of this chapter, it is only necessary to know the components of these tensors for simple shear and simple extension.

For simple shear, the components of the Cauchy and Finger tensors are shown below.

$$C_{ij}(t_1,t_2) = \begin{bmatrix} 1 & [\gamma(t_2)-\gamma(t_1)] & 0 \\ [\gamma(t_2)-\gamma(t_1)] & \left\{1+[\gamma(t_2)-\gamma(t_1)]^2\right\} & 0 \\ 0 & 0 & 1 \end{bmatrix} \tag{5.34}$$

and

$$B_{ij}(t_1,t_2) = \begin{bmatrix} \left\{1+[\gamma(t_1)-\gamma(t_2)]^2\right\} & [\gamma(t_1)-\gamma(t_2)] & 0 \\ [\gamma(t_1)-\gamma(t_2)] & 1 & 0 \\ 0 & 0 & 1 \end{bmatrix} \tag{5.35}$$

For simple extension, the components of the Cauchy and Finger tensors are:

$$C_{ij}(t_1,t_2) = \begin{bmatrix} e^{2[\varepsilon(t_2)-\varepsilon(t_1)]} & 0 & 0 \\ 0 & e^{-[\varepsilon(t_2)-\varepsilon(t_1)]} & 0 \\ 0 & 0 & e^{-[\varepsilon(t_2)-\varepsilon(t_1)]} \end{bmatrix} \tag{5.36}$$

$$B_{ij}(t_1,t_2) = \begin{bmatrix} e^{2[\varepsilon(t_1)-\varepsilon(t_2)]} & 0 & 0 \\ 0 & e^{-[\varepsilon(t_1)-\varepsilon(t_2)]} & 0 \\ 0 & 0 & e^{-[\varepsilon(t_1)-\varepsilon(t_2)]} \end{bmatrix} \quad (5.37)$$

We note that the components of the Cauchy and Finger tensors are not zero when a material is in its undeformed state, since $e^0 = 1$, and both tensors become equal to the unit tensor.

$$B_{ij} = C_{ij} = \begin{bmatrix} 1 & 0 & 0 \\ 0 & 1 & 0 \\ 0 & 0 & 1 \end{bmatrix}$$

For this reason, it is sometimes convenient to use the Cauchy *strain* tensor and the Finger *strain* tensor, defined as follows:

$$\text{Cauchy strain tensor} = C_{ij} - \delta_{ij}$$
$$\text{Finger strain tensor} = \delta_{ij} - B_{ij}$$

where δ_{ij} is the Kronecker delta defined as follows:

$$\delta_{ij} = 1 \qquad (\mathrm{i} = \mathrm{j})$$
$$\delta_{ij} = 0 \qquad (\mathrm{i} \neq \mathrm{j})$$

It is not difficult to show using equations in this chapter to show that for *infinitesimal* deformations:

$$\gamma_{ij} = C_{ij} - \delta_{ij} = \delta_{ij} - B_{ij}$$

It is comforting to find that the Cauchy and Finger strain tensors reduce to the strain tensor for very small deformations in the limit of vanishing strain.

5.6.2 Reference Configurations

In the definitions of the Finger and Cauchy tensors provided in Appendix B the meaning of the time arguments t_1 and t_2 is as follows: $B_{ij}(t_1, t_2)$ is the Finger tensor at time t_2 relative to the configuration of the fluid element at time t_1. Thus, time t_1 establishes the reference state with respect to which the strain is to be measured at any other time. In an experiment that begins when the sample has been at rest for a long period of time and is free of deforming stresses, it is convenient to let t_1 be the time t_0 at which deformation begins.

This is the obvious reference state for a laboratory test involving a homogeneous deformation starting from the rest state, but it is not useful for writing a general model capable of predicting the viscoelastic response to a complex sequence of deformations of varying kinematics, because even if an "initial" configuration could be established, it would become less and less relevant as a fluid element moved along due to the fluid's fading memory and would eventually lose its relevance altogether. We want to predict the stresses in a fluid element at time t based on deformations that occurred at earlier times t', where $t' \leq t$. For a viscoelastic fluid the only configuration that continues to have any unique significance is that at time t. This suggests that strains be relative to the configuration of a fluid element at time t. Therefore, we take t_2 to be t', the time at which a *past contribution* to the strain took place, and we let the reference time t_1 be t, the time at which the stress is to be evaluated. In other words, the fluid element is in its reference configuration at time t. The Finger tensor, evaluated at time t', relative to t is thus $B_{ij}(t, t')$.

5.6.3 Scalar Invariants of the Finger Tensor

A vector has one scalar invariant, i.e., a scalar that is determined from the components of the vector but whose value, unlike those of the components, is independent of the coordinate system used to describe the vector. The single scalar invariant of a vector **v** is its magnitude v given by:

$$v = \sqrt{v_1^2 + v_2^2 + v_3^2}$$

A second order Cartesian tensor has three scalar invariants. In this book, we will make use of the scalar invariants of the Finger tensor for an incompressible fluid. For a given deformation, these are:

$$I_1\left(B_{ij}\right) = B_{11} + B_{22} + B_{33} \tag{5.38}$$

$$I_2\left(B_{ij}\right) = C_{11} + C_{22} + C_{33} \tag{5.39}$$

$$I_3\left(B_{ij}\right) = 1 \tag{5.40}$$

Since B_{ij} is a function of the two times t_1 and t_2, the scalar invariants also depend on these two quantities. For example, the values of $I_1(B_{ii})$ and $I_2(B_{ij})$ for simple shear, using the diagonal components from Eqs. (5.34) and (5.35) are:

$$I_1(B_{ij}) = I_2(B_{ij}) = [\gamma(t) - \gamma(t')]^2 + 3 \tag{5.41}$$

5.7 Integral Constitutive Equations Based on Continuum Mechanics Concepts

A first, empirical step to develop an empirical constitutive equation to describe nonlinear viscoelastic phenomena is to replace the infinitesimal strain tensor γ_{ij} in the Boltzrnann superposition principle by a measure of finite strain that meets basic continuum mechanics criteria. If γ_{ii} in the Boltzmann superposition principle is replaced by such a strain tensor, the result is called a model of *finite linear viscoelasticity* or a *quasilinear viscoelastic* model. It will be helpful to examine some simple examples.

5.7.1 Lodge's Rubberlike Liquid Model

Either the Cauchy or the Finger tensor, or a combination of the two, can be used to formulate a model of nonlinear viscoelastic behavior, and rational mechanics provides no guidance as to which, if either, choice will prove useful. Experimental observation, however, has shown that for a viscoelastic fluid, if only one tensor is to be used, the Finger tensor gives the best indication of behavior that is just beyond the regime of linear viscoelasticity. The result of using the Finger tensor to generalize the Boltzmann superposition principle is as follows:

$$\tau_{ij}(t) = \int_{-\infty}^{t} m\,(t - t')B_{ij}(t, t')dt' \tag{5.42}$$

Lodge called Eq. (5.42) the *rubberlike liquid* and reported its predictions for many types of deformation [4].

The memory function $m(t - t')$ is related to the linear relaxation modulus $G(t)$ as follows:

$$G(t) = \int_{t}^{\infty} m(t - t')dt' \tag{5.43a}$$

$$m(t - t') = \frac{dG(t - t')}{dt'} \tag{5.43b}$$

In other words, the relaxation modulus is identical to the one that appears in the Boltzmann superposition principle, which governs the relaxation after a strain that is very slow or very small. The only new feature is the use of a strain tensor for describing large, rapid deformations.

For the generalized Maxwell model, Eq. (3.20), $m(t - t')$ is:

$$m(t - t') = \sum_{i=1}^{N} \frac{G_i}{\tau_i} \exp\left(-\frac{t - t'}{\tau_i}\right) \tag{5.44}$$

And using this in Eq. (5.44), we obtain Lodge's network model.

$$\tau_{ij}(t) = \int_{-\infty}^{t} \sum_{i=1}^{N} \frac{G_i}{\tau_i} \exp\left(-\frac{t - t'}{\tau_i}\right) B_{ij}(t, t')dt'. \tag{5.45}$$

Equation (5.45) is the integral form of a differential constitutive equation called the *upper convected Maxwell model*. It is one of a series of differential equation models that differ only in the type of time derivative involved. It is the *upper convected derivative* that appears in the differential form of Lodge's network model. There is also a *lower convected derivative* whose use leads to a differential equation that is equivalent to Eq. (5.45) with the Finger tensor replaced by the Cauchy tensor C_{ij}. In addition there is a corotational derivative that yields the corotational Maxwell model. All of these special time derivatives involve tracking a fluid element, since it is the relaxation of a fluid element that is governed by a relaxation model. For example the corotational derivative translates and rotates with a fluid element, based on the fact that simple translation and rotation do not involve deformation. One can find detailed descriptions of these models in the book by Bird et al. [3]

The predictions of the rubberlike liquid for the shear stress in any simple shear flow are the same as those of the theory of linear viscoelasticity, but the first normal stress difference is:

$$N_1(t,\gamma) = \int_{-\infty}^{t} m(t-t')[\gamma(t)-\gamma(t')]^2 \mathrm{d}t' \tag{5.46}$$

For step shear strain, the first normal stress relaxation function is:

$$N_1(t,\gamma) = \gamma^2 G(t) \tag{5.47}$$

And the *stress ratio* introduced above Eq. (4.6):

$$\frac{N_1(t,\gamma)}{\sigma(\gamma,t)} = \gamma$$

This is the *Lodge-Meissner relation*, which has often been observed to hold for molten polymers well beyond the normal regime of validity of the rubberlike liquid equation.

In steady simple shear, the first normal stress is.

$$N_1(\dot{\gamma}) = \dot{\gamma}^2 \int_0^{\infty} m(s)s^2 \mathrm{d}s = 2\dot{\gamma}^2 \int_0^{\infty} G(s)s\mathrm{d}s \tag{5.48}$$

And this implies that the fist normal stress coefficient is independent of shear rate:

$$\Psi_1 \equiv \frac{N_1}{\dot{\gamma}^2} = \int_0^{\infty} m(s)s^2 \mathrm{d}s = 2\int_0^{\infty} G(s)\mathrm{d}s \tag{5.49}$$

It has been found that in the limit of very small shear rates, N_1 for polymer melts does become proportional to $\dot{\gamma}^2$, and it was this observation that inspired the definitions of the normal stress coefficients. Thus, Eq. (5.49) gives the correct limiting low shear rate behavior of molten polymers and thus provides an accurate prediction of $\Psi_{1,0}$.

During steady simple shear, if the shear stress is suddenly eliminated while the fluid is held between the shearing surfaces it will exhibit recoil, i.e., negative strain, and the ultimate recoverable shear strain predicted by the rubberlike liquid model is given by Eq. (5.50).

$$\gamma_\infty = \frac{N_1}{2\sigma} \tag{5.50}$$

The quantity on the right hand side of Eq. (5.50) is sometimes called the "recoverable shear", but it is not equal to the actual recoverable shear except at very low shear rates.

Step shear strain in extension is not a practical experiment for a molten plastic, and the test usually used is tensile start-up flow. Using the appropriate components of the Finger tensor given in Eq. (5.37), together with Eq. (5.42) we can derive the tensile stress growth coefficient for a rubberlike liquid.

$$\eta_E^+(t,\dot{\varepsilon}) = \int_0^t G(s)(2e^{2\dot{\varepsilon}s} + e^{-\dot{\varepsilon}s})\mathrm{d}s \tag{5.51}$$

To obtain a clearer picture of the behavior of the rubberlike liquid at the start-up of extensional flow, we can use the relaxation modulus of the Maxwell fluid given by Eq. (3.19) to obtain:

$$\eta_E^+(t,\dot{\varepsilon}) = \left(\frac{2\eta}{1-2\dot{\varepsilon}\lambda}\right)\left[1 - e^{-1(1-2\dot{\varepsilon}\lambda)t/\lambda}\right] + \left(\frac{\eta}{1+\dot{\varepsilon}\lambda}\right)\left[1 - e^{-1(1+\dot{\varepsilon}\lambda)t/\lambda}\right] \tag{5.52}$$

where η is equal to $G_0\tau$ for a Maxwell fluid.

Figure 5.8 is a plot of η_E^+/η versus (t/τ) for several values of $(\dot{\varepsilon}\tau)$ As the Weissenberg number $(\dot{\varepsilon}\lambda)$approaches zero, we recover Eq. (3.36) for linear viscoelastic behavior. The curves for nonzero $(\dot{\varepsilon}\lambda)$ rise above the linear curve, and this type of behavior is called *strain hardening*, which is observed for long-chain branched polymers like LDPE. In reality of course, the stress cannot increase without limit, and either a steady stress is reached or the sample ruptures.

We have seen that by replacing the infinitesimal strain tensor by the Finger tensor we obtain a model that predicts a nonzero first normal stress difference and strain hardening in extension. However, the shear stress predicted in shear flows is the same as that of linear viscoelasticity, and the second normal stress difference is zero. We will now examine the effects or introducing a strain-dependent relaxation modulus and other measures of strain.

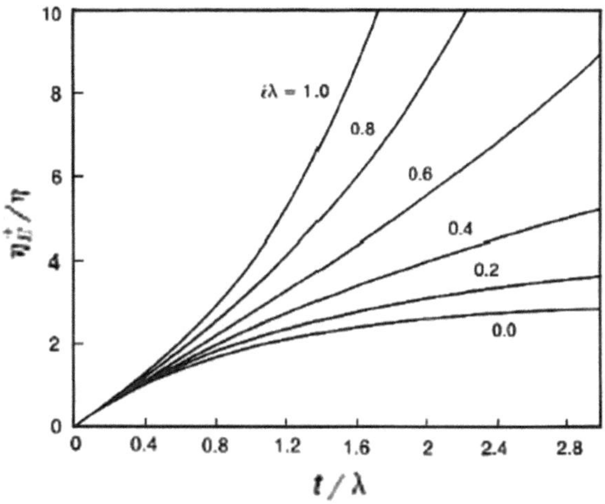

Fig. 5.8 Tensile stress growth coefficient divided by the (constant) viscosity for a rubberlike liquid with a single, exponential relaxation modulus. The lowest curve corresponds to linear viscoelastic behavior (Eq. 5.36), which is the limiting case of Eq. (5.52) as $\dot{\varepsilon}\lambda \to 0$

5.7.2 Strain Dependent Memory Function and the Cauchy Tensor

While staying with the Finger tensor in a single integral constitutive equation, there have been suggestions that the memory function be permitted to depend on strain as well as time. A simple example is to let the memory function depend on the scalar invariants of the Finger tensor, which are given by Eqs. (5.38) and (5.39). The resulting constitutive equation is:

$$\sigma(t) \equiv \sigma_{21}(t) = \int_{-\infty}^{t} M[(t-t'), I_1(B_{ij}), I_2(B_{ij})]B_{ij}(t,t')\mathrm{d}t' \tag{5.53}$$

As explained in Chap. 4, stress relaxation data for melts and for crosslinked rubbers can often be described by a relaxation modulus that is the product of a time-dependent term and a strain-dependent term. This observation suggests that the memory function for a polymeric liquid can be expressed as the product of a strain-independent function of time and a function of strain. The strain-dependent factor is called the *damping function*, which was introduced in Chap. 4, where separability was expressed by Eq. (4.1).

$$G(t,\gamma) = G(t)h(\gamma) \tag{4.1}$$

Wagner [5] made use of this idea to define the nonlinear memory function shown below.

$$M[(t-t'), I_1, I_2] = m(t-t')h(I_1, I_2)$$

where $h(I_1, I_2)$ is the damping function and $m(t-t')$ is the memory function of linear viscoelasticity. The resulting constitutive equation is:

$$\sigma_{ij}(t) = \int_{-\infty}^{t} m(t - t')h(I_1, I_2)B_{ij}(t, t')dt' \qquad (5.54)$$

This is not a complete constitutive equation, since it contains the empirical function $h(I_1, I_2)$, which must be determined by fitting experimental data In addition, it is not possible to determine the dependence of h on the two scalar invariants of B_{ij} using data from conventional experiments. This is because in simple shear and in simple extension the two invariants cannot be varied independently. However, if $h(\gamma)$ has been determined using step shear data, it is often possible to make approximate predictions of behavior in other shear flows. A constitutive equation of the form of Eq. (5.54) predicts a nonlinear extensional stress growth coefficient that does not increase without limit, but the second normal stress difference is still zero.

A non-zero second normal stress difference is obtained when the Cauchy tensor is used, but by itself this does not lead to realistic predictions. The BKZ model [6] was inspired by the theory of rubber elasticity and makes use of both the Finger and Cauchy tensors along with a strain-dependent elastic energy potential function that depends on the history of the first and second invariants of the strain tensor. The strain energy function is determined from experimental data.

5.8 Continuum Differential Constitutive Equations

For use in numerical flow simulations, differential equations are much preferred to integral equations, because they do not require the tracking of each fluid element to determine its strain history. The differential equation version of the Maxwell model was shown in tensor form as Eq. (5.30). It was also noted that (5.30) could be made into a nonlinear model equivalent to Lodge's network model, obtained by inserting (5.44) into (5.42), by replacing the derivative of the stress tensor by the "upper convected derivative" of the stress. Differential equation models that can make more realistic predictions are based on more sophisticated network ideas. A popular example is that of Phan-Thien and Tanner [7], which has been used for many flow simulations.

5.9 Constitutive Equations from Molecular Models

The Doi-Edwards tube model is much too complex for use in modeling any but the simplest flows, for example homogeneous simple shear. In order to arrive at a closed-form constitutive equation it is necessary to make several simplifying assumptions:

1. There is no chain stretch, i.e., retraction is instantaneous.
2. Each segment of the tube orients independently of the others; the *independent alignment* assumption.

The result is Eq. (5.42) with the memory function equivalent to Eq. (3.84), shown below as (5.55),

$$m(s) = \frac{8G}{5\tau_d \pi^2} \sum_{podd} \exp(-sp^2/\tau_d) \tag{5.55}$$

and the Finger tensor replaced by the Doi-Edwards strain tensor, $Q(t', t)$, whose components can be represented in terms of the components of the Finger and Cauchy tensors as follows:

$$Q_{ij}(t) = \phi_1(I_1, I_2)B_{ij}(t, t') + \phi_2(I_1, I_2)C_{ij}(t, t') \tag{5.56}$$

The two functions of the first and second invariants of the Finger tensor, given by Eqs. (5.38) and (5.39), cannot be written in a closed form, but Currie [8] showed that they can be approximated by the following analytical expressions.

$$\phi_1 = \frac{5}{J-1} \tag{5.57}$$

$$\phi_2 = -\left[\frac{5}{(J-1)(I_2 + 13/4)^{1/2}}\right] \tag{5.58}$$

where: $J \equiv I_1 + 2(I_2 + 13/4)^{1/2}$

This model predicts a damping function that is in good agreement with data for linear, high-molecular weight, monodisperse polymers, but other predictions are not accurate. In particular, the shear stress in steady simple shear has a maximum and a minimum as a function of shear rate, implying that it has the same value at three shear rates. At one time this was thought to explain the spurt and oscillatory flow phenomena sometimes observed in capillary flow and described in Chap. 6. But these phenomena are now known to result from slip.

Over the years since the appearance of the first tube model, many modifications have been proposed to deal with its deficiencies. One of these is constraint release, to deal with the fact that the tube does not stand still while the molecule inside relaxes. This is especially important for polydisperse materials, for which the *double-reptation* semi-empirical concept has been found useful. Another development was *convective constraint release*, introduced to correct the incorrect prediction of a multivalued stress in steady shear flow. Also, a retraction time was introduced to allow for time-dependent retraction.

If instead of using molecular concepts to calculate a relaxation spectrum, an experimentally determined one is used, it has been possible to develop models that have had some success in simulating the flow of linear and long-chain branched polymers. An example is the molecular stress model of Wagner et al. [9, 10].

Additional ideas have been introduced to deal with long-chain branching, which suppresses reptation so that relaxation must rely instead on fluctuations of arm length. One of these, the *pom–pom model* was developed to model a molecule with two branch points, each with multiple arms [11]. It has been used to model the flow of LDPE into and out of a slit [12]. A modified form called the "extended pom–pom" was developed for the simulation of plastics forming operations [13]. However, the connection between the detailed structure of the polymer and the model is now completely lost. And there is also very little to link the model to the classical concept of reptation. What is left is an empirical model that is an improvement on earlier ones based on continuum concepts alone.

5.10 Numerical Simulation of Melt Flows

Simulations of melt forming processes have proven useful for flows in which viscoelasticity does not play an important role. These are mainly internal flows in dies, extruders and molds. Even in these flows the presence of a sharp corner, for example when the diameter of a channel suddenly decreases, poses a major challenge because of mathematical singularities that arise at such a point. And when melt exits a die into air, its behavior is governed primarily by the viscoelastic response to the stresses generated during die flow. Here the lack of a reliable constitutive equation limits the usefulness of the flow model.

References

1. Larson RG (1988) Constitutive equations for polymer melts and solutions. Butterworths, Boston
2. Dealy JM, Larson RG (2006) Structure and rheology of molten polymers. Hanser Publishers, Munich
3. Bird RB, Armstrong RC, Hassager O (1987) Dynamics of polymeric liquids, vol 1, Fluid mechanics, 2nd edn. Wiley, New York
4. Lodge AS (1964) Elastic liquids. Academic Press, New York
5. Wagner MH (1979) Zur Netzwerktheorie von Polymer-Schmelzen. Rheol Acta 18:33–50
6. Tanner RI (1988) From A to (BK) Z in constitutive relations. J Rheol 32:673–702
7. Phan-Thien N, Tanner RI (1977) A new constitutive equation derived from network theory. J Non-Newt Fl Mech 2:353–365
8. Currie PK (1982) Constitutive equations for polymer melts: predictions by the Doi-Edwards and Curtiss-Bird kinetic theory models. J Non-Newt Fl Mech 11:53–68
9. Wagner MH, Rubio P, Bastian H (2001) The molecular stress function model for polydisperse polymer melts with dissipative convective constraint release. J Rheol 45:1387–1412
10. Wagner MH, Yamaguchi M, Takahashi M (2003) Quantitative assessment of strain hardening of low-density polyethylene melts by the molecular stress function model. J Rheol 47:779–793

11. McLeish TCB, Larson RG (1998) Molecular constitutive equations for a class of branched polymers, the pom–pom polymer. J Rheol 42:81–110
12. Lee K, Mackley MR, McLeish TCB, Nicholson TM, Harlen OG (2001) Experimental observation and numerical simulation of transient "stress fangs" with flow of molten polyethylene. J Rheol 45:1261–1276
13. Boltussen MGHM, Hulsen MA, Peters GWM (2010) Numerical simulation of the fountain flow instability in injection molding. J Non-Newt Fl Mech 165:11–12

Chapter 6
Measurement Techniques

Abstract The measurement of the rheological behavior of viscoelastic melts is much more challenging than determining the viscosity of a Newtonian fluid, but reliable data on melts are essential for resin characterization, quality control, evaluating processability and for use in process simulation. This chapter describes the instruments and techniques used to learn about the rheological behavior of molten polymers. Rotational rheometers are used to obtain linear behavior but are of limited utility in exploring nonlinear behavior. Capillary rheometers are used to determine viscosity at moderate to high shear rates and to study the occurrence of flow instabilities and melt fracture. The *extrusion plastometer*, commonly called a melt indexer and widely used in the plastics industry, is a useful tool although not a true rheometer. The use of rheometers as process instruments and in high-throughput resin development is described. Finally, extensional rheometers for melts are described, and applications are discussed.

6.1 Introduction

In the plastics industry, data from rheological measurements are used for the following purposes:

1. Polymer characterization

Data provide information about the structure of the melt, including average molecular weight, molecular weight distribution, and branching structure.

2. Quality control—statistical process control

This is a specialized type of characterization in which a particular rheological characteristic is used to indicate the degree to which a batch of resin adheres to product specifications. Current practice is to use statistical process control rather than product quality control to avoid producing off-spec product.

J. M. Dealy and J. Wang, *Melt Rheology and its Applications in the Plastics Industry*, Engineering Materials and Processes, DOI: 10.1007/978-94-007-6395-1_6,

3. Measure of processability

In the development of new polymers or new grades of existing polymers, it is highly desirable to be able to evaluate the potential of a candidate resin for use in a melt forming process using a small sample. We will see in later chapters that for processes in which viscoelasticity plays an important role, there is no fully reliable way to do this, and it is necessary at some point to evaluate the material using production scale equipment. However, it is often possible to do a preliminary screening of candidate materials based on laboratory tests in order to limit the number of resins to be evaluated using process machinery.

4. Data for process simulation

The simulation of melt forming operations requires a model describing rheological behavior, and data are required to develop such models. However, advances in process simulation are currently limited by the availability of reliable constitutive models for nonlinear viscoelastic flow behavior.

It is important to distinguish between two types of rheological characterization. One is the measurement of a well-defined rheological property, such as viscosity or storage and loss moduli. The second type is an empirical test that does not yield a well-defined physical property but is used to compare materials based on the result of a simple test. The principal example is the *extrusion plastometer*, commonly called a *melt indexer*. Such testers are used primarily for quality control and statistical process control.

The most-used tools used to make rheological measurements on melts are capillary rheometers, rotational rheometers and extensional rheometers. In order to use these instruments effectively the user must understand how the raw data are related to the property to be measured and must have the skill required to obtain reliable data. Rheological measurements on melts are challenging in several ways. Issues of concern include residual stresses, flow instabilities, thermo-oxidative degradation, and wall slip. Additional issues in the case of rotational rheometers are fixture alignment and sample rim trimming. These issues are discussed in this chapter.

6.2 Rotational and Other Drag-Flow Rheometers

In a drag-flow rheometer, one surface is moved relative to another to generate shear in the fluid between them. The simplest example is the flow between two flat plates, one of which translates with respect to the other, as was illustrated in Fig. 2.1. There are rheometers based on this principle, and Fig. 6.1 is a sketch of a *sliding plate rheometer* developed by Dealy et al. [1, 2]. Such instruments are used for the study of wall slip and non-linear viscoelasticity but are not convenient for linear viscoelastic characterization.

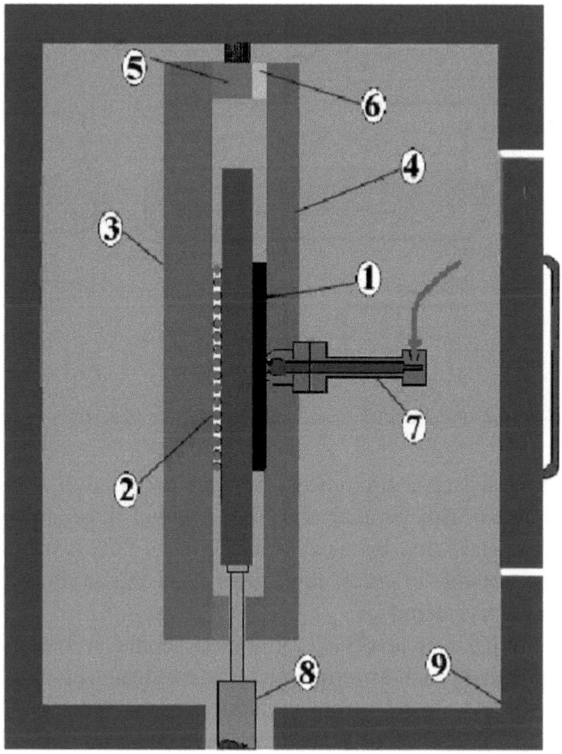

Fig. 6.1 Sketch showing basic features of the sliding plate rheometer for melts. 1 = sample, 2 = bearing table, 3 = back-plate, 4 = stationary plate, 5 = Support for frame, 6 = gap spacing shim, 7 = shear stress transducer, 8 = linear actuator, 9 = oven

In preparing samples for the rheometers described in this section, it is important to avoid residual stresses arising from the sample molding process. A hot press is usually used for sample molding, with sample particles placed in the cavity formed by a circular hole in a sheet of metal placed between the platens. If an excess of polymer is placed in the cavity, significant compression can result, which can compromise data. It is also important to ensure that samples are free of moisture, monomer, and dissolved gases. This usually requires vacuum drying at an elevated temperature. Polystyrene is especially sensitive to dissolved gas, which plasticizes the polymer. It is recommended that samples be conditioned for 8 h at room temperature or 2 h at 80 °C. Absorbed moisture is a very common problem with polar polymers. The special difficulties caused by reactions involving water in the case of polyesters and polyamides are dealt with in Sect. 6.9 at the end of this chapter.

For the determination of viscoelastic behavior in shear, rotational rheometers are most commonly used. In such devices the melt is placed between to circular surfaces, one of which rotates with respect to the other. Two fixture geometries are used, cone-and-plate and parallel disk, also called plate–plate, which are shown in Fig. 6.2. Cone-plate fixtures have the advantage that the shear rate is very nearly uniform throughout the sample, and this is essential for tests such as large-strain stress relaxation and start-up of steady shear and for measuring viscosity and

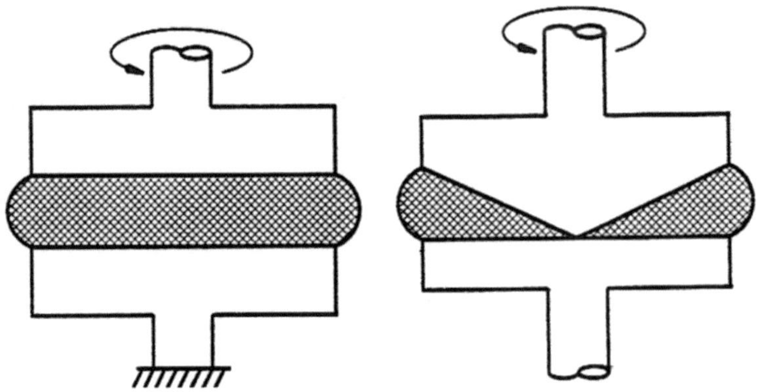

Fig. 6.2 Parallel disk (plate–plate) and cone-plate fixtures for use in rotational rheometers

normal stress differences. Sample preparation and loading, however, are much simpler with parallel disk fixtures, and these are preferred for studies of linear viscoelasticity. Because of the onset of flow instabilities [3] rotational rheometers are not able to access high shear rates and are thus limited in their ability to probe nonlinear behavior.

Whenever possible, calibration should be based on the measurement of basic quantities such as torque and angular displacement. Scatter in results from several tests arise mainly from variations in gap spacing, sample trimming and temperature. Round-robin tests have revealed that variability from one laboratory to another is much higher than that within a laboratory, probably due to differences in test procedure and equipment. When establishing the limiting strain amplitude for linear behavior, it is important to realize that this depends on frequency.

Two types of motor-drive are available. In *controlled-strain* or *controlled rate* (CR) instruments a motor rotates one fixture, while torque and normal force transducers are coupled to the stationary fixture. In *controlled-stress* (CS) instruments, the rotating fixture is driven by a motor designed to generate a prescribed torque. Thus, torque determination and motion generation are combined. A detailed discussion of the differences between these two types of instrument is available [4].

For many years, controlled-strain (CR) units made use of DC motors in which commutation was by means of brushes, while brushless (electrically commutated, EC) motors were later introduced. An optical encoder is generally used to track rotary displacement, the signal from which goes to a closed-loop control motor control system. The compliance of the body and rotor shaft causes a mismatch between the encoder output and the deformation experienced by the sample, and the *force balance transducer* was developed to deal with this problem. A description and sketch of the FBT can be found in [5, p. 344]. A drag cup motor applies a torque to balance that driving compliance, and the motor current is proportional to the torque. This device adds its own time constant to the system.

With drag cup motors, the time required for the fields to build up limits their use at high frequencies and for generating large amplitude oscillatory motion. Air-bearing supported EC (brushless DC) motors are now used in some CS rheometers, and this means that the stator torque follows the stator current almost instantaneously. Using an optical encoder indication of the strain or rate in a feedback loop makes it possible to operate in a controlled strain or rate mode.

CR units are suitable for oscillatory shear at moderate to high frequencies and steady shear, including start-up of steady shear, and for the measurement of viscosity and first normal stress difference. CS instruments are advantageous for creep and recoil, especially when very low strain rates are needed to probe the terminal region.

For creep and recovery tests, it is very important to minimize friction, as this introduces an error in the torque signal. This is especially crucial for creep recovery (elastic recoil) as this requires that the torque be suddenly reduced to zero, and this is virtually impossible to achieve with an air bearing due to the small torque resulting from the air flow. Magnetic bearings can eliminate this residual torque.

The equations used to calculate stress and strain in cone-plate and parallel plate rheometers are based on the following assumptions:

1. Inertia can be neglected because of the high viscosity of melts.
2. Surface tension at the exposed edges of the sample does not affect the torque.
3. The free surface at the edge of the sample is spherical in shape.
4. The flow is uniform out to the edge.
5. In the case of cone-plate geometry, the cone angle is sufficiently small that certain trigonometric identities can be used.

Assumptions 1 and 2 are generally valid for melts, and number five is valid for the cone angles normally used. Assumption 4 is not always valid, particularly in cone-plate fixtures, for reasons that are given in the next section. Care is required to ensure that assumption 3 is valid. The usual procedure is to use a sample somewhat larger than needed to fill the gap, bring the surfaces together until the gap is a bit larger than its final value, trim the extruded polymer to give a smooth surface, and then set the final gap spacing. It has been reported [6, p. 810] that the edge shape changed during repeated tests with the same sample of a polyisoprene, and this resulted in a large reduction in stress. A concave surface must be avoided as it initiates unstable flow [7]. There is always a small residual uncertainty due to the edge shape issue. Overall, even with the best possible practice using commercial rheometers, the best accuracy that can be achieved for molten polymer in rotational rheometers is about $\pm 3\ \%$.

6.2.1 Cone-Plate Rheometry

The fixture parameters are the radius R and the cone angle β. Either the torque $M(t)$ or the angular displacement $\phi(t)$ is programmed to vary with time in a prescribed manner. For steady shear, M and Ω, the angular velocity in rad/s, are constant. If the cone angle is less than 0.1 rad, the error related to the small-cone-angle approximation is less than 1 %, and for $\beta < 0.2$ rad the error is about 2 %. The resulting approximate values of the shear strain, strain rate, shear stress and normal stress differences are given by the following expressions. Derivations of these can be found in the book by Macosko [5].

$$\gamma = \phi/\beta \tag{6.1}$$

$$\dot{\gamma} = \Omega/\beta \tag{6.2}$$

$$\sigma = \sigma_{\phi\theta} = \frac{3M}{2\pi R^3} \tag{6.3}$$

$$N_1 = \frac{2F}{\pi R^2} \tag{6.4}$$

The modulus amplitude in small-amplitude oscillatory shear is:

$$G_d = \frac{3M_0\beta}{2\pi R^3 \phi_0} \tag{6.5}$$

where M_0 and ϕ_0 are the torque and angular displacement amplitudes respectively. At small strain amplitudes the storage and loss moduli are calculated by rheometer software using Eqs. (3.41) and (3.42).

Controlled strain is the preferred mode of operation for nonlinear studies. In step-strain experiments, an important source of experimental error is the deviation of the actual strain history from a perfect step [8]. Methods to deal with this problem have been reported [9, 10].

One uncertainty in cone-plate data arises from the precision with which the cone was fabricated. In order to avoid problems arising from a sharp apex, cones are truncated, and in setting the gap, the virtual height of the missing cone tip must be calculated. A comparison of data from many laboratories indicated that the error in stress due to uncertainties in cone angle and truncation height was ±2.5 % [11].

Instrument compliance is another cause of concern in making mechanical measurements, since the forces generated by the fluid in response to a deformation will tend to twist and compress the rheometer components that also experience these forces. Compliant components include shear and normal force transducers and the rheometer frame. This problem can be dealt with by use of a force rebalance transducer that corrects for the deformation by means of a servo. Most rheometer manufacturers now provide such devices to compensate for twisting deformations due to torque, and at least one also offers a normal force rebalance transducer. The performance of an early torque rebalance transducer has been described [12–14].

Wall slip becomes prominent in the shear flow of entangled, linear polymers when the shear stress is sufficiently large over a period of time, and this represents a serious limitation to the use of such flows to study nonlinear viscoelasticity, i.e., when the strain and strain rate are large [8]. When the substrate is steel or another high-energy surface, it has been shown [15] that this phenomenon involves partial disentanglement between molecules strongly adsorbed at the wall and those in the bulk rather than cohesive failure at the polymer-metal interface. Thus, even in the absence of flow instabilities, the actual strain in a rotational rheometer may not be equal to the *nominal* strain calculated from the angular displacement and the gap. But the bulk deformation may still be homogeneous in the presence of slip, and in step strain or steady shear, if the true strain or strain rate is known, useful data can be obtained by correcting the strain data for slip. But in transient flows, a time-dependent slip transient is superposed on the viscoelastic response, and rheologically useful information cannot be obtained.

Several types of flow irregularity occur in cone-plate flow of molten polymers, and these limit these fixtures to use at low shear rates [3]. Some irregularities manifest themselves as visible distortions at the edge of the sample. Edge fracture starts as an axisymmetric indentation in the meniscus [16] like that sketched in Fig. 6.3, and as the shear rate increases, the indentation develops into a crack that penetrates into the sample, leading eventually to severe breakup and debonding [17]. An advanced stage of fracture is sketched in Fig. 6.4 [18]. A second irregularity that affects the edge of the sample is the occurrence of *edge vortices*, which are the ends of cone-shaped eddies that extend to the flow axis [19, 20] and are thought to be associated with the first normal stress difference [10]. This type of irregularity has also been observed in large-amplitude oscillatory shear [21, 22]. Finally, for polyethylenes in both steady shear [23] and step strain at strains greater than two [24] a distorted edge has been observed. Larson [24] found that it was necessary to wait several hours after this occurred to reproduce the original data in an LDPE. It is generally not advisable to use the same sample for a series of experiments in the nonlinear regime.

The effects of edge distortions can be avoided by use of a cone-partitioned-plate fixture [25], but to make use of this fixture it is necessary to modify the oven design. It was reported that this fixture could be used to measure shear stress at shear rates up to 70 s^{-1} for a linear polystyrene but that it was not useful for normal stress measurement because of instrument compliance.

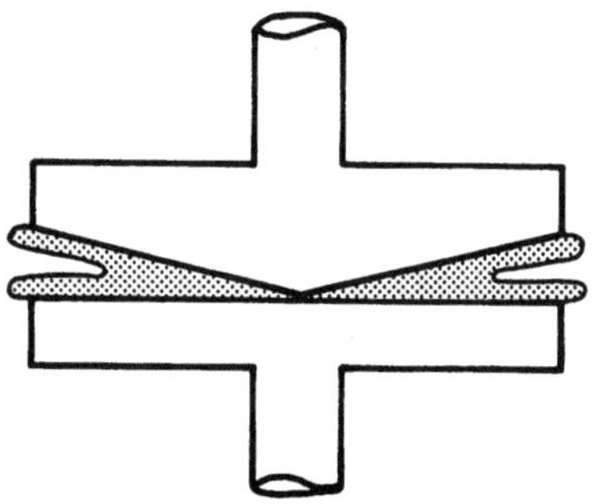

Fig. 6.3 Edge indentation that is the first stage of edge fracture

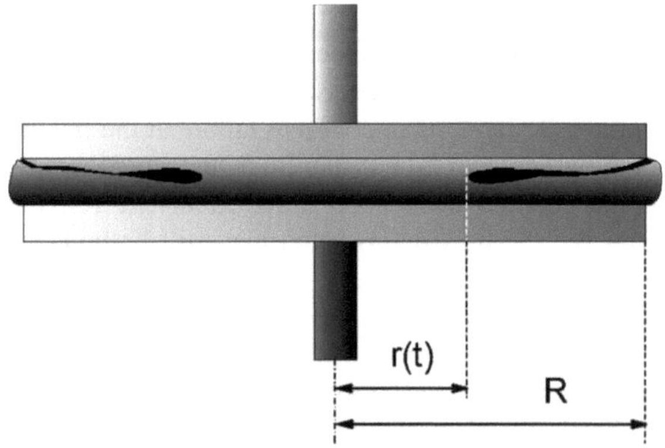

Fig. 6.4 Edge fracture as observed by Mattes et al. [18]

The first normal stress difference is usually determined by measuring the normal thrust F exerted on cone-plate fixtures by the melt, using Eq. (6.4). The normal thrust is strongly affected by changes in the gap due to axial instrument compliance as these cause squeezing or stretching of the sample. It also varies with temperature much more than does the shear stress, so temperature control and measurement are critical. And any variation in temperature due to control system imperfections or viscous heating will change the melt density, contributing to the squeezing and stretching. A servo-positioner can be used to keep the gap constant, but the dynamics of the servo system may interfere with transient experiments.

The second normal stress difference is more difficult to measure than the first. Techniques developed to accomplish this include the use of micro-miniature pressure sensors [26] and cone-partitioned plate fixtures [27].

6.2.2 Parallel Disk Rheometry

Parallel disk fixtures in rotational rheometers are used primarily to determine the linear viscoelastic behavior of melts. For cone-plate fixtures, disk-shaped samples must be squeezed substantially between the cone and the plate, and this introduces a normal thrust that can take a long time to relax. In addition, the molten sample may not be centered with respect to the cone and plate after squeezing. While methods for using parallel-disk fixtures to determine the viscosity [28] and the nonlinear relaxation modulus [29] have been proposed, these are not recommended, and cone-plate fixtures are preferred because of the approximate uniformity of strain and stresses throughout the sample.

The equations for calculating the storage and loss moduli from the torque amplitude M_0, the phase angle δ, the disk radius and gap R and h, and the angular amplitude ϕ_0 are given below.

$$G' = \frac{2M_0 h}{\pi R^4 \phi_0} \cos \delta \tag{6.6}$$

$$G'' = \frac{2M_0 h}{\pi R^4 \phi_0} \sin \delta \tag{6.7}$$

Most of the flow irregularities that occur in cone-plate rheometry can also happen in parallel disk fixtures, but are generally much less troublesome because of the small strains used to probe linear behavior.

Parallel disk geometry is also useful for creep tests within the regime of linear viscoelasticy. Calculating the compliance from the torque and angular displacement is straightforward, because the compliance is independent of radius. The torque is fixed at M_0, and the angular displacement $\phi(t)$ is measured as a function of time. The shear strain at a radius r is:

$$\gamma(r,t) = \frac{\phi(t)\, r}{h} \tag{6.8}$$

The creep compliance is independent of strain and thus of radius, which implies that:

$$J(t) = \frac{\gamma(r,t)}{\sigma(r)} = \frac{\phi(t)\, r}{h\sigma(r)} \tag{6.9}$$

This relationship implies that the stress is also linear in r and can be represented as follows:

$$\sigma(r) = \frac{\phi(t) r}{J(t) h} \tag{6.10}$$

The torque M_0 is related to the local shear stress by integrating from $r = 0$ to R.

$$M_0 = \int_0^R \sigma(r) 2\pi r^3 dr = \frac{\phi(t)}{J(t)h} \frac{2\pi R^4}{4} \tag{6.11}$$

Solving for the creep compliance, we obtain:

$$J(t) = \frac{\pi R^4 \phi(t)}{2hM_0} \tag{6.12}$$

6.2.3 *Accessing the Terminal Zone Using Creep and Creep Recovery*

The plateau and terminal zones are of principal interest, because it is in these ranges of frequency that viscoelastic behavior is most closely related to molecular structure. For example, the zero-shear viscosity, which describes behavior in the limit of zero frequency, is very sensitive to molecular weight. However, it may not be possible to reach the terminal zone using small-amplitude oscillatory shear, because the time required is excessive or the torque signal is too low. In particular, this arises when the polymer of interest has a broad molecular weight distribution or a high level of long-chain branching. Creep measurements using a torque-controlled rheometer can often be used to probe the terminal zone, but for very polydisperse materials, particularly those with even a small amount of high-molecular-weight polymer, creep measurements may also be problematic because of the need to measure extremely slow strains while maintaining the stress at the very low, constant value required to ensure that the behavior is linear. In addition, thermo-oxidative degradation may take place if the experiment lasts many hours. And a difficulty that arises when an air bearing is used is that it is not possible to maintain the torque at precisely zero, although this problem can be avoided by the use of a magnetic bearing.

If the torque level is increased, strain data are more precise, but if this is allowed to continue to steady state the linear elastic regime will be vacated. Kraft et al. [30] proposed a technique for determining the creep compliance up to the steady-state without moving outside the regime of linear behavior. During creep at a relatively high stress σ_0, they reduce the stress to zero at a time t_1, when the deformation is still within the linear range, and monitor the resulting recoil. This experiment can be analyzed by use of Eq. (6.13), which is an expression of the Boltzmann superposition principle for an experiment in which the stress, rather than the strain, is the controlled (independent) variable.

$$\gamma(t) = \int J(t - t')d\sigma(t') \tag{6.13}$$

The unloading that takes place at t_1 represents a second creep experiment commencing at this time and driven by a negative stress of $-\sigma_0$. From Eq. (6.13), the resulting shear deformation, $\gamma(t)$, is related to the creep compliance as follows:

$$\gamma(t) = J(t)\sigma_0 + J(t - t_1)(-\sigma_0) = \sigma_0[J(t) - J(t - t_1)] \tag{6.14}$$

The creep compliance is only measured directly from $t = 0$ to t_1. However, solving (6.14) for $J(t)$, we find that we can use the recovery data to extend $J(t)$ up to $t = 2t_1$ as follows:

$$J(t) = \gamma(t)/\sigma_0 + J(t - t_1)(t_1 < t < 2t_1) \tag{6.15}$$

Now that $J(t)$ has been determined up to $t = 2t_1$, this information can be used in combination with the next portion of the recoil curve to determine the compliance at times up to $3t_1$. This procedure is repeated until the terminal zone is reached, i.e., until $J(t)$ becomes linear with time. Finally, the zero-shear viscosity is calculated as the reciprocal of the slope of this line.

There remains the problem of combining creep data with storage and loss modulus data to extend the characterization into the terminal zone. To accomplish this, He et al. [31] calculated continuous retardation spectra using both modulus and creep data and plotted these together. The resulting graph shows the zones in which each technique provides reliable data and the zone of overlap. They argue that this procedure is to be preferred to the use of a discrete spectrum to combine the data from the two types of experiment. To infer a continuous spectrum from storage and loss modulus data, one never has data over the frequency from 0 to ∞, as are ideally required, so there is uncertainty in any spectrum inferred from experimental data. A traditional guideline is that the inferred spectrum is valid between values of t equal to the reciprocals of the maximum and minimum frequencies at which data are available. However, Davies and Anderssen [32] showed that the range of validity of the spectrum is substantially less than this.

6.3 Pressure-Driven Rheometers

6.3.1 Capillary and Slit Rheometers

Pressure-driven capillary rheometers are used to determine melt viscosity at high shear rates and have been the rheological work-horses of the plastics industry for many years. Their use has recently decreased somewhat as rotational rheometers have become more widely used, and the complex viscosity is now often used in place of viscosity. But capillary instruments are also used to observe extrudate swell, sharkskin melt fracture, and die build-up (die drool).

The flow is generated either by a piston moving in a cylindrical reservoir or by application of gas pressure to the reservoir. Downstream from an entrance region the flow in the capillary becomes fully-developed, i.e. the velocity and shear stress profiles become independent of distance from the entrance, if the flow is assumed isothermal and we neglect the effect of pressure on viscosity. Raw data consist of the reservoir pressures P_d corresponding to various volumetric flow rates Q, but to calculate the viscosity one needs the wall shear stress σ_w and the wall shear rate $\dot{\gamma}_w$:

$$\eta = \frac{\sigma_w}{\dot{\gamma}_w}$$

Data are sometimes reported in terms of a *flow curve* of apparent wall shear stress σ_A, which ignores the entrance correction, versus apparent wall shear rate $\dot{\gamma}_A$, which would be the shear rate for a Newtonian fluid:

$$\sigma_A \equiv \frac{P_d R}{2L} \tag{6.16}$$

$$\dot{\gamma}_A \equiv \frac{4Q}{\pi R^3} \tag{6.17}$$

The equation for the apparent shear stress, (6.16), ignores the entrance pressure drop, and Eq. (6.17) for the apparent shear rate is only correct for a Newtonian fluid in the absence of wall slip. If the true viscosity–shear rate curve is needed, corrections must be applied to both quantities.

Figure 6.5 shows pressure data obtained by wall-mounted pressure transducers in the reservoir and capillary [33]. The exit pressure indicated in the figure is generally neglected. It is clear that use of the driving pressure in Eq. (6.16) yields a wall shear stress that is significantly larger than that in the fully-developed flow zone, where the pressure gradient is constant. The ΔP_{ent} shown in the figure is the pressure drop that must be subtracted from P_{d} to give the pressure drop for fully-developed flow throughout a capillary of length L.

$$P_d = \Delta P_{\text{ent}} + \Delta P_{\text{cap}} \tag{6.18}$$

If the driving pressure in Eq. (6.16) is replaced by ΔP_{cap} we obtain an expression for the true wall shear stress.

$$\sigma_w = \frac{\Delta P_{\text{cap}} R}{2L} \tag{6.19}$$

Bagley [34] proposed a method for obtaining the entrance correction that involves the use of capillaries of several diameters. The procedure is illustrated in Fig. 6.6. Plotting the driving pressure versus *L*/*R* for data obtained for given values of the apparent wall shear rate, one extrapolates to zero P_{d} to get an *end correction*, *e* that can then be used to correct Eq. (6.16).

$$\sigma_{\text{w}} = \frac{P_{\text{d}}}{2(L/R + e)} \tag{6.20}$$

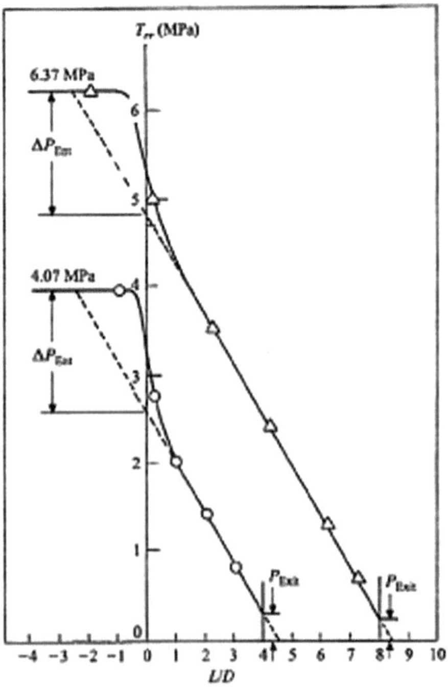

Fig. 6.5 Wall pressures measured in the reservoir and along the length of two capillaries for the flow of a HDPE. There is a sharp pressure drop in the immediate neighborhood of the entrance and a linear pressure gradient after that. If P_{exit} is neglected, P_{ent} is the quantity to be subtracted from the driving pressure (e.g. 6.37 MPa) to obtain the pressure drop for fully-developed flow in a capillary of length L. From Han [33]

The factor e is the hypothetical extra length divided by R that would yield the true wall stress. It can also be related to the ΔP_{ent}

$$e = \Delta P_{\text{ent}} / 2\sigma_{\text{w}} \tag{6.21}$$

A problem with this method is that the points on the plot often do not fall on a straight line due to the effect of pressure on viscosity. In addition it is quite time-consuming, requiring the use of at least three capillaries. A simpler method of correcting for the entrance loss is the use of an orifice die [35]. Kim and Dealy [36] reported that the most accurate determination of entrance pressure drop is obtained when L/D is less than 0.5, and the entrance angle is greater than 90°. It is very important to design the exit region to avoid its being wetted by the melt [36]. Otherwise the measured value of ΔP_{ent} will be too high [37, 38].

Another quantity sometimes used to report data is the apparent viscosity:

$$\eta_{\text{A}} \equiv \sigma_{\text{w}} / \dot{\gamma}_{\text{A}} = \frac{\pi R^4 P_{\text{cap}}}{8QL} \tag{6.22}$$

The true shear rate at the wall can be determined by use of the Rabinowich correction, which requires the differentiation of data obtained using capillaries having several radii [5, p. 240]. But a much simpler approximate method is based on the power-law viscosity model and does not require the use of several dies [39, 40]. According to the power-law model the wall shear rate in a capillary is:

$$\dot{\gamma}_W = \left(\frac{3n+1}{4n}\right)^{\frac{n}{n-1}} \frac{4Q}{\pi R^3} = \left(\frac{3n+1}{4n}\right)^{\frac{n}{n-1}} \dot{\gamma}_A \tag{6.23}$$

But over the range of values of n from 0.2 to 1.3, a value of 0.83 for the shift factor yields a viscosity error that is less than 2 %.

$$\dot{\gamma}_W \approx (0.83)\,\dot{\gamma}_A \quad (0.2 < n < 1.3) \tag{6.24}$$

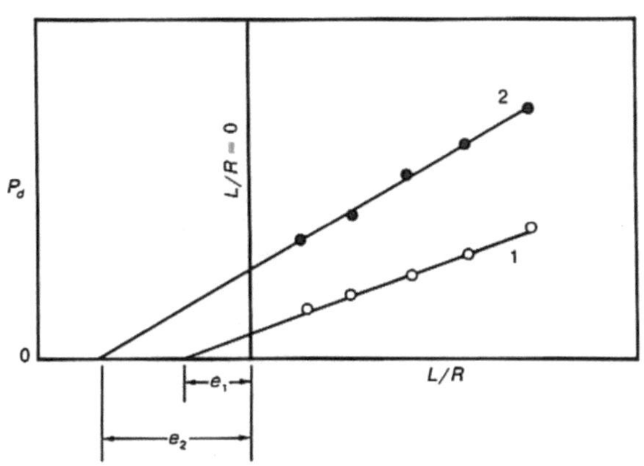

Fig. 6.6 Sketch showing the use of a Bagley plot to determine the end correction e for capillary flow. This is added to L/R to obtain the effective capillary dimensions that can be used with the driving pressure to obtain the true wall shear stress using Eq. (6.20)

This implies that points on a log–log plot of shear stress versus apparent shear rate can be shifted by 0.83 units to the left to obtain a reasonable approximation of the shear stress versus shear rate curve.

Slit rheometers are more difficult to build and use but are preferred for research studies, because the flat flow channel makes it possible to mount pressure sensors and to make optical measurements. As in the case of capillary rheometers, there are established methods for calculating the true wall shear stress and shear rate from experimental data [5].

The standard methods for interpreting capillary and slit rheometer data to determine the viscosity are based on the assumptions that the temperature and density are uniform and that the axial variation in pressure has no effect on the viscosity. However, it is known that the effects of pressure-dependent viscosity and viscous heating are significant and should be taken into account in analyzing capillary flow data [41, 42], and the same is true for slit flow [43]. There is also a variation in melt density, but it has been estimated that its effect on calculated pressure and temperature coefficients is less than 5 %, even at high driving pressures [41].

A final source of uncertainty in the analysis of data from pressure-driven rheometers is the possibility of wall slip [44]. In fact, well-entangled, linear polymers nearly always slip at a sufficiently high wall shear stress. A large slip velocity often announces itself in the flow curve ($\sigma_W$vs..$\dot{\gamma}_A$) by the occurrence of an oscillatory shear regime in constant-piston-speed rheometers, or a sudden large jump in flow rate (*spurt*) in pressure-controlled instruments. However, the presence of low slip velocities at pressures below those at which these phenomena occur may not be apparent from an inspection of data. More information on these phenomena is provided in the following section.

Capillary [45] and slit [46] rheometers have been used to determine the effect of pressure on viscosity, either by using a piston-driven instrument with a throttling valve at the exit and operating at fixed flow rates [47–50] or by controlling the pressure at the entrance and the exit and operating at a fixed pressure drop [51]. Park et al. [42] compared these methods and discussed methods for interpreting data in terms of a pressure coefficient for viscosity. Issues that need to be addressed are the variation of pressure along the capillary and viscous heating.

We note in conclusion that capillary rheometers are useful for the determination of melt viscosity at shear rates far above those accessible in rotational rheometers. However, at high shear rates pressure and temperature variations become important and must be taken into account in the analysis of data. Also, the occurrence of slip may limit the shear rate at which reliable data can be obtained. The shear rate at which these phenomena appear depends on the polymer and its viscosity. For example, a high viscosity HDPE can exhibit gross melt fracture at a shear rate below 100 s^{-1}, while a high-flow (low viscosity) melt might produce reliable results at shear rates several times this high.

6.3.2 Extrudate Distortion: Gross Melt Fracture and Sharkskin

Several polymers, particularly polyolefins and polystyrene, are subject to several types of instability in capillary flow that occur at the entrance or the exit from the die, and some of these also affect melt forming processes. A comprehensive treatment of all these phenomena can be found in the book by Koopmans et al. [52]. At the entrance to the capillary streamlines converge, and this results in stretching along streamlines and high tensile stresses, especially when the entrance has a sharp corner. One manifestation of these stresses is a swirling motion of the melt entering the capillary, which results in extrudate in the form of a spiral.

A more dramatic instability is *gross melt fracture* (GMF), which occurs when the tensile stress at the entrance is sufficient to cause the melt to rupture [53]. This is shown in Fig. 6.7, which displays photos of cross-sections of extruded melt at three flow rates [54]. The extrudate distortions were made visible by placing layers of carbon-black-filled resin in the barrel. The entrance acceleration stretches these layers out into cylinders. At the lowest shear-rate (top photo) the exterior of the extrudate, as well as the black cylinders, are smooth, and there is no distortion. At the next highest shear rate (center) there is an exterior roughness called sharksin, which is discussed below, while the interior flow remains smooth, showing that there is no entrance instability. The third photo (bottom) shows the result of melt at the rupture at the entrance where the melt has been broken into pieces. Following the rupture of the melt in the highly stretched region near the center, melt flows into the capillary from the surrounding region. Then continuous flow reestablishes itself, only to break once more.

Kim and Dealy [54] estimated the critical tensile stress for the onset of GMF. They reported that this stress was insensitive to molecular weight but increased with polydispersity and level of long-chain branching [55]. Furthermore, it decreased somewhat with entrance angle at high flow rates at angles below 90°.

Gross melt fracture is not thought to have an important effect on melt forming processes with the possible exception of high-speed wire coating [53]. However, viscous dissipation plays an important role in wire coating due to the very high shear rate near the wall, and the resulting large increase in temperature greatly reduces the stresses in the melt.

Surface melt fracture, usually called sharkskin, results from a tensile failure of melt at the exit of the die. The high tensile stress results from the sudden acceleration of melt from the wall region when it is abruptly freed from the restraining influence of the wall and is pulled forward by shear stress from the faster-moving melt closer to the axis. This results in small grooves that sometimes spiral around the extrudate. Figure 6.8 is a photo of extrudate showing this phenomenon [56], and Fig. 6.9 is a sketch based on the detailed observations of Migler et al. [57]. A crack forms at the die exit, then separates and moves downstream to be followed by another crack. Sharkskin can be a problem in film blowing, sheet extrusion and blow molding. It is most often addressed by the use of a processing aid, usually a

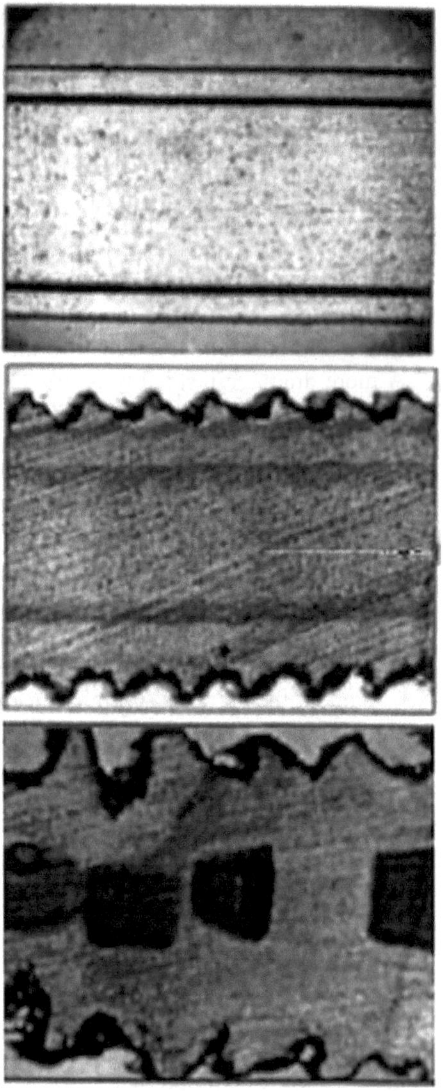

Fig. 6.7 Cross-sections of melt extruded from a capillary. Flow patterns were made visible by loading layers of carbon-black-filled resin into the reservoir. At the lowest flow rate (*top*) the exterior is smooth, and there is no interior distortion of the streamlines. At the next highest shear rate sharkskin fracture is present on the surface, but the interior flow is till smooth. At the highest flow rate, stretching of melt at the entrance to the capillary has broken the melt into pieces. Research of the late Sing Oh Kim [54]

fluoropolymer that promotes slip at the die wall. Slip greatly reduces the wall shear stress so that the melt near the wall is much closer to that of the bulk flow when it reaches the exit. It has been claimed that there is a *change of slope* in a flow curve at the onset of sharkskin, but there is no objective basis for this [58]. In any event, the "onset" reported is the point at which roughness becomes readily visible to the naked eye, while closer inspection reveals that it is already present at lower flow rates.

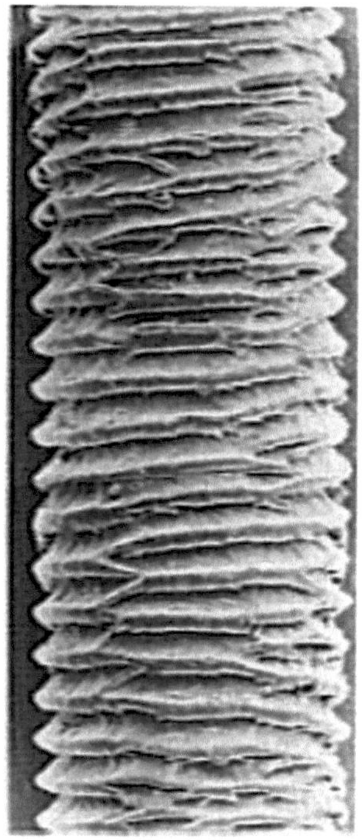

Fig. 6.8 Photo of sharkskin exit fracture. From Bergen [56]

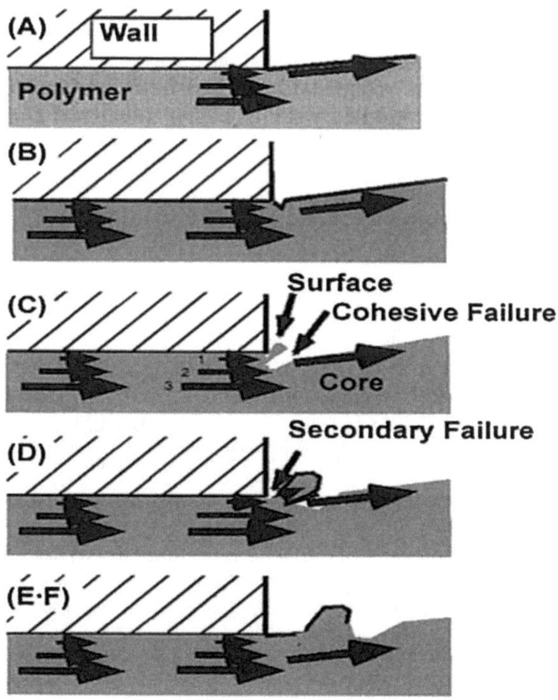

Fig. 6.9 Sketch based on detailed observations of sharkskin melt fracture. From Migler et al. [57]

6.3.3 The Spurt Effect, Oscillating Flow, and Wall Slip

In the capillary flow of linear polymers at controlled driving pressures, a curious thing happens at a particular shear stress. This is shown in Fig. 6.10, which is the flow curve of a linear polyethylene [59]. At the shear stress (proportional to the apparent shear rate) indicated by A, the flow rate suddenly increases by a factor of about twenty! This is called the spurt effect. When the flow is driven by a piston moving at successively increasing speeds, at the constant speed corresponding to point A the flow rate increases suddenly to point B but then decreases toward point C. At C it suddenly shifts back to point B, only to start the process over again. This phenomenon is called a stick–slip instability or oscillating flow. One may wonder how the flow rate can vary while the piston speed is constant. While melts are taken to be incompressible for many purposes, their densities do indeed depend on pressure. At point A, wall slip suddenly occurs. This drastically decreases the wall shear rate and the wall shear stress and thus the pressure drop in the capillary. The melt in the reservoir expands in response to the reduced pressure, and the melt released floods into the capillary. If the pressure is increased to compensate for this, the flow rate must increase until the pressure drop for slip flow is high enough to match that for non-slip flow at the same total flow rate. In the case of constant piston speed, the sudden release of compressed melt leads to a reduction of pressure and a decrease in flow rate back to point C, where pressure starts to rebuild and the process repeats itself.

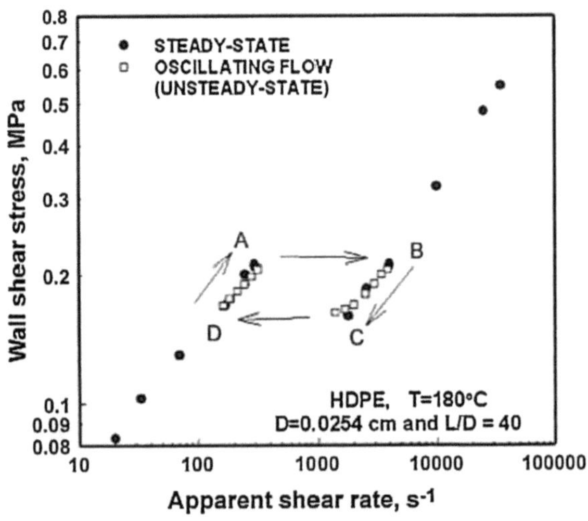

Fig. 6.10 Flow curve of a HDPE. As the flow rate increases from zero, steady flow is interrupted at point *A* by oscillations in which the system moves around the cycle A–B–C–D until the feed flow rate reaches that for point *B*, after which it increases steadily. The oscillating flow regime results from periodic wall slip (*low pressure*) and re-adhesion (*increasing pressure*), and starting when the driving pressure is compatible with steady slip flow, the flow rate increases steadily again. Data of Hatzikiriakos [59]

Wall slip occurs in rheometers and melt forming processes and has been investigated extensively [60]. It is now understood that melts do not actually slip on a high-energy surface such as steel. The molecules in contact with the wall are strongly adsorbed at multiple sites, leaving only loops and chain ends to entangle with bulk material. This entanglement is relatively weak, allowing the bulk to slip past the adsorbed molecules. True wall slip can be promoted by coating the wall with a polymer that is incompatible with the bulk so that none of its molecules are adsorbed at the wall [61]. For example, fluorocarbons promote slip in polyolefins, and polyolefins promote the slip of fluorocarbons.

Early studies made use of capillary flow and often reported exponential relationships between slip velocity and shear stress:

$$v_s = k\sigma^n \tag{6.25}$$

More reliable data can be obtained using a sliding plate instrument in which the shear rate and stress are uniform, and the metal surfaces can be easily exposed for cleaning and coating. If a nominal shear rate is defined as V/h, where V is the velocity of the moving plate, and h is the gap between the two plates, the true shear rate is given by:

$$\dot{\gamma} = \frac{V - v_s}{h} = \dot{\gamma}_n - \frac{v_s}{h} \tag{6.26}$$

Figure 6.11 shows sliding plate data for a polybutadiene, which shows slip velocity as a function of shear stress and pressure [62]. The presence of two maxima is consistent with the observation of spurt and oscillatory flow in capillary flow, and we see that there is a relatively weak slip regime at low stresses with a sudden switch to a high-slip regime when the stress reaches a maximum. But on reducing the stress, the switch back to the low-stress regime occurs at a lower stress giving rise to the hysteresis observed in capillary data.

6.3.4 The Extrusion Plastometer (Melt Indexer)

The extrusion plastometer was developed around 1950 in response to the need for a standard way of specifying the fluidity of molten polymers for use by resin manufacturers as a product characteristic. Widely referred to as a *melt indexer*, this is a primitive capillary instrument in which the driving force is applied to the piston by placing a weight on a platform attached to its upper end. A specific protocol for the use of this device was subsequently adopted by the American society for testing and materials (ASTM) as Standard Test Method Dl238. ASTM test methods are revised from time to time, and the version published in a given year indicates, after the number of the test method, the year of the most recent revision; at time of publication the most recent version was ASTM D1238-10. The test method prescribes the exact geometry of the barrel and die and the detailed procedure to be used for the measurement. Figure 6.12 shows the essential

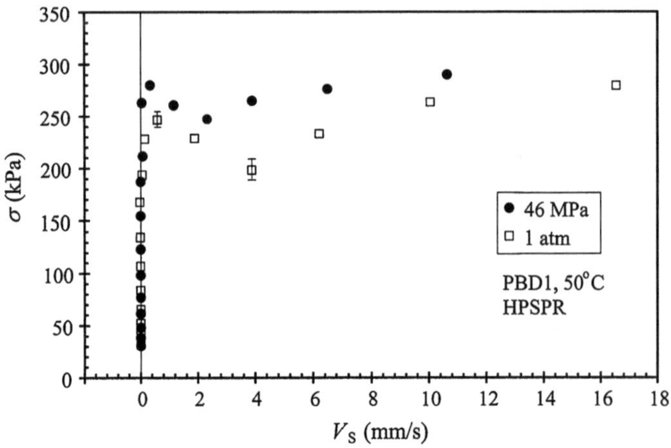

Fig. 6.11 Slip velocity versus wall shear stress for a polybutadiene at two pressures as measured in a sliding plate rheometer. Between the maximum and minimum stresses there is a zone of transition from weak slip to strong slip. Due to melt compressibility, this zone cannot be probed using a capillary rheometer, because the pressure, density and shear stress are all related. From Lim et al. [62]

elements of the extrusion plastometer described in ASTM D1238. Polymer pellets are loaded into the barrel (6) equipped with a thermometer (3) and surrounded by an electrical heater (4) and an insulating jacket (5). The weight (1) drives the piston (2), which forces the polymer through the die (7). "Procedure A" involves cutting and weighing the length of extrudate produced during a time interval indicated by a stop watch. Alternatives to manual operation, "Procedures B and C" involve automatic measurement of the distance the piston travels during a specified time interval. To calculate the mass of melt extruded it is necessary to know the melt density.

A table in the standard test method lists a number of standard combinations of temperatures and loads. For example, the conditions usually used for polyethylenes are indicated as 190/2.16, which implies a measurement at 190 °C with a total load 2.16 kg. The mass of the weight is 2 kg, and that of the piston is 0.16 kg. The recorded datum is the *flow rate* (FR), which is the mass of polymer extruded during a particular time interval, in units of grams per ten minutes. Other names commonly used for this quantity are melt flow index (MFI), melt index (MI) and melt flow rate (MFR). In reporting an MI value it should be accompanied by the temperature and load used, for example 190/2.16. The mass of the sample and the time interval for collecting extrudate are selected based on the expected value of the Flow Rate. For example, for FR values less than 1.0 g/10 min, 2.5–3 g of sample and a measurement time of 6 min are recommended, while for a high-flow polymer with FR greater than 25, 4–8 g of sample and a time of 0.25 min are recommended. Obviously FR or MI increases as viscosity decreases, i.e., a *high MI* resin has a low viscosity.

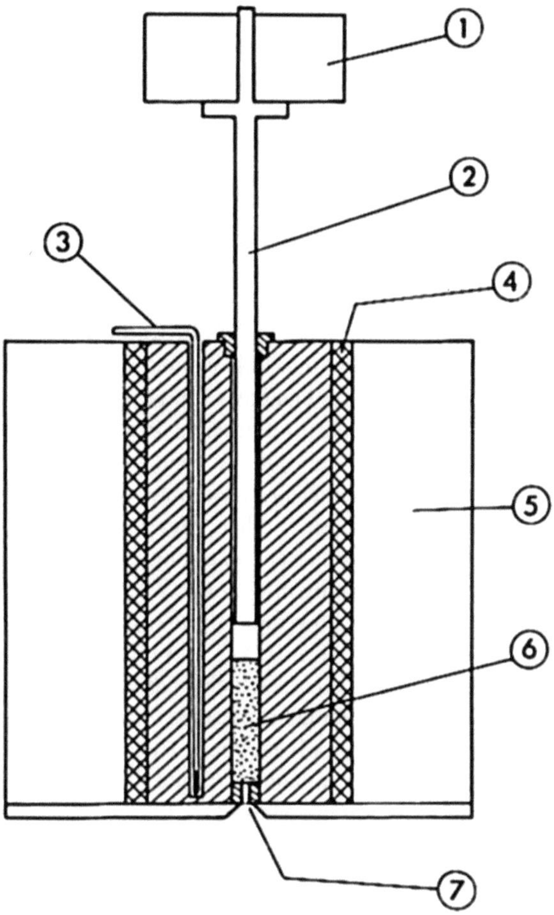

Fig. 6.12 Cross-section of the extrusion plastometer (melt indexer) described in ASTM Standard Test Method D1238. Drawing courtesy of Jean Leblanc

The Flow Rate, or Melt Index, is now used globally, and standard test methods equivalent to ASTM D1238 have been adopted by virtually every industrial nation. As a result MI is nearly always the only melt flow property listed among the specifications of commercial thermoplastic resins. Typical values for polyethylenes under condition 190/2.16 are: blow molding 0.2–0.8 (called *fractional MI* grades); extrusion and film blowing 1.0; rotational molding 2–10, and injection molding 1–30. *High-flow* resins for injection molding of thin parts have values ranging up to 80 or higher.

While one can calculate the driving pressure P_d and then use Eqs. (6.16) and (6.17) to calculate an apparent shear stress and an apparent shear rate, these have little relationship to the true values of these quantities, because the L/D is only four, and the region of fully-developed flow, if any, is very short. Therefore, FR cannot be used to calculate a well-defined physical property. A variation in resin formulation can affect both the entrance pressure drop and the viscosity, and the change in FR will reflect both of these effects. In brief, the extrusion plastometer is not a rheometer but an empirically-based melt characterizing tool. The ASTM Standard Test Method warns the user regarding the fundamental significance of the results.

The temperature history of a sample has an effect on the measured MI; ASTM D 1238 specifies a preheat time of 7 min, starting from the loading of the solid polymer into the barrel, and the solid polymer loaded normally comes from a pelletizing operation that introduces its own thermal history. Shekhtmeyster et al. [63] showed that the addition of a nitrogen blanket for laboratory polypropylene measurements can reduce the MI by 20 % and that variations in the additive package for a high-density polyethylene (HDPE) can change the MI by a factor of 3.

Even if the test could be relied upon to yield a good approximation of the viscosity at one shear stress, it still would have the limitation of being a single-point characterization that tells us nothing about the shape of the viscosity curve. As a result, two resins can have similar values of MI but quite different behaviors at shear rates relevant to extrusion. Figure 6.13 shows viscosity curves of three polymers that have very similar melt index values. As can be seen, this reflects their behavior at a rather low shear rate, while the viscosities at high shear rates are quite different from each other. Measuring the flow rates at two or more loads provides some information regarding the shape of the viscosity curve, i.e. the degree of shear thinning of the melt. The most commonly used second total load is 10 kg, yielding the flow rate I10, and the ratio of this to flow rate I2 for a 2 kg weight (total load of 2.16 kg) is reported as I10/I2. Another flow rate ratio sometimes used is I21/I2, which involves the use of a total load of 21.6 kg. Of course both measurements must be made at the same temperature, and this is normally 190 °C for polyethylene. The official name of this ratio is the Flow Rate Ratio FRR, but it is sometimes called the melt flow ratio, MFR, which should not to be confused with the Melt Flow Rate. In general, increasing the polydispersity of a linear polymer increases I10/I2. And increasing long-chain branching while keeping polydispersity constant also increases I10/I2. This is illustrated in

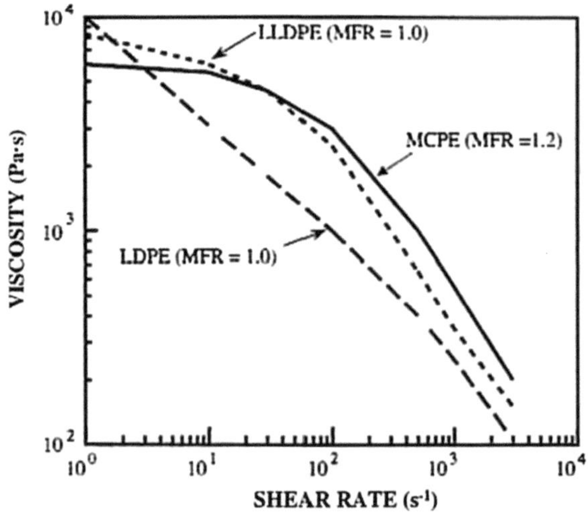

Fig. 6.13 Viscosity versus shear rate plots for three polyethylenes having very similar melt index values but whose viscosity curves have different shapes. The conditions approximating those in the melt indexer are in the *upper left corner*, while many melt forming processes involve much higher shear rates where the viscosities of the three polymers are significantly different. Figure courtesy of Don Delaney, Dynisco Polymer Test

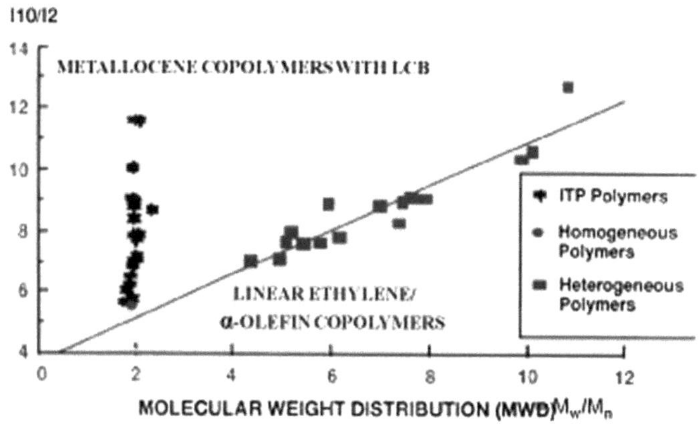

Fig. 6.14 The ratio of high-load to low-load flow rates in an extrusion plastometer used to indicate degree of shear-thinning of metallocene polyethylenes. In traditional HDPEs the polyidispersity index (PI $\equiv M_w/M_n$) is increased to achieve stronger shear-thinning. But all the samples made using the CGC technology mentioned in Sect. 7.2.3.1 have about the same polydispersity index (PI $\approx$ 2) but vary greatly in viscosity behavior due to the presence of low levels of long-chain branching. From Lai and Knight [64]

Fig. 6.14, which shows the effect of adding small, gradually increasing levels of LCB to a metallocene polyethylene while keeping the polydispersity index (M_w/M_n) equal to about two [64].

6.4 On-Line Rheometers

Rheometers that can be mounted directly on a process stream provide a nearly real-time indication of the state of the melt. Although it is not possible to exercise the same control over experimental conditions as that in the laboratory, on-line monitoring can yield improved precision, especially in the case of high viscosity resins, owing to the reproducible sample history and measuring procedure. On-line monitoring may also be advantageous in the case of polymers that are highly sensitive to environmental conditions, such as polyethylene terephthalate (PET), which is very sensitive to moisture, and polypropylene, which is very sensitive to oxygen. A bypass rheometer consists of a sample line, a gear pump, a capillary, and a pressure transducer, as shown schematically in Fig. 6.15. Compact units use an annular, combined feed and return line, and some can sample melt directly from the barrel of an extruder.

We note that the flow rate in the feed line is governed by the shear rate that one wishes to generate in the capillary or slit, and because the flow channel has a small diameter this flow rate is quite low. The long delay between the time that a change occurs in the process and the time that the output signal reflects this change can be greatly reduced by use of two gear pumps, one that draws melt from the process

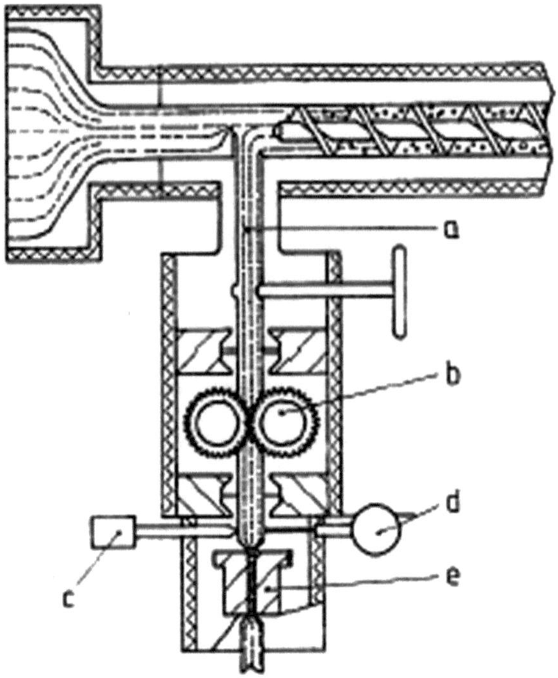

Fig. 6.15 Cross-section of an on-line melt rheometer showing: **a** melt side stream from extruder; **b** gear pump; **c** pressure transducer; **d** thermocouple; **e** capillary. Drawing courtesy of Göttfert

stream and operates at a high rate, and a second that feeds a small side stream to the capillary.

Other aspects of the sampling circuit are also important. A side stream takes melt from the region near the wall of the main process channel, and this is slow moving material that may not be representative of the average properties of the entire flow. The degree to which this is a problem can be determined by comparing on-line measurements with those made in the laboratory using pellets made from extrudate. Of course this procedure assumes that no change takes place in the sample between the point where the rheometer is installed and the pelletizer. Finally, the temperature of the feed line is important. If there are hot spots, these will promote degradation, while cold spots will restrict the sampling flow.

The simplest mode of operation is to select the apparent shear rate, calculate the corresponding flow rate using Eq. (6.17) and set the gear pump to generate this flow rate. Uncertainties arise from the measurement of the pressure drop and the pump speed. A pressure transducer should not be used in the lowest 10 % of its full-scale reading. Instrument error is normally specified as a percentage of the full-scale reading.

On-line instruments are subject to large temperature variations in the plant setting that affect the output signal. In the case of bypass units, the exposure of the bottom of the capillary to ambient air causes the polymer near the exit to be cooler. The sensitivity and zero of pressure transducers depend on their temperature and that of the housing of their electronic components. These characteristics also drift with time, and periodic recalibration is a necessary part of routine maintenance.

6.4.1 On-Line Detection of Melt Index

The flow rate (FR) or melt index (MI) is useful for tracking variations in molecular weight when the shape of the molecular weight distribution does not vary and there is no long-chain branching. Because this quantity is widely used in quality control, it is sometimes desirable to have an on-line rheometer that is capable of providing a continuous indication of the melt index. Of course, it is not possible to reproduce exactly the procedure described in ASTM Standard Test Method D 1238, which involves the use of a piston driven by a dead weight. A simulation of the standard method is achieved by making the entrance and capillary dimensions identical to those in D 1238. A feedback loop is used to provide the gear pump speed required to generate the nominal driving pressure of the test method. A pressure of 43.2 psi is generally used to simulate the melt index. What remains is to convert the volumetric flow rate, Q, to the MI. This requires the density of the melt, ρ_m, since MI is a mass flow rate and is related to the volumetric rate as shown by Eq. (6.27).

$$\mathrm{MI}(\mathrm{g}/10\,\mathrm{min.}) = 10\rho_\mathrm{m}(\mathrm{g}/\mathrm{cm}^3)Q(\mathrm{cm}^3/\mathrm{min.}) \tag{6.27}$$

This procedure is said to work best for materials having MI values between 1 and 20.

Shekhtmeyster et al. [63] demonstrated that the thermal stability of the polymer is an important factor that is often neglected in comparing on-line and laboratory measurements, because it is never possible to reproduce, on-line, the temperature history of the sample of the standard test method.

6.4.2 On-Line Determination of Viscosity

A single-point characterization, such as the melt index, or the viscosity at a single shear rate, may be adequate to track quality when the shape of the molecular weight distribution is not changing and there is no long-chain branching. When the shape of the molecular weight distribution may change, or when long-chain branching is present, a single-point measurement may not be adequate to track resin variation, because several materials may have the same viscosity at one shear rate or stress but quite different viscosity curves. There are on-line units that employ two capillaries to permit measurements at two shear rates.

Viscosity is very sensitive to temperature, and if it is not possible to maintain the temperature of the melt at a standard test temperature, one must compensate for the deviation of the melt temperature from this value. Temperature compensation makes use of concepts introduced in Chap. 3. If the vertical shift factor is taken to be one, the stress versus shear rate curve can be shifted horizontally as indicated by Eq. (6.28).

$$\sigma(\dot{\gamma}, t_\mathrm{o}) = \sigma\{[a_\mathrm{T}(T)\dot{\gamma}],\ t\} \tag{6.28}$$

6.5 High-Throughput Rheometry

High-throughput techniques have been used with success for the rapid evaluation of the potential value of many small experimental samples, particularly in the development of new pharmaceuticals and polymer catalysts. The analytical tools used to characterize samples probe molecules, usually in solution, but rheological evaluation of polymers requires bulk, fully-entangled samples, so samples of at least 5–10 g are required. However, melt rheology is useful at a second stage of the development process after initial screening using traditional high-throughput techniques. While there have been a number of reports of instruments for use with small samples, there remains the challenge of the speed with which a series of materials can be evaluated. A group at the National Institute of Standards and Technology (NIST) has developed a multi-sample micro-slit rheometer (MMR) capable of measuring the viscosity of four samples simultaneously [65, 66].

There are two commercially-available apparatuses for the rapid evaluation polymers, both based on oscillatory shear. The VTM rheometer made by Dynisco is based on an instrument originally developed for the rubber industry. Pre-molded samples are clamped between two sheets of film in a closed-cavity shearing chamber. There is thus some uncertainty as to the exact sample size, and since the cavity is closed around the rim of the sample, and the shape of the cavity conforms to neither cone-plate nor parallel disk geometry, the data cannot be reliably interpreted in terms of storage and loss moduli. The High-Throughput Rheometer (HTR) from Anton Paar [67] is a standard rotational rheometer with parallel-disk fixtures that operates in conjunction with a multi-axis robot that carries out all the steps normally done by a technician to determine the storage and loss moduli. This instrument provides high-quality rheological data, but as one might expect it is large and expensive.

6.6 Extensional Rheometers

6.6.1 Introduction

It is more difficult to measure precisely the response of a melt to a stretching deformation than to a shearing deformation. However, without a reliable constitutive equation based on a molecular model, there is no way to predict how a melt will behave in a large, rapid extensional flow, based only on knowledge of its response to shearing deformations. In other words, extensional tests reveal aspects of the nonlinear viscoelasticity that cannot be predicted from shear data. These aspects are particularly important for polymers with significant levels of long-chain branching. Therefore, the essential criterion for the success of an extensional flow technique is the degree of departure from linear viscoelastic behavior that can be generated. Melt behavior has been studied using uniaxial (also called simple or tensile), biaxial, and planar extensional flows [5, Chap. 7]. These are described in Chap. 4 and shown in Figs. 4.8–4.10. However, only uniaxial extension is in general use and discussed here.

The extensional rheometers described in the next section are limited to use at strain rates well below 10 s^{-1}. In order to reach higher rates, drawdown of an extruded filament (melt spinning) and converging flow at the entrance to a capillary have been used to determine *apparent* extensional viscosities, but they cannot be used to infer well-defined material properties such as $\eta_E^+(t, \dot{\varepsilon})$. These methods are described in Sect. 6.6.3.

6.6.2 Rheometers for Uniaxial (Simple) Extension

A uniaxial extensional rheometer generates a deformation in which the Hencky strain rate $\dot{\varepsilon}$ (defined by Eq. 4.18) is maintained constant. The material function that is determined is the tensile stress growth coefficient $\eta_E^+(t, \dot{\varepsilon})$ defined by Eq. (4.21). In the early stages of either a start-up of steady simple extension or when the strain rate is very low, the behavior should follow the prediction of the Boltzmann superposition principle. This can be calculated using the linear relaxation modulus (Eq. 4.22) thus providing a valuable criterion for the accuracy of data from an extensional rheometer, at short times and low rates.

The first reliable extensional rheometer for operation at fixed strain rates was that of Meissner [68, 69] which made use of a *rotary clamp*. In this device two gear-like rotors turn at the same speed but in different directions at the end of a cylindrical sample to generating a constant velocity at a fixed point in space. The Hencky strain rate is the velocity of the melt passing through the rotors divided by the distance from the rotors to a vertical plane where the axial velocity is zero. The sample floats on the surface of an oil bath. A second set of slowly turning rotors at the other end of the sample provides a balancing tensile force, and a later embodiment [70] made use of adhesive to fasten this end of the sample to a fixed end-plate submerged in the oil. In later designs a sample is clamped at both ends, with one clamp fixed in position and the other displaced by means of a servomotor. The most successful configuration is that of Münstedt [71] the essential features of which are shown in Fig. 6.16. The sample is mounted in a vertical oil bath. The bottom end of the small sample is glued to an end plate coupled to a load cell, while the top end plate is connected to the end of a flexible band that winds onto a drum driven by a DC servomotor. This instrument has been used in many important studies of melt rheology. But while it is a fine tool for scientific work, it is not convenient for routine use.

An extensional rheometer suitable for routine use was designed as a fixture to be installed in a standard rotational instrument; this is the Sentmanat Extensional Rheometer (SER) [72, 73] shown in Fig. 6.17. Two drums are rotated by the motor of the rotational rheometer, and the force in the sample is calculated from the torque. A similar device is the Extensional Viscosity Fixture (EVF) developed at

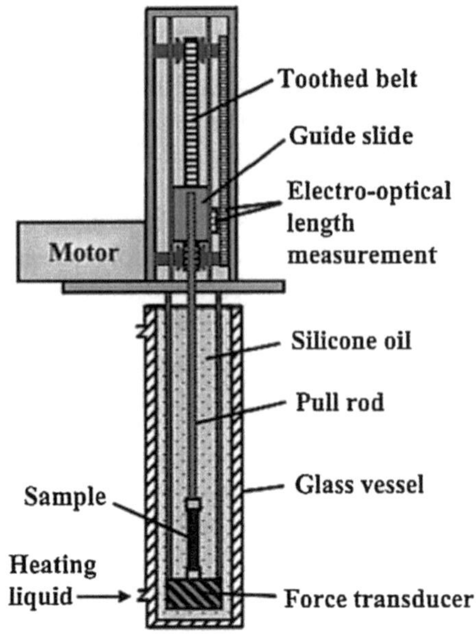

Fig. 6.16 Extensional rheometer for molten plastics developed by Münstedt. The sample is fastened to metal end plates by an adhesive

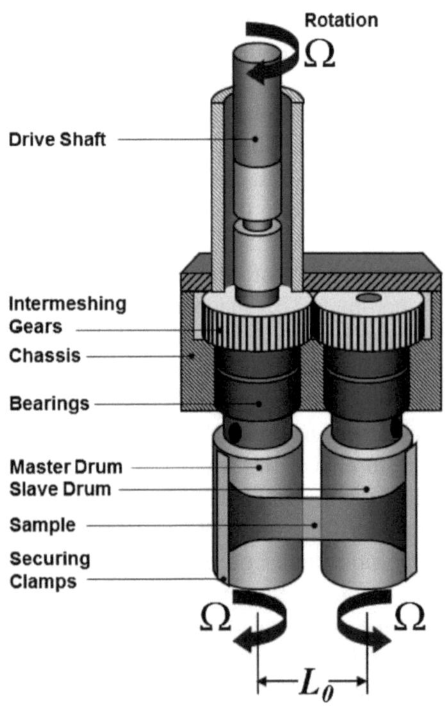

Fig. 6.17 Sentmanat extensional rheometer (SER) fixture for use in a standard rotational rheometer. Drawing courtesy of Xpansion instruments

TA Instruments [74]. There is no bearing friction contribution to the torque signal, because the two drums are decoupled in the EVF, but one of the drums rotates around the other, so the sample does not stay in view during stretching. In both devices, the sample is a small molded rectangular plaque that is clamped at each end to a rotating drum. Because the sample is small and thin, and all its surfaces are exposed to the oven environment, heating is very fast. Sample changing is also fast, taking only about a minute. In addition to start-up of steady, simple extension, creep and stress relaxation experiments can be carried out.

The control of sample temperature requires special care, since sample surfaces are entirely exposed to the oven atmosphere. A recommended procedure is to preheat the oven for fifteen minutes before loading a sample and to position the temperature sensor as close to the sample as possible. If the sample loading time can be kept to 15–20 s, 30–60 s should suffice for heating the sample in an oven of modern design.

Sample preparation is another issue of concern because of the need to avoid residual stresses. This is not likely to be a problem with resins for which stresses relax rapidly, thermal expansion is small, and crystallinity is low. But in general it is wise to compression mold a flat sheet between non-stick surfaces, with no edge constraints, to the desired thickness and then cut samples from the center of the sheet using a dual-blade knife. For thermoplastic melts sample widths from 3 to 12.7 mm and thicknesses from 0.2 to 0.8 mm are commonly used. Lower molecular weight, linear polymer samples generate lower torques and sag more, and use of a wider sample maximizes the torque and reduces sagging.

Nielsen et al. [75] used a model to determine the effect of sample aspect ratio (thickness H over width W) on data obtained using the fixtures mentioned above. They found that when this ratio is much less than one the deformation deviates from uniaxial extension and starts to approach planar extension. However, data comparisons have indicated that while this may be a problem with elastomers and at high strains, it is not a problem with most thermoplastics. For very elastic materials, when this is a concern, a sample thickness less than 0.3 mm and a width less than 3.5 mm are recommended.

Equation (6.27) corrects the sample cross-sectional area for the decrease in density on melting and for the increase in length [72].

$$A(t) = A_0 \left(\frac{\rho_S}{\rho_M}\right)^{2/3} \exp(-\dot{\varepsilon}\, t) \tag{6.29}$$

The density correction is from 10 to 20 % and need not be made if one is only comparing similar polymers, but it is necessary to obtain true values of rheological properties.

The sample sags a bit after being clamped, while being heated and before stretching begins, and there are two ways to deal with this. One approach is to pre-stretch the sample at a very low rate until there is a small torque value, wait for the force to relax, and then start the experiment. Another approach that avoids introducing a pre-test strain history is to use a time offset when plotting the data.

The SER has been found to yield data that are in reasonable agreement with those from the sophisticated Münstedt tensile rheometer [76]. Aho et al. [77] proposed a protocol for use of the SER that gives good results; by not using the sample clamps they were able to achieve higher strains, since the sample does not have to wrap on top of the clamp after one revolution. They also corrected for thermal expansion of the sample and ensured good temperature control and uniformity by mounting the temperature sensor next to the SER frame and leaving the fixture in the oven for up to two hours before use to ensure that it is at the set temperature.

Figure 6.18 [76] compares data obtained using two SER fixtures, one installed in an ARES rheometer (TA Instruments) and the other in an MCR501 rheometer (Anton Paar). Also shown is the linear viscoelasticity prediction based on storage and loss moduli data. The agreement between the two instruments is very good, as is the agreement with the linear prediction at low strain rates or short times.

Figure 6.19 is a photo of the EVF fixture. Stadler et al. reported on the use of the EVF to study the behavior of HDPE [78] and LDPE [78, 79]. Jackson et al. [80] used an EVF to evaluate PVC additives for foam applications and polylactic acid (PLA) for use in film blowing. For comparison of materials they plot $\eta_E^+(t, \dot{\varepsilon})$ versus time using linear scales at a single rate. Figure 6.20 shows an example of the use of this technique to study the effect of resin additives [81].

A problem that arises in all the methods described in this section is the necking of the sample that always occurs at some point during a stretching test. For strain-hardening melts, there is a maximum in the transient property $\eta_E^+(t, \dot{\varepsilon})$, and using available instruments it was impossible to reach strains significantly beyond the maximum before necking and rupture occurred. Figure 4.9 showed data for LDPE that exhibited maxima at all strain rates. Published values of the uniaxial extensional viscosity $\eta_E(\dot{\varepsilon})$ defined by Eq. (4.23) were based on the assumption that the maximum was the steady-state value [81]. However, data obtained using a filament stretching rheometer (FSR) [82] suggest that that in strain-hardening melts the

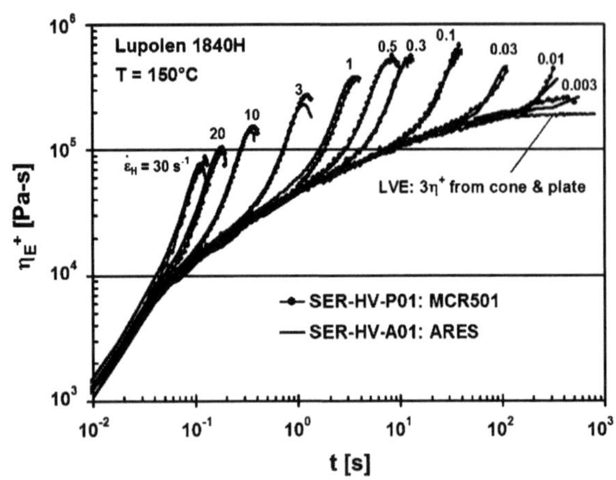

Fig. 6.18 Tensile stress growth coefficient data for an LDPE obtained using an SER installed in rotational rheometers made by TA Instruments (ARES) and by Anton Paar (MCR 501). From Sentmanat et al. [76]

Fig. 6.19 Photo of extensional viscosity fixture (EVF) made by TA instruments. Courtesy of instruments courtesy of Catherine Jackson [80]

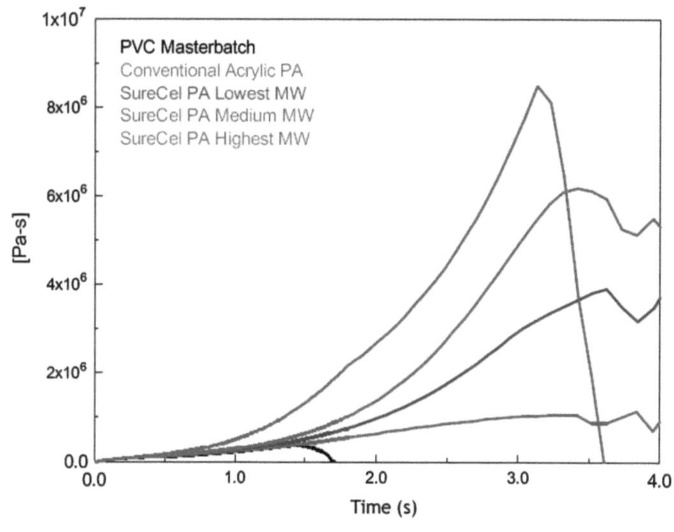

Fig. 6.20 Use of an EVF to compare the effectiveness of processing aids to make PVC strain hardening for use in foam. Extensional stress growth coefficient $\eta_{\mathrm{E}}^{+}(t)$ as function of time at 1 s^{-1} and 185 °C. From C. L. Jackson [80]

maximum in $\eta_E^+(t, \dot{\varepsilon})$ is followed by a decrease to a true steady state, as shown in Fig. 6.21 [83, 84]. Because this sophisticated research tool is not suitable for routine industrial use, the devices now widely used to determine the response of melts to stretching are the rotating cylinder fixtures used with standard rotational rheometers. Thus, for practical applications, one relies on the transient portion of

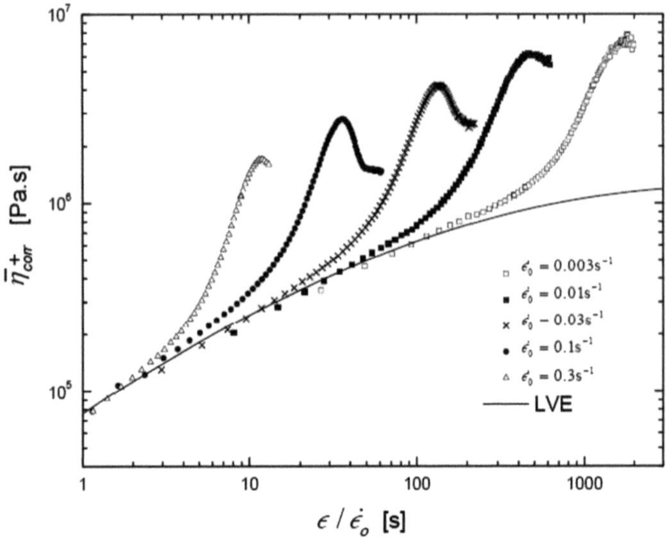

Fig. 6.21 Extensional stress growth coefficient of a LDPE measured using a filament stretching rheometer. At some strain rates, after reaching a peak the coefficient decreases to reach a plateau. From Huang et al. [84]

the stress-growth data obtained prior to necking. In any event, the conditions under which steady extensional strain can be achieved are very far removed from those that arise in melt forming processes. To determine when necking starts to compromise data, one can eliminate the torque near its maximum at several strain rates and observe stress relaxation data. An atypically, rapid initial relaxation at a higher shear rate is a sign that necking has begun.

6.6.3 Converging Flow and Melt Strength

We have seen that rheometers capable of accurate measurements of extensional flow properties are limited to use at low Hencky strain rates, usually well below 10 s^{-1}. In order to reach higher strain rates, the drawdown of an extruded filament (melt spinning) and the converging flow into an orifice die or capillary have been used to determine an *apparent extensional viscosity*. Since the stress and strain are not uniform in these flows, it is necessary to model the flow in order to interpret data in terms of material functions or constants.

In converging flow, one knows only the overall pressure drop and the flow rate, and the results are very sensitive to die exit shape [36, 85]. The first attempt to interpret these data in terms of an apparent extensional viscosity was that of Cogswell [86] based on power laws. Genieser et al. [87] concluded from their detailed simulation of entrance flow that it is not possible to generate a significant

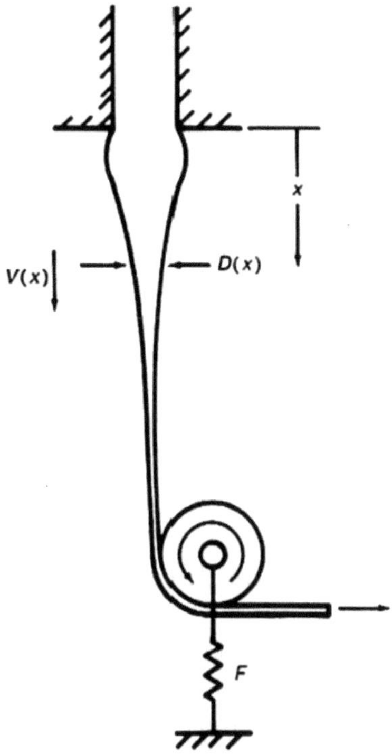

Fig. 6.22 Extrudate drawing test. Melt is extruded from a capillary and drawn down by a rotating element while the tensile stress is monitored

degree of chain stretch using this technique. And data comparisons [35, 88] indicate that differences in extensional flow behavior between resins cannot be reliably correlated with entrance pressure losses.

In the extrudate drawing test, melt is forced through a capillary, usually by means of a capillary rheometer, as shown in Fig. 6.22. The extruded filament is cooled by exposure to ambient air, drawn down by means of a rotary clamp, and the tensile force in the filament is determined by measuring the vertical force on the rotary clamps. In the test procedure usually used, the draw-down is steadily increased until the filament breaks and the stress at break is called the *melt strength*. The Goettfert Rheotens is a commercial device based on this concept.

While the melt strength test has found application in the evaluation of polyethylenes for film blowing, sheet extrusion, and foaming. It does not yield a well-defined rheological property, because neither the strain nor the temperature is uniform in the filament. The melt first experiences shear deformation in the capillary, and the deformation in the filament is primarily uniaxial extension, but the strain rate and temperature vary with distance from the capillary exit up to the point where the melt solidifies.

Laun and Schuch [35] found that filament breakage occurs at a stress of about 0.8 MPa for all the low density polyethylenes they studied, whereas the melt strength, i.e., the force at break, varied strongly with drawdown ratio. This implies that the tensile stress at break is a basic physical property of a polymer. While the stress at break is not a rheological property, the drawdown required to produce this stress depends strongly on the extensional flow properties. Laun and Schuch [35] also examined the effect of die geometry and other parameters on the reliability of melt strength measurements and made the following recommendations. A capillary die with a large *L/D* will minimize the effect of entrance flow history and swell. A high flow rate minimizes the effects of heat transfer on the filament temperature. And a high nominal drawdown ratio (v_L/v_0) and a low value of $\dot{\gamma}_A$ in the die will minimize the effects of swell. These guidelines are not always compatible, and compromise is usually necessary in designing a test protocol for a given polymer.

Wagner et al. [89] measured the extensional flow behavior of several polymers using both an extensional rheometer and a Rheotens melt strength apparatus and compared the data. A melt strength test can be carried out isothermally [90], but this makes the test more complicated and does not make it possible to determine any basic rheological property [91].

6.7 Torque Rheometers

A torque rheometer consists of a powerful motor drive, in the range of 1–7 horsepower, a torque sensor, and a mixing head or lab-scale extruder. The mixing heads are miniaturized batch mixers of various designs, from low-intensity open mixers to high-intensity internal mixers. Within dual cavities, rotors of several types generate a complex mixing pattern. These range from sigma blades, for low shear, through cams for medium shear, to rollers for high shear; also available are planetary and Banbury mixing heads. The recipe of interest is loaded into the cavities through a port equipped with a weight-driven piston to keep the mix under pressure. As mixing proceeds the torque on the drive shaft is monitored to produce a characteristic pattern of mechanical energy input.

Although there have been attempts to develop relationships between the torque and the viscosity of the mixture, the complexity of the flow in the mixing head renders this a highly unreliable procedure. However, for certain applications torque rheometer data are reasonably repeatable and are sometimes used for process control. The advantage of the torque rheometer is its ability to simulate in the laboratory industrial mixing processes in which rheology plays an important role. In particular, the time evolution of a mixing or reaction process can be monitored.

Torque rheometers have been found useful for PVC compounds, crosslinking polymers, and elastomers. For example, ASTM Standard Test Method D 2396 [92] deals with plasticizer absorption by PVC, while D 2538 [93] describes a test for the fusion rate of PVC. And D 3795 [94] deals with the cure properties of thermo sets as determined using a torque rheometer.

Also available as accessories for use with the basic drive unit of torque rheometers are small single- and twin-screw extruders and other continuous mixing systems. Among the various dies offered for use with the extruders are capillary and slit dies designed for the measurement of melt viscosity.

6.8 Using Rheology for Statistical Process Control

This topic is treated in detail in the book by Dealy and Saucier [95], and we present here only a brief summary. Quality control is a long-established method of ensuring that all products shipped to or received by a resin customer meet contract specifications. One or more samples of product are taken from each batch or at regular time intervals from a process stream and subjected to laboratory tests. By the time a resin manufacturer discovers that off-spec material has been produced, a substantial amount of this material has already been produced and cannot be sold as is. The more modern approach to quality is statistical process control.

The melt index or an apparent viscosity value is often used for statistical process control. First, a record is kept of test data obtained in successive tests, and these are plotted versus time to create a run chart. This shows the normal variation of the melt index from one sample to the next, and based on experience with a given resin upper and lower *control limits* are established that are well within the *specification limits*. These are shown as horizontal lines on the run chart to form a control chart. Figure 6.23 is an example of a control chart that shows the viscosity of each sample and the moving range (MR) which is the absolute value of the difference between successive values. The latter highlights test-to test variations. Also shown are the long-term average line and upper and lower control limits.

Instead of rejecting material that goes outside the specification limits the objective is to watch the control chart for signs that the process may be out of control, before any off-spec material has been produced. A commonly used set of rules for deciding when there has been a significant process upset the cause of which needs to be sought was developed by Western Electric, and one example is when two out of three consecutive points are on the same side of the average and more than two standard deviations above or below the long-term average.

6.9 Sample Stability: Thermo-Oxidative Degradation and Hydrolysis

At elevated temperatures, the rheological properties of polymer are subject to change with time due to reactions of several types. Polyolefins are prone to thermo-oxidative degradation in the presence of oxygen; polypropylene is especially sensitive to this problem due to its many tertiary carbons. In rotational rheometers the oven containing the test fixtures should be flooded with nitrogen, but technical grade nitrogen contains a small amount of oxygen, and the sample itself is may contain a bit. Commercial resins are generally stabilized by means of an additive such as the Irganox® line of antioxidants (BASF). The additive effectiveness decreases with time at temperature, and the amount added is generally sufficient to stabilize the resin through a typical melt processing operation. It is always wise when studying a new material to verify the time available for

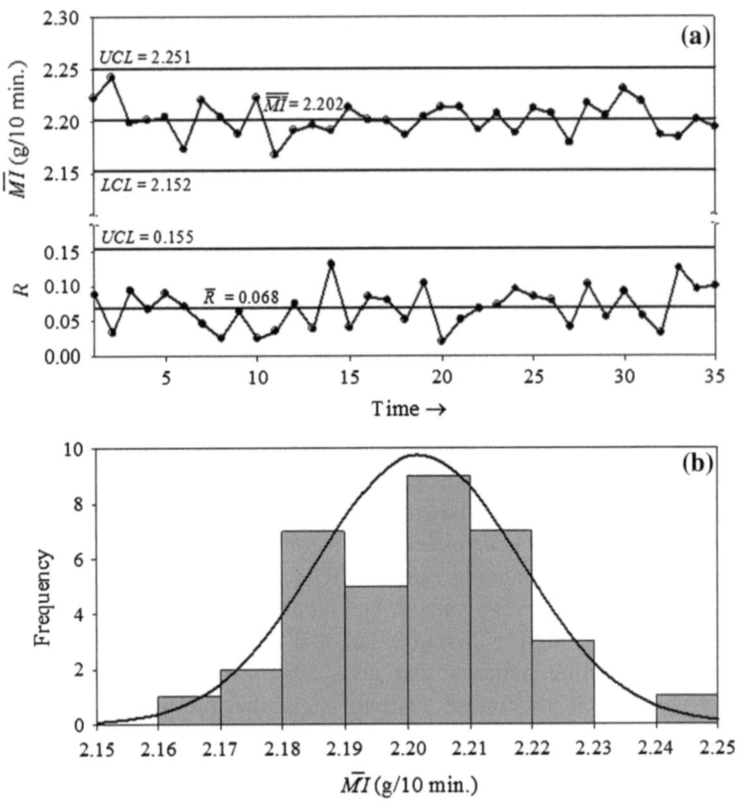

Fig. 6.23 Upper figure is a control chart of subgroup average melt index test results over time. Upper and lower control limits are shown as well as the long-term average value (2.202 gm/10 min). Points at bottom show the range of values within each subgroup. The bar chart shows the frequency distribution of MI subgroup averages

measurements by doing a time sweep in a rotational rheometer at a fixed frequency. A fairly low frequency should be used as the stress amplitude will be more sensitive to the presence of the longer molecules most likely to undergo chain scission. Samples synthesized for research should have stabilizer added before experiments are undertake. This is especially important when the amount of material available is small.

Condensation polymers such as polyamides (Nylons), polycarbonates and polyesters, e.g., polyethylene terephthalate (PET), are very sensitive to moisture, and this poses a problem in both processing and laboratory studies. If the amount of water present is less than the equilibrium concentration, polymerization takes place, and the molecular weight and the viscosity increase, whereas if the water concentration is more than the equilibrium concentration, hydrolysis takes place, and the molecular weight and the viscosity decrease. It is crucial to control the humidity in both situations. Pezzin [96] and Laun [97] studied the effect of moisture on capillary rheometer data for a polyamide (PA) 6. And Khanna et al. [98] reported that reducing moisture in a PA6 from 0.1 to 0.02 % doubled the zero-shear viscosity. Although PA11 absorbs less moisture than PA6 or PA 6–6, reversible hydrolysis is still a concern. Figure 6.24 shows the variation of water concentration with time for a PA11 [99], as determined by Karl Fischer titration. Vacuum drying of samples is essential; exposure of dried samples to air prior to testing was avoided by squeezing pellets in the rheometer rather than forming disks in an earlier step. Figure 6.25 shows the evolution with time of the viscosity of the PA11 of Fig. 6.24 at three temperatures for an initial water content of 0.095 %. As moisture is absorbed, the molecular weight and viscosity increase.

A study of the variation of viscosity with time of polyethylene terephthalate [100] revealed two effects: a fast hydrolysis reaction followed by slow thermo-oxidative degradation. The data were interpreted in terms of rate constants for these two processes.

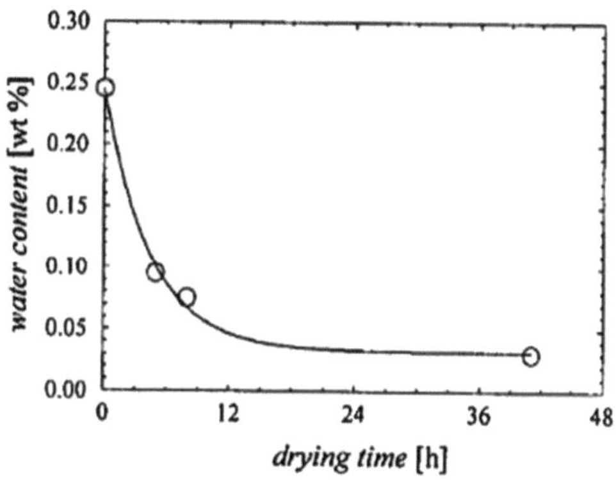

Fig. 6.24 Water concentration in a polyamide as a function of drying time. From Acierno and Van Puyvelde [99]

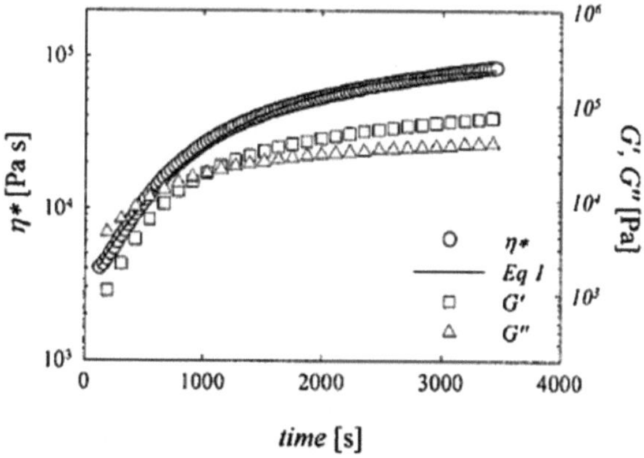

Fig. 6.25 Evolution of viscoelastic properties with time of the same polyamide as in Fig. 6.24. From Acierno and Van Puyvelde [99]

References

1. Giacomin AJ, Samurkas T, Dealy JM (1989) A novel sliding plate rheometer for molten polymers. Polym Eng Sci 29:499–504
2. Dealy JM, Giacomin AJ (1998) Sliding plate and sliding cylinder rheometers. In: Collyer AA, Clegg DW (eds) Rheological measurement, 2nd edn. Chapan and Hall, London
3. Larson RG (1992) Instabilities in viscoelastic flows. Rheol Acta 31:213–226
4. Läuger J, Stettin H (2010) Differences between stress and strain control in the non-linear behavior of complex fluids. Rheol Acta 49:909–930
5. Macosko CW (1994) Rheology: principles, measurements and applications. VCH Publishers, New York
6. Auhl D, Ramirez J, Likhtman AE, Chambon P, Fernyhough C (2008) Linear and nonlinear shear flow behavior of monodisperse polyisoprene melts with a large range of molecular weights. J Rheol 52:801–835
7. Stadler FJ, Münstedt H (2008) Terminal viscous and elastic properties of linear ethane/α-olefin copolymers. J Rheol 52:697–712
8. Gevgilili H, Kalyon DM (2001) Step strain flow: Wall slip effects and other error sources. J Rheol 45:467–475
9. Laun HMD (1978) Description of the non-linear shear behavior of a low density polyethylene melt by means of an experimentally determined strain dependent memory function. Rheol Acta 17:1–15
10. Venerus DC, Kahvand H (1994) Normal stress relaxation in reversing double-step flows. J Rheol 38:1297–1315
11. Mackay ME, Dick GK (1995) A comparison of the dimensions of a truncated cone measured with various techniques: The cone measurement project. J Rheol 39:673–677
12. Vermant J, Moldenaers P, Mewis J, Ellis M, Garritano R (1987) Orthogonal superposition measurements using a rheometer equipped with a force rebalanced transducer. Rev Sci Instrum 68:4090–4096
13. Mackay ME, Halley PJ (1991) Angular compliance error in force rebalance torque transducers. J Rheol 35:1609–1614
14. Niemiec JM, Pesce JJ, McKenna GB (1996) Anomalies in the normal force measurement when using a force rebalance transducer. J Rheol 40:323–334
15. Archer LA, Chen YL, Larson RG (1995) Delayed slip after step strains in highly entangled polystyrene solutions. J Rheol 39:519–525
16. Hutton JF (1969) Fracture and secondary flow of elastic liquids. Rheol Acta 8:54–59
17. Gleissle W (1974) Schub-und Normalspannungsmessungen an Silikonölen bei hohen Schergefällen mit einem neuen Kegel-Platte-Rheometer. Coll Polym Sci 252:848–853
18. Mattes KM, Vogt R, Friedrich C (2008) Analysis of the edge fracture process in oscillation for polystyrene melts. Rheol Acta 47:929–942
19. Kulicke WM, Porter RS (1979) Irregularities in steady flow of non-Newtonian fluids between cone and plate. J Appl Polym Sci 23:953–965
20. Kulicke WM, Jeberien HE, Kiss H, Porter RS (1979) Visual observation of flow irregularities in polymer solutions. Rheol Acta 18:711–716
21. Chen YL, Larson RG, Patel SS (1994) Shear fracture of polystyrene melts and solutions. Rheol Acta 33:243–256
22. Pearson DS, Rochefort WE (1982) Behavior of concentrated polystyrene solutions in large-amplitude oscillatory shear. J Polym Sci: Polym Phys 20:83–98
23. Quinzani LM, Vallés EM (1986) The Use of a modified cone-and-plate geometry (MCP) in a Rotational rheometer for the measurement of material functions. J Rheol 30(S):S1–S21
24. Larson RG (1985) Nonlinear shear relaxation modulus for a linear low-density polyethylene. J Rheol 29:823–831
25. Snijkers F, Vlassopoulos D (2011) Cone-partitioned-plate geometry for the ARES rheometer with temperature control. J Rheol 55:1167–1186

26. Baek SG, Magda JJ (2003) Monolithic rheometer plate fabricated using silicon micromachining technology and containing miniature pressure sensors for N_1 and N_2 measurements. J Rheol 47:1249–1260
27. Schweizer T (2003) Comparing cone-partitioned plate and cone-standard plate shear rheometry of a polystyrene melt. J Rheol 47:1071–1085
28. Geiger K (1988) Weissenberg-Rabinowitsch-Auswertung der mit dem Platte–Platte-Rotationsrheometer gemessenen fließkurve mittels eines fließgesetzes vom carreauschen Typ. Rheol Acta 27:209–211
29. Soskey PR, Winter HH (1984) Large step shear strain experiments with parallel-disk rotational rheometers. J Rheol 28:625–645
30. Kraft M, Meissner J, Kaschta J (1995) Linear viscoelastic characterization of polymer melts with long relaxation times. Macromolecules 32:751–757
31. He C, Wood-Adams P, Dealy JM (2004) Broad frequency characterization of molten polymers. J Rheol 48:711–724
32. Davies AR, Anderssen RS (1997) Sampling localization in determining the relaxation spectrum. J Non-Newt Fl Mech 73:163–179
33. Han CD (2007) Rheology and processing of polymeric materials, Vol 1. Polymer rheology. Oxford University Press, New York
34. Bagley EB (1957) End Corrections in the capillary flow of polyethylene. J Appl Phys 28:624–627
35. Laun HM, Schuch H (1989) Transient elongational viscosities and drawability of polymer melts. J Rheol 33:119–175
36. Kim S, Dealy JM (2001) Design of an orifice die to measure entrance pressure drop. J Rheol 45:1413–1419
37. Hatzikiriakos SG, Mitsoulis E (1996) Excess pressure losses in the capillary flow of molten polymers. Rheol Acta 35:545–555
38. Aho J, Syrjälä S (2008) Evaluation of different methods for determining the entrance pressure drop in capillary rheometry. Appl Rheol 18:63258-1–63258-5
39. Schummer P, Worthoff RH (1978) An elementary method for the evaluation of a flow curve. Chem Eng Sci 33:759–763
40. Laun HM (1983) Polymer melt rheology with a slit die. Rheol Acta 22:171–185
41. Laun HM (2003) Pressure dependent viscosity and dissipative heating in capillary rheometry of polymer melts. Rheol Acta 42:295–308
42. Park HE, Lim ST, Laun MH, Dealy JM (2008) Measurement of pressure coefficient of polymer viscosity: drag flow versus capillary flow. Rheol Acta 47:1023–1038
43. Hay G, Mackay ME, Awati KM, Park Y (1999) Pressure and temperature effects in slit rheometry. J Rheol 43:1099–1116
44. Hatzikiriakos SG, Dealy JM (1992) Wall slip of molten high density polyethylenes, II, Capillary rheometer studies. J Rheol 36:703–741
45. Kadijk S, Van den Brule BHAA (1994) On the pressure dependency of the viscosity of molten polymers. Polym Eng Sci 34:1535–1546
46. Aho J, Syrjl S (2011) Shear viscosity measurement of polymer melts using injection molding machine with adjustable slit die. Polym Testing 30:595–601
47. Couch MA, Binding DM (2000) High pressure capillary rheometory of polymeric fluids. Polym 41:6323–6334
48. Cardinaels R, Van Puyvelde P, Moldenaers P (2007) Evaluation and comparison of routes to obtain pressure coefficients from high-pressure capillary rheometer data. Rheol Acta 46:495–505
49. Carreras ES, El Kissi N, Piau JM, Toussaint F, Nigen S (2006) Pressure effects on viscosity and flow stability of polyethylene melts during extrusion. Rheol Acta 45:209–222
50. Aho J, Syrjälä S (2010) Measurement of the pressure dependence of viscosity of polymer melts using a back pressure-regulated capillary rheometer. J Appl Polym Sci 117:1076–1084
51. Laun HM (2004) Capillary rheometry for polymer melts revisited. Rheol Acta 43:509–528

52. Koopmans R, Doelder J, Molenaar J (2011) Polymer melt fracture. CRC Press, New York
53. Dealy JM, Kim S (2005) Gross melt fracture in extrusion. In: Hatzikiriakos SG, Migler KB (eds) Polymer processing instabilities. Marcel Dekker, New York
54. Kim S, Dealy JM (2002) Gross melt fracture of polyethylene. I: a criterion based on tensile stress. Polym Eng Sci 42:482–494
55. Kim S, Dealy JM (2002) Gross melt fracture of polyethylene. II: Effects of molecular structure. Polym Eng Sci 42:495–503
56. Bergem N (1976) Visualization studies of polymer melt flow anomalies in extrusion. Proc 7th Intern Cong Rheol 50–54
57. Migler KB, Son Y, Qiao F, Flynn K (2002) Extensional deformation, cohesive failure, and boundary conditions during sharkskin melt fracture. J Rheol 46:383–400
58. Shaw MT (2007) Detection of multiple flow regimes in capillary flow at low shear rates. J Rheol 51:1303–1318
59. Hatzikiriakos SG, Dealy JM (1992) Role of slip and fracture in the oscillating flow of HDPE in a capillary. J Rheol 36:845–884
60. Archer LA (2005) Wall slip: measurement and modeling issues. In: Hatzikiriakos SG, Migler KB (eds) Polymer processing instabilities. Marcel Dekker, New York
61. Kharchenko SB, Migler KB, Hatzikiriakos SG (2005) Conventional polymer processing additives. In: Hatzikiriakos SG, Migler KB (eds) Polymer processing instabilities. Marcel Dekker, New York
62. Park HE, Lim ST, Smillo F, Dealy JM, Robertson CG (2008) Wall slip and spurt of polybutadiene. J Rheol 52:1201–1239
63. Shekhtmeyster J, Garritano R (1997) On-line flow index measurement. SPE ANTEC Tech Papers. 43:1113
64. Lai S, Knight GW (1993) Dow constrained geometry catalyst technology (CGCT): New rules for ethylene α-olefin interpolymers—Controlled rheology polyolefins. Soc Plast Engrs Ann Tech Conf (ANTEC) :1188
65. Moon D, Rur AJ, Migler KB (2008) Multi-sample micro-slit rheometry. J Rheol 52:1131–1142
66. Moon D, Migler KB (2009) Measurement of dynamic capillary pressure and viscosity via multi-sample micro-slit rheometer. Chem Eng Sci 64:4537–4542
67. Laeuger J (2008) Automation and high throughput in rotational rheometry for polymers, SPE ANTEC Tech Papers
68. Meissner J (1969) Rheometer zur Untersuchung der deformationsmechanischen Eigenschaften von Kunstoff-Schmelzen unter definierter Zugbeansprunchung. Rheol Acta 8:78–88
69. Meissner J (1971) Dehnungsverhalten von Polyäthylen-Schmelzen. Rheol Acta 10:230–242
70. Laun HM, Münstedt H (1978) New universal extensional rheometer for polymer melts. measurements on a polystyrene sample. J Rheol 17:421–435
71. Münstedt H (1979) New universal extensional rheometer for polymer melts. J Rheol 24:847–867
72. Sentmanat ML (2004) Miniature universal testing platform from extensional melt rheology to solid-state deformation behavior. Rheol Acta 43:657–669
73. Xpansion instruments. www.xinst.com
74. TA instruments. www.tainst.com
75. Nielsen JK, Hassager O, Rasmussen HK, McKinely GH (2009) Observing the chain stretch transition in a highly-entangled polyisoprene melt using transient extensional rheometry. J Rheol 53:1327–1346
76. Sentmanat M, Wang BN, McKinley GH (2005) Measuring the transient extensional Rheology of polyethylene melts using the SER universal testing platform. J Rheol 49:585–606
77. Aho J, Rolón-Garrido VH, Syrjälä S, Wagner MH (2010) Measurement technique and data analysis of extensional viscosity for polymer melts by Sentmanat extensional rheometer. Rheol Acta 49:359–370

78. Stadler FJ, Nishioka A, Stange J, Koyama K, Münstedt H (2007) Comparison of the elongational behavior of various polyolefins in uniaxial and equibiaxial flows. Rheol Acta 46:1003–1012
79. Stadler FJ, Kaschta J, Münstedt H, Becker F, Buback M (2009) Influence of molar mass distribution and long-chain branching on strain hardening of low density polyethylene. Rheol Acta 68:479–490
80. Jackson CL, VanRheenan PD, Calzia KJ (2010) Melt extensional viscosity for polymer processing of PVC and PLA. Ann Mtg of Society of Rheology
81. Münstedt H, Laun HM (1981) Elongational properties and molecular structure of polyethylene melts. Rheol Acta 20:211–221
82. Anna SL, McKinley GH, Nguyen DA, Sridhar T, Muller SJ, James DF, Huang J (2001) An interlaboratory comparison of measurements from filament-stretching rheometers using common test fluids. J Rheol 45:83–114
83. Rasmussen HK, Nielsen JK, Bach A, Hassager O (2005) Viscosity overshoot in the start-up uniaxial elongation of low density polyethylene melts. J Rheol 49:369–381
84. Huang Q, Skov AL, Rasmussen HK, Hassager O, Harlen O, Hoyle DM, McLeish TC, Hassel D, Lord TD, Mackley MR (2011) Stress maximum and steady extensional flow of branched polymer melts. Society of rheology 83rd Ann. Mtg, Cleveland, Ohio
85. Rajagopalan D (2000) Computational analysis of techniques to determine extensional viscosity from entrance flows. Rheol Acta 39:138–151
86. Cogswell FN (1972) Converging flow of polymer melts in extrusion dies. Polym Eng Sci 64:64–73
87. Genieser LH, Brown RA, Armstrong RC (2003) Comparison of measured centerline stress and velocity fields with predictions of viscoelastic constitutive models. J Rheol 47:1331–1350
88. Hingmann R, Marczinke BL (1994) Shear and elongational flow properties of polypropylene melts. J Rheol 38:573–587
89. Wagner MH, Bastian H, Bernnat A, Kurzbeck S, Chai CK (2002) Determination of elongational viscosity of polymer melts by RME and Rheotens experiments. Rheol Acta 41:316–325
90. Sampers J, Leblans PJR (1988) An experimental and theoretical study of the effect of elongational history on the dynamics of isothermal melt spinning. J Non-Newt Fl Mech 30:325–342
91. Revenu P, Guillet JG, Carrot C (1993) Elongational flow of polyethylenes in isothermal melt spinning. J Rheol 37:1041–1056
92. ASTM D 2396. Standard Test Method for Powder-Mix Time of Poly (Vinyl Chloride) (PVC) Resins Using a Torque Rheometer
93. ASTM D 2538. Standard practice for fusion of Poly (Vinyl Chloride) (PVC) compounds using a torque rheometer
94. ASTM D 3795. Standard test method for thermal flow and cure properties of thermosetting plastics by torque rheometer
95. Dealy JM, Saucier PC (2000) Rheology in plastics quality control. Hanser publishers, Munich
96. Pezzin G (1964) Capillary viscometry of molten polycaprolactam. J Appl Polym Sci 8:2195–2212
97. Laun HM (1979) Das viskoelastische Verhalten von Polyamide-6-Schmelzen. Rheol Acta 18:478–491
98. Khanna PP, Han PK, Day ED (1996) New developments in the melt rheology of nylons 1: Effect of moisture and molecular weight. Polym Eng Sci 36:1745–1754
99. Acierno S, Van Puyvelde P (2005) Rheological behaviour of polyamide 11 with varying initial moisture content. J Appl Polym Sci 97:666–670
100. Seo KS, Cloyd JD (1991) Kinetics of hydrolysis and thermal degradation of polyester melts. J Appl Polym Sci 42:845–850

Chapter 7
Rheology and Molecular Structure

Abstract The sensitivity of rheological properties to molecular structure makes rheological information valuable for characterizing polymers and for understanding their behavior in melt processing operations. Structural factors explored include molecular weight distribution, tacticity, comonomer content, and long-chain branching (LCB). The effects of LCB on melt behavior are complex but very important, and this subject is explored in detail.

The relationship between the rheological properties of polymers and their molecular structure is dealt with in detail in a book devoted entirely to this subject [1]. This chapter contains information the authors believe to be useful to those interested in the development and processing of commercial polymers. In general, we can say that linear viscoelastic behavior, e.g., the storage and loss moduli, is the richest source of information about the molecular structure of linear polymers, while extensional flow behavior provides important additional information in the case of long-chain branched polymers.

7.1 Rheology and Structure of Linear Polymers

7.1.1 Effect of Molecular Weight on Zero-Shear Viscosity

Section 2.4 explains how molecular structure affects viscosity, and this information is summarized briefly here. The zero-shear viscosity of linear, monodisperse polymers is governed by the following relationships:

$$M < M_C \quad \eta \propto M \tag{2.17}$$

$$M > M_C \quad \eta_0 = KM^{\alpha} \quad (\alpha = 3.5 \pm 0.2) \tag{2.18}$$

J. M. Dealy and J. Wang, *Melt Rheology and its Applications in the Plastics Industry*, Engineering Materials and Processes, DOI: 10.1007/978-94-007-6395-1_7,

For polydisperse, linear polymers, Eq. (2.18) is valid with M replaced by the weight-average molecular weight, as long as there are few unentangled molecules present.

$$\eta_0 = KM_w^\alpha \tag{2.19}$$

This implies a blending law for a binary blend of two monodisperse samples of the same polymer having molecular weights M_1 and M_2; according to (2.19) the weight-average molecular weight of the blend is:

$$M_{wb} = w_1 M_1 + w_2 M_2 \tag{2.20}$$

where w, and w, are the weight fractions of the blend components.

Using (2.18) and (2.19) we obtain:

$$\eta_{0,b} = KM_w^\alpha = \left(w_1 \eta_{0,1}^{1/\alpha} + w_2 \eta_{0,2}^{1/\alpha}\right)^\alpha \tag{2.21}$$

In Sect. 6.2.3 it was pointed out that the determination of the zero-shear viscosity may not possible using oscillatory shear, especially in the case of a broad molecular weight distribution. And it was shown there how creep or creep together with creep recovery can be used to probe very long-time (low-frequency) behavior.

7.1.2 *Effect of Molecular Weight Distribution on Shear Behavior*

Molecular weight distribution affects both shear-thinning and end-product performance. While broadening the MWD enhances shear thinning, which is usually desirable, it may not yield optimum product properties, and in resin development it is often necessary to compromise on these objectives. We will see in this chapter how some modern technologies permit independent control of viscosity and MWD.

An example of an early technique to control rheology involved efforts to broaden the market for polypropylene. Traditional Ziegler–Natta catalysts do not permit precise control of MWD, and this proved a barrier to the use of polypropylene for some applications. This motivated the development of *controlled rheology* polypropylene resins having narrower MWD than reactor product that were made by chemical or thermal degradation, a process called *visbreaking*. Narrow distribution polypropylenes can now be polymerized using metallocene catalysts. And we will see in this chapter that certain metallocene catalyst systems can be used to achieve independent control of MWD and shear thinning.

7.1.2.1 Determining MWD from Viscosity

The technique generally used to determine molecular weight distribution (MWD) is gel permeation chromatography (GPC) also called size exclusion chromatography

(SEC). But rheological data are sometimes advantageous for this purpose. First, many polymers dissolve either with great difficulty or not at all. Second, rheological data are more sensitive to high molecular weight fractions than GPC elution curves, and these fractions have an important effect on the ease with which a polymer can be processed in the molten state, i.e., its processability. Also, the number average is significantly less reproducible than the weight average, which leads to uncertainty in the polydispersity index (M_w/M_n). And finally, standard rheological properties such as the viscosity and the storage and loss moduli are easier to determine than GPC elution curves. Several methods have been proposed to infer MWD from rheological data, and these are reviewed in detail by Dealy and Larson [1]. We provide here only a brief summary.

Bruce Bersted [2, 3] in 1975 suggested a technique to estimate the viscosity of a linear polyethylene from its MWD. This was based on the idea that each molecule makes a contribution to the viscosity equal to its zero-shear viscosity but that as the shear rate increases, the maximum length of molecules that make such a contribution decreases. In other words, as the shear rate increases, the effects of progressively shorter molecules on the viscosity are eliminated. Bersted and Slee [4] used this idea as the basis for a method to determine weight distribution from viscosity data. This idea was developed further [5, 6] and improved [7–16]. The method requires the second derivative of (log η) with respect to (log $\dot{\gamma}$) and at least seven precise data points per decade. Wood-Adams and Dealy [8] demonstrated its successful use with a polydisperse polyethylene. They used $|\eta^*|$ versus frequency rather than viscosity versus shear rate data, because precision, accuracy and range are all enhanced in this way. The most important portions of the viscosity curve are those where changes in shape occur [8, 11]. Peaks in the MWD show up in the data as minima in the second derivative. They used a plot of the second derivative versus reduced frequency (ω/ω_c) to evaluate the reliability of the MWD calculation. The region around a minimum is most important. If the measured points, as opposed to the extrapolated points, include this minimum, then the peak in the MWD will be within the experimental window. Figure 7.1 is an example of such a curve for a polyethylene [8].

Wu [14] proposed a method to determine MWD from $G'(\omega)$ data in the terminal and plateau zones. His basic assumption was that the cumulative molecular weight distribution curve has the same shape as the $G'(\omega)$ curve. Turninello [15] reviewed work on relations between $G'(\omega)$ and MWD and proposed an improvement of Wu's method that is better for polymers having a bimodal MWD.

7.1.2.2 Polydispersity Correlations

Zeichner and coworkers [16, 17] developed a measure of the breadth of the molecular weight distribution for a particular set of polypropylenes that is based on the curves of storage and loss moduli versus frequency. Using data for a series of polypropylenes made by Ziegler–Natta catalysts and degraded by random chain scission, they defined a polydispersity parameter PI(Z), as shown by Eq. (7.1),

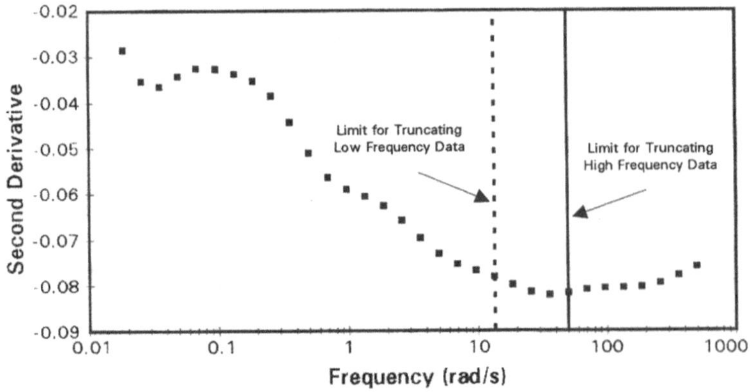

Fig. 7.1 Second derivative of the complex viscosity with respect to frequency for a HDPE. The key features for MWD determination are the minima, as these are related to peaks on the MWD. From Wood-Adams and Dealy [8]

where G_c is the crossover modulus, and found that it was approximately equal to the polydispersity index defined as M_w/M_n.

$$\mathrm{PI(Z)} = 10^6(\mathrm{dynes/cm^2})/G_\mathrm{C} \tag{7.1}$$

The crossover modulus is the value of G' (and G'') at the frequency where the two moduli are equal. An objective identification of the crossover point can be made by fitting the points on both curves in the vicinity of this point with cubic splines. Bafna [18] pointed out that this empirical correlation is based on data for a single family of polypropylenes, i.e. a group of polymers having similar molecular weight distributions, and that it cannot be expected to be valid for other groups of polymers.

Yoo [19] reported that the above correlation based on the crossover modulus is not useful for polypropylenes with melt flow rates above about 40 gm/10 min. For very high MFR resins, he proposed a correlation based on the frequencies at which the storage and loss moduli equal a particular value in dynes/cm^2. For example, for a value of 10,000 dynes/cm^2 the *modulus separation* (Modsep) is defined as follows:

$$\mathrm{Modsep(10{,}000)} \equiv \frac{\omega(\mathrm{G}' = 10{,}000\ \mathrm{dynes/cm^2})}{\omega(\mathrm{G}'' = 10{,}000\ \mathrm{dynes/cm^2})} \tag{7.2}$$

For a large set of MFR polypropylenes, this modulus value was found to provide a useful correlation as follows:

$$\mathrm{PI} = 11.5 - 1.92[\mathrm{Modsep}\ (10{,}000)] \tag{7.3}$$

Yoo showed that the Modsep correlation agreed well with the crossover modulus method based on PI(Z) over the range of MFR values in which both were useful and that Modsep was useful over the range from 30 to 1,000 gm/10 min^{-1}.

Shroff and Mavridis [20] compared several methods for inferring the polydispersity index from rheological data. They point out that some of the parameters that have been proposed depend on temperature and that highly asymmetric MWDs and the presence of even a small amount of LCB make these techniques unreliable.

7.1.3 Effect of Molecular Weight Distribution on Steady State Compliance

In Chap. 3 it is shown that the Rouse-Bueche model for unentangled melts predicts that the steady-state compliance is linear with molecular weight and that the tube model predicts that for a perfectly monodisperse sample it is independent of molecular weight. Data have shown that these predictions are correct, although the molecular weight M_C' at which J_s^0 becomes independent of M is larger than M_C, the critical value for η_0. However, even a small amount of polydispersity in an entangled polymer increases J_s^0 dramatically [21, 22]. This is illustrated in Fig. 7.2, which shows data for blends of nearly monodisperse polydimethyl siloxanes A and B, where B has an M_w about ten times that of A [23]. The values of J_s^0 are normalized by dividing by the value for polymer B, and this quantity is thus unity for pure B at the right end of the scale. We see that the viscosity of the blend can be over twenty times that of either of the components.

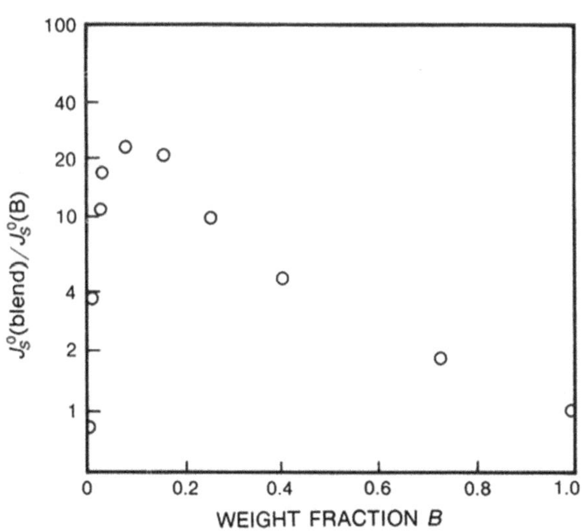

Fig. 7.2 Steady-state compliance of blends of two polydisperse silicones. For component A, M_w is 5.9×10^4, while for B it is 6.0×10^5, but the two have very similar values of J_s^0. The compliance is divided by that of component B, so that this ratio is nearly unity for A and exactly unity for 100 % B. Broadening MWD has a dramatic effect on J_s^0, with the ratio rising to nearly 30 at about 10 wt % B. From Graessley [23]

7.1.4 Effect of Tacticity on Viscosity

Vinyl polymers can exist in a variety of molecular forms depending on the relationship between the orientations of neighboring side groups. Many commercial polymerization processes produce atactic products, in which the distribution of orientations is random; an important example is polystyrene. In isotactic polymers, alternate side groups have the same orientation, while in syndiotactic products they have opposite orientations. An atactic polymer is one in which the orientations are randomly distributed along the chain.

Using metallocene catalysts it is possible to produce materials that are nearly perfectly iso- or syndiotactic. Studies of polystyrene [24], polypropylene [25], poly(methyl-methacrylate) [26, 27] and polybutadiene [27] have revealed that the plateau moduli of the iso- and atactic polymers are similar (0.4–0.5 MPa for PP) but that of the syndiotactic form is significantly higher (0.9–1.3 MPa for PP), with M_e following the opposite trend. In making such comparisons it is important to use values at the same temperature. Figure 7.3 shows $|\eta^*|$ versus frequency data reported by Huang et al. [24] for three polystyrenes having very similar molecular weights (2.5×10^5) but different tacticities. The zero-shear viscosities were found to follow the following relationship with molecular weight when M_w was normalized by M_e.

$$\eta_0\,(\text{Pa s}) = 0.029 \left(\frac{M_w}{M_e}\right)^{3.6} \quad (280\,^\circ\text{C}) \tag{7.4}$$

As with the dependence of viscosity on molecular weight, at high shear rates viscosity becomes independent of tacticity.

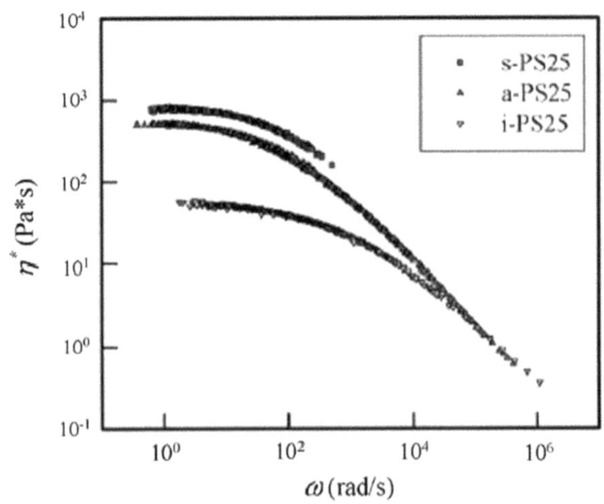

Fig. 7.3 Complex viscosity versus frequency of atactic, syndiotactic and isotactic polystyrenes at 280 °C. From Huang et al. [24]

7.1.5 Effect of Comonomer on Linear Shear Behavior

The use of an alpha-olefin comonomer in the polymerization of linear polyethylene is a widely-used method of controlling the crystallinity and thus density of polyethylene. For many years this has been carried out using Ziegler–Natta catalysts, and the product is called *linear low density polyethylene* (LLDPE). Metallocene catalysts were later found to make more efficient use of more costly comonomer by distributing it more uniformly among the molecules. LLDPEs made using Ziegler-Natta catalysts have more comonomer in the lower molecular weight fractions [28]. In addition to having a more uniform chemical composition distribution (CCD), linear products of a single reactor using a single catalyst of this type have molecular weight distributions described by the *most-probable* or Schultz-Flory distribution. This distribution has a polydispersity index (M_w/M_n) of two, and metallocene polymers have values very close to this theoretical value. We will refer to such a copolymer as an mLLDPE.

The principal effects of comonomer on rheology can be described in terms of its impact on the zero-shear viscosity and the plateau modulus. Fetters et al. [29] found that the average molecular weight per backbone bond m_b was a useful correlating parameter for use with ethylene/α-olefin copolymers. García-Franco et al. [30] used this parameter to analyze data for a series of ethylene/butane copolymers synthesized using a metallocene catalyst. They found that the zero-shear viscosities of all the samples were well described by Eq. (7.5).

$$\eta_0\ (\text{Pa s}) = 4.743 \times 10^{-10} \left(\frac{M_w}{m_b}\right)^{3.3} \quad (190\ {}^\circ\text{C}) \tag{7.5}$$

Data for hydrogenated polybutadiene that had been anionically synthesized ($m_b = 14$) were also found to obey this relationship.

The plateau modulus for copolymers having a comonomer fraction greater than about 0.2 was found to be related to the crossover modulus by Eq. (3.71), which had been proposed by Wu [31]. Inserting the polydispersity ratio for metallocene polymers, 2.0, the result is $G_N^0/G_C = 6.85$.

Using the backbone equivalence concept, Fetters et al. [29] developed a relationship between the plateau modulus of a copolymer and that of polyethylene homopolymer.

$$G_N^0 = \left(\frac{14}{m_b}\right)^3 \left(G_N^0\right)_{PE} \tag{7.6}$$

Figure 7.4 compares their data with the predictions of this model. The dashed line is Eq. (7.6), while the dotted and solid lines are empirical power laws that fitted the data better than the backbone equivalence model. While the model does not fit the date quantitatively, it does describe the general trend quite well. It was possible to place the storage and loss modulus data for all the copolymers on master curves by dividing the moduli by the plateau modulus and multiplying the frequency by η_0/G_N^0.

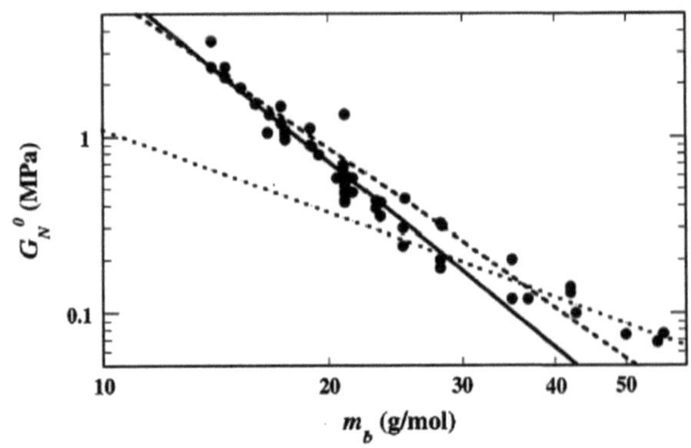

Fig. 7.4 Plateau modulus versus molecular weight per backbone bond for a large number of polyolefin copolymers. The *dashed line* is Eq. (7.6), and the *dotted* and *solid lines* are empirical power laws fitted to two regions of the plot. The backbone equivalent model alone does not lead to an accurate correlation, but it obviously plays an important role. From Fetters et al. [29]

The van Gurp and Palmen plot, a graph of loss angle δ versus the logarithm of $|G^*|$, was described in Sect. 3.16, and it was pointed out there that such a plot is invariant with temperature if the polymer is thermorheologically simple, and to average molecular weight for linear homopolymers. This type of representation is quite sensitive to comonomer content, but García-Fanco et al. [32] found they could describe this effect by use of a van Gurp and Palmen plot in which the complex modulus is normalized by the plateau modulusG^*/G_N^0. This is illustrated in Fig. 7.5 for a series of ethylene/octene copolymers. The number shown is wt % octene, and the upper data set is for monodisperse, hydrogenated polybutadiene.

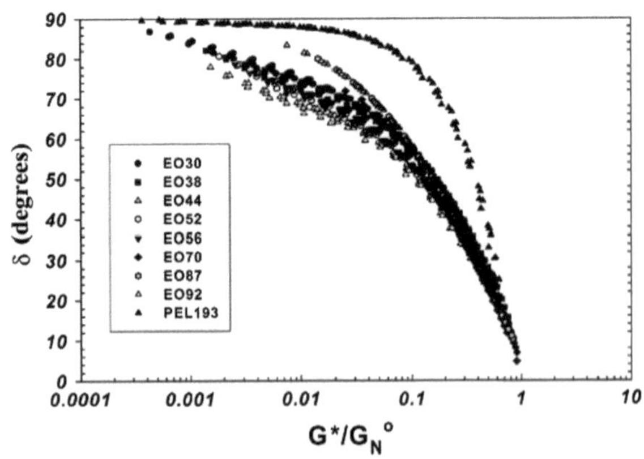

Fig. 7.5 A modified van Gurp and Palmen plot of loss angle versus $|G^*|/G_N^0$ for eight ethylene/octene copolymers. The number shown is the wt % octene, and the upper data set is for monodisperse, hydrogenated polybutadiene, which is expected to be very similar to monodisperse polyethylene. Normalizing using G_N^0 tends to bring data for all the copolymers close together. From García-Franco et al. [32]

7.1.6 Extensional Flow Behavior of Linear Polymers

Tube models predict that chain stretch and strain hardening can occur in an entangled, linear, monodisperse melt only when it is well entangled (M greater than about $3M_e$) and the rate of deformation exceeds the reciprocal of the longest Rouse time. For polyethylene M_e is relatively low ($\approx$1000), which means that a very high value of M would be required for significant stretch at practically achievable strain rates. But for polystyrene ($M_e \approx$ 17,000) one might be able to generate chain stretch and thus strain hardening in a practical experiment. Studies of highly-entangled, monodisperse polystyrene melts [33] using a filament stretching rheometer revealed slight strain softening at the lowest strain rates and marked strain hardening at higher strain rates. But such samples are not typical of commercial resins.

While we do not expect to see strain hardening in linear polymers at accessible strain rates unless the molecular weight is extremely high, it has been reported that a very small amount of relatively high molecular weight material in a sample consisting primarily of molecules having moderate molecular weights can result in strain hardening at very low strain rates. This has been observed in polystyrene [34–36], polypropylene [37] and polyethylene [38]. For example, Münstedt [34] compared the behavior of four linear polystyrenes, one of which exhibited strong strain hardening. This sample had an M_w of only 39,000 but also a much smaller MWD peak at about 70,000. However, neither a sample with M_w = 74,000 and a very small high molecular weight tail nor a commercial polymer with M_w = 219,000 and M_w/M_n = 2.3 exhibited significant strain hardening. For HDPE a very high MW tail can result in strong strain hardening around strain rate of 0.01 s^{-1} or lower, and this is advantageous for extrusion blow molding, because strain hardening at low strain rates suppresses parison sag.

7.2 Long-Chain Branching and Melt Rheology

7.2.1 Introduction

The effects of long-chain branching on linear viscoelastic properties have been extensively studied. When the detailed branching structure is known, quantitative relationships can often be developed between structure and rheology, but when the structure is nonhomogeneous and not well characterized, quantitative relationships cannot be made.

When is a branch "long"? Data from NMR measurements give the same response for all side-chains longer than five monomer units and this might be classified as long. But the zero-shear viscosity undergoes a dramatic change in its dependence on molecular weight around the critical molecular weight M_C, which involves a lot more than five carbons. And the steady-state compliance has a somewhat different value of M for the onset of entanglement effects. For our purposes, a long-chain is one that has a significant effect on rheological properties.

A great variety of branching structures are possible, and we will see that it is not possible to establish quantitative relationships between rheological data and branching structure unless something is known about the type of branching structure involved. This information can only come from information about the way the polymer was produced. Without any knowledge of the origin of a sample, it is not possible to use any combination of characterization techniques to determine the detailed branching structure.

In general, long-chain branched polymers are thermorheologically complex [39, 40]. However, branched systems are sometimes characterized in terms of an activation energy of flow valid over a specific, narrow range of times, often in the terminal zone [41]. Various ways have been proposed to infer values of such a parameter from data, and care must be taken in interpreting reported flow activation energies.

Strain hardening in extension is the most prominent indicator of long-chain branching. Star polymers do not exhibit strain hardening, because the arms are not stretched. But inner chain segments with branch points at each end promote chain stretch. Gabriel and Münstedt [42] compared a number of polyolefins having various branching structures. They concluded that both the number of branches per molecule and the number of entanglements per branch play crucial roles in strain hardening. In general, except for a polymer whose structure is known with some certainty, for example a metallocene polyethylene, shear behavior is of limited utility for the detection of the important aspects of long-chain branching structure. What one often wants to know from a practical point of view is the degree to which LCB enhances strain hardening, and extensional flow is the most direct way to discern this.

7.2.2 Branched Polymers with Monodisperse Structures

Polymers having well-defined structures can be prepared by means of anionic polymerization, and this technique has been widely used to prepare samples for rheological study. This has been a fruitful approach to the study of the effects of long-chain branching on rheological behavior, and we begin our discussion of this subject by describing the behavior of such "model polymers".

For star polymer with arms of increasing length, when the arms are still too short to entangle there is a decrease in the viscosity compared with a linear polymer of the same molecular weight, because the size of the molecule decreases as the branch length increases. But when the arm molecular weight M_a reaches two or three times the molecular weight between entanglements M_e, η_0 starts to increase approximately exponentially with molecular weight. This is illustrated in Fig. 7.6, which shows data for linear, three-arm and four-arm polybutadiene stars [43]. At moderate molecular weights the data for the stars lie below, but parallel

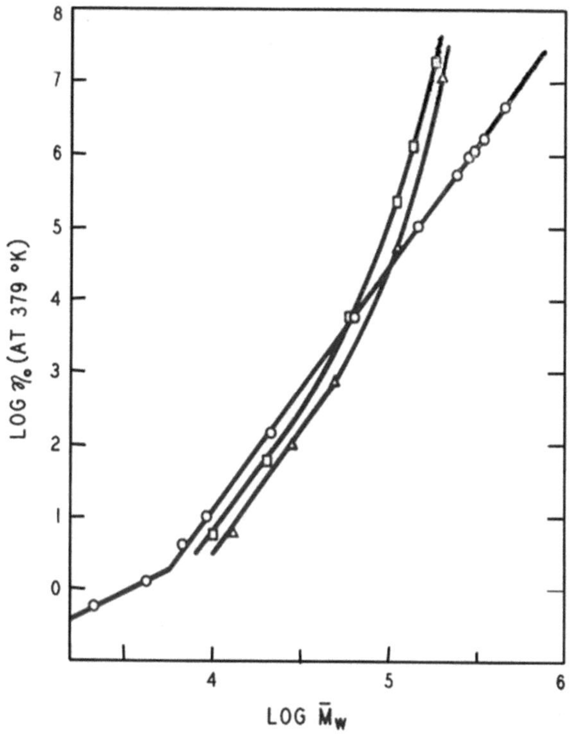

Fig. 7.6 Zero-shear viscosity of three series of polybutadienes: linear (*circles*), three-arm stars (*squares*) and four-arm stars (*triangles*). At low MW the data are below but parallel to the line for the unbranched polymer, and when the branch length rises to three or four times M_e the star data rise above the line approaching exponential growth. From Kraus and Gruver [43]

to, the line for linear, entangled polymer with $\alpha = 3.4$, in accord with Eq. (2.20), but when the branch length reaches three to four times M_e the data for the stars rise sharply and cross the line for linear polymers. The exponential increase of viscosity seen in Fig. 7.6 is consistent a the picture in which relaxation occurs primarily by means of arm retraction, since reptation is suppressed by the branch point. Gell et al. [44] studied asymmetric stars made by adding arms of varying length at the midpoint of a monodisperse highly entangled backbone. They found that even very short arms could lead to a marked increase in η_0.

For multi-armed stars, when M_a is above M'_C we expect the onset of entanglement to cause marked deviations from the linear increase of J_s^0 with M that is in line with the Rouse-Bueche model for linear, unentangled polymers. However, for stars J_s^0 continues to increase with M_a beyond M'_C.

The storage and loss moduli of a number of model branched polymers have been reported including stars [45–52] and combs of various types [53–55]. Branching leads to deviations from the behavior shown in Chap. 3 for linear polymers, but the details depend strongly on branching structure. Trinkle et al. [56, 57] used the van Gurp-Palmen plot to reveal the presence of long chain branching in polyethylene. Lohse et al. [58] found that this type of plot for hydrogenated polybutadienes (model PEs) having well-defined structures was useful to bring out the distinctive effects of branching on linear viscoelastic behavior. Figure 7.7 shows such a plot of

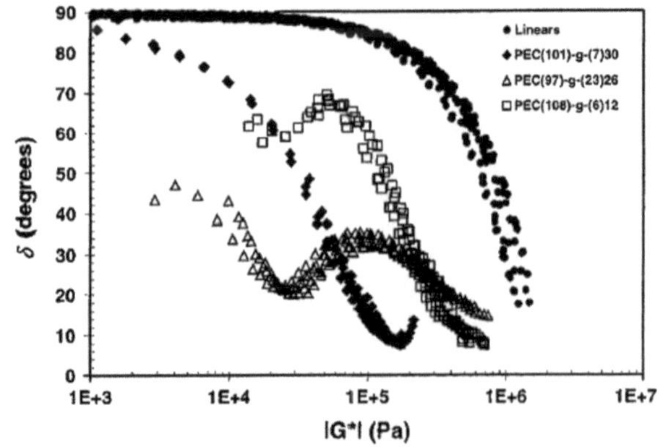

Fig. 7.7 A van-Gurp and Palmen plot of data for comb polymers. Each structure has distinctive features. Sample PEC(101) has 30 arms of molecular weight 6.5, PEC(97) has 26 arms with MW of 23.5, and PEC(108) has 12 arms with MW of 5.8. From Lohse et al. [58]

their data for several comb polymers, it reveals the features of the curve associated with each branching structure. Sample PEC(101) had 30 arms of molecular weight 6.5, PEC(97) had 26 arms with MW of 23.5, and PEC(108) had 12 arms with MW of 5.8. One should keep in mind that frequency-dependence information, which is essential for characterizing viscoelastic behavior, is lost in this type of data presentation, so it is not complete characterization of linear behavior.

Star polymers do not exhibit strain hardening in extension [59], because there is no chain stretch; we will see below that data on branched metallocene polymers suggest that at least two branch points on some molecules are required to produce strain hardening.

7.2.3 Long-Chain Branching in Metallocene Polymers

7.2.3.1 Introduction

Polyethylene copolymers (LLDPE), in which comonomer content governs crystallinity and thus density, were described in Sect. 7.1.5, and it was pointed out that copolymers made using single-site catalysts (mLLDPE) have a more uniform distribution of co-monomer among the chains (CCD) than LLDPEs made using Ziegler-Natta catalysts. Films made from LLDPE have improved mechanical properties but are more difficult to process, because of their lower degree of shear thinning, as compared with more polydisperse or long-chain-branched polymers.

This deficiency was corrected when the Constrained Geometry Catalyst (CGC) [60] was developed. Using this catalyst in a particular type of reactor, it was possible to produce homopolymers having long chain branching frequencies λ, the average number of branch points per 1,000 carbon atoms, in the range of 0.01–3 [61]. The introduction of LCB has an important effect on the viscosity, substantially increasing the degree of shear thinning. It thus became possible to have reproducible MWD and controlled levels of crystallinity as well as strong shear thinning. Aside from their commercial importance, these materials are of fundamental interest, because they make it possible to carry out detailed studies of the relationships between branching structure and rheological behavior.

A key difference between branched metallocene polyolefins and other branched systems is that mPEs contain a large fraction of linear chains, and all chain segments have the same size distribution as the linear chains. For polyethylene the level of branching λ, the average number of branch points per thousand carbon atoms, is related to β, the average number of branch points per molecule through Eq. (7.7).

$$\beta = \frac{M_N \lambda}{14 \times 10^3} \tag{7.7}$$

Costeux [62] derived the following useful expressions relating these parameters to polydispersity.

$$\frac{M_W}{M_N} = 2(\beta + 1) \tag{7.8}$$

$$\lambda = \frac{14 \times 10^3 (2\beta)(\beta + 1)}{M_W} \tag{7.9}$$

Costeux [62] also derived equations for the distributions of structures among the molecules.

The key feature from the point of view of rheological behavior is the distribution of segment types among the molecules rather than simply the number of branches, because extensional flow behavior is primarily sensitive to the presence of segments with branch points at both ends. The simplest such molecule has an H shape. Costeux et al. [63] calculated the branching distributions of all polymers that can be made using a single, constrained-geometry catalyst in a single reactor. They represented these systems using a triangular diagram in which the vertexes correspond to linear molecules, free arms, and inner backbone segments. Seven simulated systems are shown on Fig. 7.8. The point nearest the lower-right-hand vertex corresponds to a system made up mostly of linear molecules with a few three-armed stars. We will see that the rheological behavior of such materials can be quantitatively related to these features of their structure.

The CGC technology used to make the copolymers discussed above is a solution polymerization process, and the vinyl-terminated macromonomers formed in solution compete with comonomer for insertion to form long branches, and as a result the branches have the same molecular weight distribution as the backbone chain. Since there are very few long-chain branches, if the reincorporation is random there should be no significant effect of comomoner concentration on the level of long-chain branching. This implies that the rheological effects of the short and long branches should be independent. This may not be case with very long comonomers [64], but these are not of commercial interest. However, reincorporation of macromonomers in mLLDPEs polymerized in slurry or gas-phase reactors may occur by a different mechanism [65], and it is not clear how this might alter the effect of comonomer on structure and rheology.

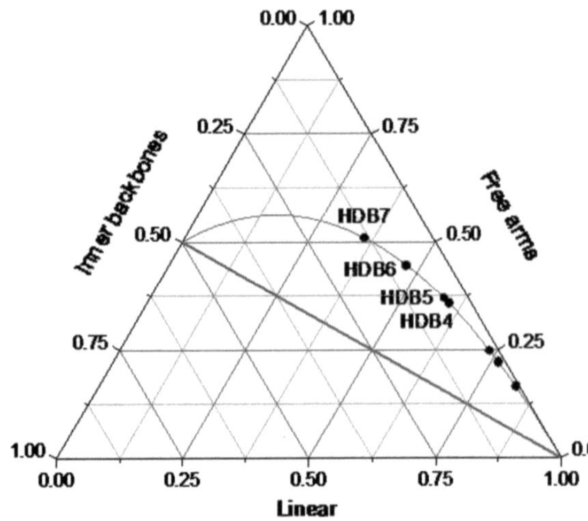

Fig. 7.8 Triangular diagram showing the segment-type compositions of all long-chain branched polyethylenes that can be made by CGC technology. The lower-right apex is linear molecules, the lower *left* corner is inner backbones, and the *top* apex is free arms. From Costeux et al. [63]. The *points* indicated refer to samples whose rheological behavior has been studied [67]

7.2.3.2 Shear Behavior of Long-Chain Branched Metallocene Polyolefins

The zero-shear viscosity of branched metallocene polyethylenes at fixed M_w increases with branching level; the ratio of η_0 of a branched polymer to that of its linear analog having the same M_w rises from 10 to 40 as the branching level increases from 0.4 to 2.0 LCB per 10,000 carbon atoms at 150 °C [66, 67],

Methods have been proposed to infer branching level in branched metallocene polyethylenes from viscosity data. These require calibration using LCB information from multi-detector GPC or NMR [61]. As explained above, all polymers made using a single metallocene catalyst in a single reactor are members of a well-defined set, and any rheological property that differentiates between the members of this set can indicate the branching level, although some techniques may be more sensitive than others. One method [68, 69] compares the MWD of the polymer with that predicted from viscosity data by means of the method described in Sect. 7.1.2. The difference between the actual and predicted peak in the MWD was found to be quantitatively related to the branching level. A simpler approach is that of Lai et al. [70], who found that for strictly linear polyethylenes prepared using one single-site catalyst, the Cross equation (Eq. 2.9) provides a good fit to viscosity data, and they related departures from the Cross equation with levels of long-chain branching. Garcia-Franco et al. [71] studied lightly long-chain branched metallocene copolymers made by both the CGC solution process and a gas-phase process. They found that van Gurp-Palmen plots were sensitive to the low levels of long-chain branching in these polymers and correlated the branching level with the value of the loss angle when $|G^*| = 10$ kPa.

Karjala et al. [72] studied the combined effects of LCB and polydispersity on η_0 by blending metallocene polyethylenes to yield groups of samples having very similar values of M_w/M_n but varying LCB levels. While it was not possible to infer quantitative relationships, the data indicated that LCB has a much stronger effect on the departure from the line for linear PE and that increasing the level of LCB clearly increases this departure.

There has been very little study of the combined effects of comonomer and LCB in metallocene copolymers. Based on a small set of samples, Karjala et al. [73] reported that regardless of comonomer content, levels of branching as low as 0.01 LCB/1000C had values of η_0 that were above the 95 % confidence limits of the line describing linear homopolymers. And the effect of LCB was significantly more important than that of comonomer.

7.2.3.3 LCB in Metallocene Polyethylenes: Extensional Flow Behavior

While the levels of LCB in mPEs are not sufficient to give the strong strain hardening desired for film applications, these materials made it possible to learn a great deal about what branching structures are important in generating high stretching stresses. To study this connection, Wood-Adams et al. [66] studied seven homopolymers made using one CGC catalyst (described in Sect. 7.2.3.1) and having various levels of branching. The values of M_w, λ and β for these materials are given in Table 7.1. For comparison with their data, they also determined the η_0 of linear homopolymers made using the same catalyst and reported the following relationship with molecular weight:

$$\eta_0\,(Pa\cdot s) = 3.9\times 10^{-15} M_w^{3.65}\ (150\ ^\circ\mathrm{C}) \tag{7.10}$$

The structural makeup of the samples in this set can be located on the triangular diagram shown in Fig. 7.8 [63]. The lower right vertex is linear chains, the top vertex is arms, and the lower left vertex is inner backbones. We note that samples 1–4 consist mostly of linear chains, with some stars and very few inner backbones, and these polymers exhibited very little strain hardening [66]. The extensional flow behavior of samples 5 to 7 in Table 7.1 were later measured by Torres [67]

Table 7.1 Values of parameters defined by Eqs. (7.7–7.9) for the seven long-chain branched metallocene polyethylenes indicated in Fig. 7.8

Sample	M_w	λ	β
HDB1	77,000	0.026	0.067
HDB2	82,000	0.037	0.099
HDB3	86,000	0.042	0.116
HDB4	96,000	0.080	0.224
HDB5	79,000	0.090	0.210
HDB6	68,000	0.190	0.343
HDB7	70,000	0.330	0.537

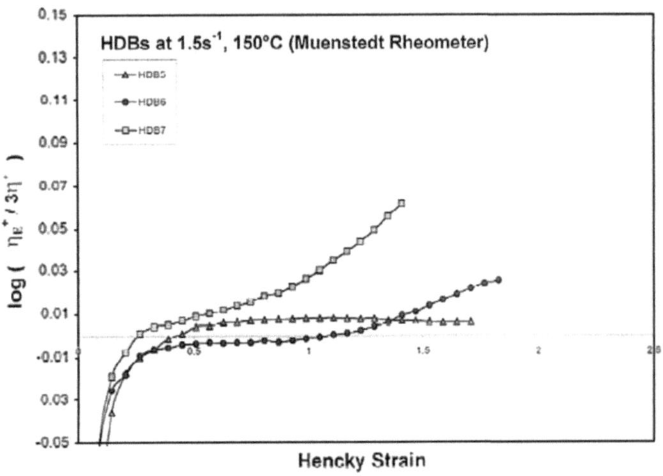

Fig. 7.9 Extensional stress growth function of three materials corresponding to points in Fig. 7.8. Equation (4.23) was used to normalize the data to emphasize departures from linear behavior. Sample HDB 5, which contains some free arms but almost no inner backbones, is a dilute blend of *stars* in linear polymer and exhibits no strain hardening. But as inner backbones start to appear, there is an increasing degree of strain hardening. From Torres [67]

using a Münstedt rheometer, and the results are shown in Fig. 7.9 where the tensile stress growth coefficient has been normalized using Eq. (4.23). Sample HDB 5 exhibits no strain hardening, but starting with HDB 6 we see significant hardening, which increases markedly as we move on to HDB7, Again referring to Fig. 7.8, we see that these samples contain growing numbers of inner backbones, which appear to be essential for strain hardening. These materials were also characterized using the much simpler SER extensional rheometer described in Chap. 6, and while the data lacked the precision of those from the Münstedt rheometer used by Torres, the effect of increasing the number of inner backbones was very clear.

Malmberg et al. [74] studied branched metallocene copolymers made using several catalysts and found marked differences between the products. They proposed that extensional flow behavior depends more on the distribution of long chain branches than the number of branches. In particular the presence of multiple branch points on a few molecules is much more effective than many branches evenly distributed among many molecules. On the other hand, the evenly distributed branches had a larger effect on the zero-shear viscosity. This is consistent with observations of the difference between star and comb polymers.

7.2.4 Random Branching Introduced by Post-Polymerization Reactions

There have been reports that some polymers classified as HDPE contain low levels of long-chain branching, for example those made using chromium oxide catalysts [75, 76]. It has been reported [77] that many variables affect branch formation in these polymers, and it is not clear whether the branching occurs during polymerization or subsequently, as the molecules are vinyl terminated and can react during melt processing. Another source of long-chain branching in a linear polyethylene is the cross-linking that can occur whenever it is heated above its melting point, particularly in the presence of air. This is a potential source of uncertainty in laboratory measurements.

Linear polymers can be made strain-hardening by the introduction of long-chain branches. Techniques used to introduce branching in already-polymerized samples yield distributions of structures that often include tree-like molecules with short segments, similar in some ways to LDPE. This complexity can make it difficult to distinguish between the effects of branching and polydispersity on the basis of viscosity alone. Treating linear polyethylene with a peroxide [78–80] leads to branching and eventually to cross-linking although the precise branching structure produced by the peroxide is somewhat uncertain and difficult to quantify. This process is used commercially to enhance the properties of HDPE or LLDPE for particular applications.

Polypropylene is another polymer whose processability in certain applications requires long-chain branching. *High melt strength polypropylene* (HMSPP) is made by treatment of reactor product using a peroxide crosslinking agent, electron radiation of macro-monomer grafting [81, 82]. Kurzbeck et al. [83] used electron-beam radiation to introduce branching into a polypropylene to produce strong strain-hardening, and the same result was obtained by Hingmann and Marczinke [84] using a chemical cross-linking agent. They estimated that fewer than three branches per molecule are required to produce this pronounced strain hardening. A problem in drawing quantitative conclusions from such studies is that polydispersity increases when monodisperse linear polymers are treated to introduce branching, and the average molecular weight decreases.

Auhl et al. [85] studied the behavior of a series of polypropylenes that had been subjected to electron beam radiation to introduce various levels of long-chain branching. A branching factor, defined as the ratio of zero-shear viscosity of a branched sample to that of its linear precursor, is shown in Fig. 7.10 as a function of radiation dose d. If little chain scission occurred, the branched samples should have the same molecular weight as the linear precursor, e.g., the point for $d = 0$ corresponds to a branching factor of unity. The viscosity increases at low branching levels, reaching a peak, and then falls to values a bit below unity. Based on all the analytical data available for these samples, as well as the rheological

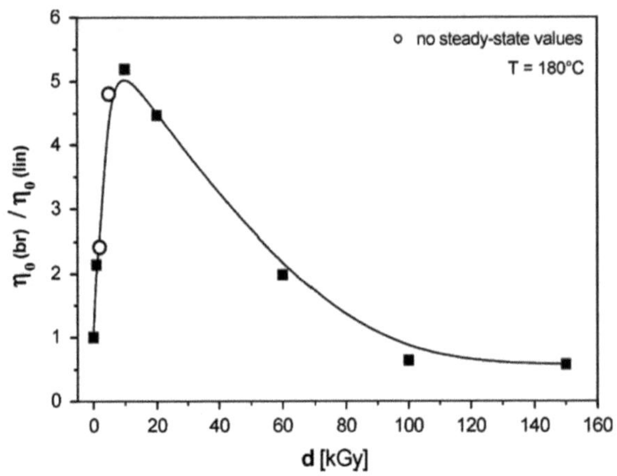

Fig. 7.10 Branching factor g, the ratio of η_0 of radiated polypropylene to that of linear PP precursor, versus radiation dose d. If little chain scission occurred all samples have the same molecular weight; i.e. $d = 0$ implies a branching factor of unity. Electron beam radiation has resulted in long-chain branching. The viscosity first increases sharply with branching level but then falls continuously to values below that of the linear precursor. From Auhl et al. [85]

data, the authors concluded that at low radiation doses the branched molecules were mainly stars, and the increase in viscosity was thus similar to that shown in Fig. 7.6 whereas the higher doses produced more treelike structures.

7.2.5 Low-Density Polyethylene

Low-density polyethylene (LDPE) is strongly shear-thinning, which makes it easy to extrude, and it is strongly strain hardening, which gives it high melt strength. Branched metallocene polyolefin copolymers also exhibit strong shear thinning but their strain hardening is significantly weaker than LDPE. LDPE therefore remains the resin of choice for fabrication processes in which strain hardening is crucial, such as extrusion-coating and extrusion foaming, and as a melt strength enhancer when blended with LLDPE or HDPE to improve bubble stability in film blowing or reduce neck-in in extrusion coating. Figure 4.12 showed data for stress growth in extension for a tubular LDPE [86]. [It was explained in Sect. 4.7 that whether the value of the extensional viscosity $\eta_E(\dot{\varepsilon})$ can be inferred from the maximum in the curve of $\eta_E^+(t,\dot{\gamma})$ is now debatable.]

There is a broad range of branching structures in an LDPE resin, with many short branches as well as complex, tree-like molecules. The structure depends on how the polymer was manufactured. LDPE is made by high-pressure, free-radical polymerization in either a tubular reactor or an autoclave, and autoclave LDPE typically has a much higher branching level and a more tree-like structure than LDPE produced in a tubular reactor [87].

As noted earlier, the zero-shear viscosities of linear homo- and copolymers with molecular weights well above M_C obey Eq. (2.21), which is repeated below.

$$\eta_0 = KM_w^a \tag{2.21}$$

The introduction of long-chain branches (LCB) leads to important deviations from the line on a log–log graph defined by this equation, and such a deviation is often cited as evidence of LCB. In general, we know that a small or modest level of entangled branches results in a significant viscosity increase, in comparison with a linear polymer with the same M_w. This was shown for star polymers in Fig. 7.6. While LDPEs have zero-shear viscosities that are usually well below those of linear polymers having the same M_w, it is sometimes reported that an LDPE has a viscosity greater than its linear analog, particularly when the sample has a relatively high molecular weight.

There are several issues that make this type of comparison questionable. First, the temperature dependence of η_0 is much stronger for LDPE than for HDPE, so changing the temperature will alter η_0 by different amounts for the two polymers. In addition, the temperature dependencies for autoclave and tubular LDPEs have also been reported to be different [88]. But a more serious issue is the value of M_w used for the comparison. The "conventional GPC" value (gpc-M_w) is obtained using an infrared detector and a calibration curve based on linear standards. This value is said to be a "backbone" M_w for any PE sample. On the other hand, the "absolute value" (abs-M_w) is obtained using a light scattering detector and includes the entire molecule: backbone, LCBs and SCBs. It was noted in Sect. 7.1.5 that SCBs have a significant effect on η_0 in copolymers having no LCBs. In addition, it is not possible to discern whether a change in viscosity is due simply to the increase in molecular weight or to the entanglement of branches. Wang et al. [88] used both molecular weight values to correlate data for a number of LDPEs made in both types of reactor, and their plots are shown in Fig. 7.11a, using abs-M_w, and in Fig. 7.11b using gpc-M_w. First we note that some points move from one side of the linear PE line to the other when the M_w method is changed. Moreover, using the gpc-M_w results in a clearer trend, especially for the samples from a tubular reactor, for which there is some indication of an exponential dependence.

7.2.6 Branching Level from Zero-Shear Viscosity and Molecular Weight

A number of correlations have been proposed to relate level of long-chain, random branching to zero-shear viscosity together with molecular weight [89–96] and/or intrinsic viscosity. Larson [91] compared some of these with model predictions and concluded that "…the zero-shear viscosity of a branched melt is extremely sensitive not only to degree of branching but also to type." and that no single type

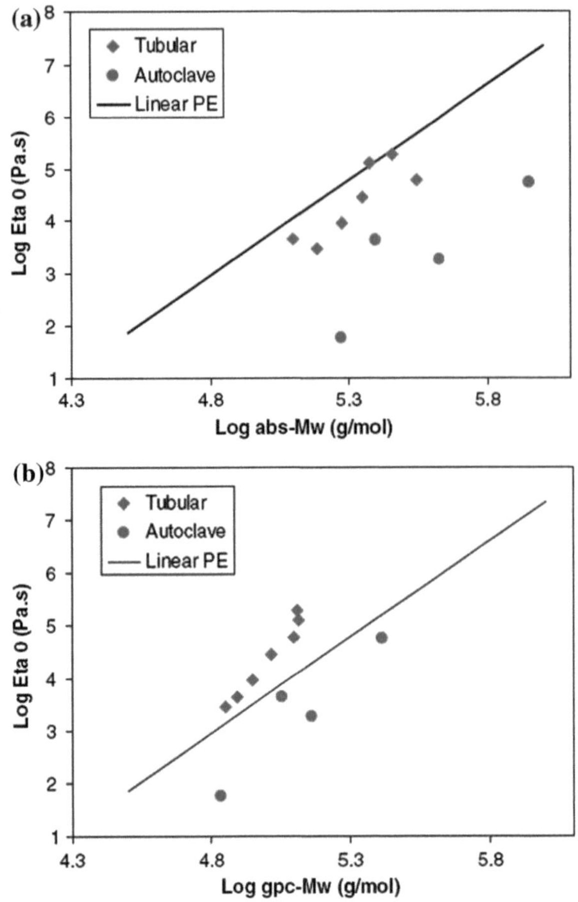

Fig. 7.11 **a** Zero-shear viscosity versus "absolute" M_w of LDPEs determined using a light-scattering detector, which includes the entire molecule. Samples from both tubular and autoclave reactors are shown. From Wang et al. [88]. **b** Zero-shear viscosity versus "conventional GPC" M_w of LDPEs determined using an infrared detector based on linear standards and called a "backbone value." Samples from both tubular and autoclave reactors are shown. The use of this M_w value yields a clearer trend. From Wang et al. [88]

of correlation could describe all branched systems. Vega et al. [96] evaluated several such correlations and concluded that a correlation for number of branches based on a few rheological and molecular parameters is at best approximately valid only for one specific type of branching structure.

References

1. Dealy JM, Larson RG (2006) Structure and rheology of molten polymers. Hanser, Munich
2. Bersted BH (1975) An empirical model relating the molecular weight distribution of high-density to the shear dependence of the steady shear viscosity. J Appl Polym Sci 19:2167–2177
3. Bersted BH (1976) A model relating the elastic properties of high-density polyethylene melts to the molecular weight distribution. J App Polym Sci 20:2705–2714

4. Bersted BH, Slee JD (1977) A relationship between steady-state shear melt viscosity and molecular weight distribution in polystyrene. J Appl Poly Sci 21:2631–2644
5. Malkin AY, Teishev AY (1988) Can the MWD of polymers be determined uniquely from the flow curve of its melt? Polym Sci USSR 29:2449–2455
6. Malkin AY, Teishev AY (1991) Flow curve-molecular weight distribution: is the solution of the inverse problem possible? Polym Eng Sci 31:1590–1596
7. Shaw MT, Tuminello WH (1994) A closer look at the MWD-viscosity transform. Polym Eng Sci 34:159–165
8. Wood-Adams PM, Dealy JM (1996) Use of rheological measurements to estimate molecular weight distribution of linear polyethylene. J Rheol 40:761–778
9. Berker A, Driscoll JJ (1998) Comment on 'Obtaining molecular-weight distribution from the viscosity data of linear polymer melts'. J Rheol 42:1555–1562
10. Liu Y, Shaw MT, Tuminello WH (1998) Response to Comment on 'Obtaining molecular-weight distribution information from the viscosity data of linear polymer melts'. J Rheol 42:1563–1564
11. Liu Y, Shaw MT, Tuminello WH (1996) Optimized data collection for determination of the MWD from the viscosity data of polymer melts. Poly Eng Sci 38:169–176
12. Liu Y, Shaw MT, Tuminello WH (1998) Obtaining molecular-weight distribution information from the viscosity data of linear polymer melts. J Rheol 42:453–476
13. Nobile MR, Cocchini F, Lawler JV (1996) On the stability of molecular weight distributions as computed from the flow curves of polymer melts. J Rheol 40:363–382
14. Wu S (1985) Polymer molecular-weight distribution from dynamic melt viscoelasticity. Polym Sci Eng 25:122–128
15. Tuminello WH (1986) Molecular weight and molecular weight distribution from dynamic measurements of polymer melts. Polym Eng Sci 26:1339–1347
16. Zeichner GR, Patel PD (1981) A comprehensive evaluation of polypropylene melt rheology. Proceedings 2nd world congress of chem eng, Montreal
17. Zeichner GR, Macosko CW (1982) On-line viscoelastic measurements for polymer melt processes. SPE ANTEC Tech Papers 28:79–81
18. Bafna SS (1997) Is the cross-over modulus a reliable measure of polymeric polydispersity? J Appl Polym Sci 63:111–113
19. Yoo HJ (1994) MWD determination of ultra high MFR polypropylene by melt rheology. Adv Polym Technol 13:030201–030205
20. Shroff R, Mavridis H (1995) New measures of polydispersity from rheological data on polymer melts. J Appl Polym Sci 57:1605–1626
21. Struglinski MJ, Graessley WW (1985) Effects of polydispersity on the linear viscoelastic properties of entangled polymers, 1. Experimental observations for binary mixtures of linear polybutadiene. Macromol 18:2630–2643
22. Fuchs K, Friedrich C, Weese J (1996) Viscoelastic properties of narrow-distribution poly(methyl methacrylates). Macromol 29:593–5901
23. Graessley WW (1971) Linear viscoelasticity in entangled polymer systems. J Chem Phys 54:5143–5157
24. Huang CL, Chen YC, Hsiao TJ, Tsai JC, Wang C (2011) Effect of tacticity on viscoelastic properties of polystyrene. Macromol 44:6155–6161
25. Eckstein A, Suhm J, Friedrich C, Maier RD, Sassamannshausen J, Bochmann M, Mülhaupt R (1998) Determination of plateau moduli and entanglement molecular weights of isotactic, syndiotactic, and atactic polypropylenes synthesized with metallocene catalysts. Macromol 31:1335–1340
26. Fuchs K, Friedrich C, Weese J (1996) Viscoelastic properties of narrow-distribution poly (methyl methacrylates). Macromol 29:593–5901
27. Wu S (1987) Entanglements between dissimilar chains in compatible blends: poly (methyl methracrylate) and poly (vinylidine flouride). J Polym Sci Phys 25:527–566
28. Zhang M, Lynch DT, Wanke SE (2001) Effect of molecular structure distribution on melting and crystallization behavior of 1-butene/ethylene copolymers. Polymer 42:3067–3075

29. Fetters LJ, Lohse DJ, García-Franco CA, Brant P, Richter D (2002) Prediction of melt state poly (α-olefin) rheological properties: the unsuspected role of the average molecular weight per backbone bond. Macromol 35:10096–10101
30. García-Franco CA, Harrington BA, Lohse DJ (2006) Effect of short-chain branching on the rheology of polyolefins. Macromol 39:2710–2717
31. Wu S (1989) Chain structure and entanglements. J Polym Sci Polym Phys 27:723–741
32. García-Franco CA, Harrington BA, Lohse DJ (2004) On the rheology of ethylene-octene copolymers. Rheol Acta 44:591–599
33. Bach A, Almdal K, Rasmussen HK, Hassager O (2003) Elongational viscosity of narrow molar mass distribution polystyrene. Macromol 36:5174–5179
34. Münstedt H (1980) Dependence of the elongational behavior of polystyrene melts on molecular weight and molecular weight distribution. J Rheol 24:847–867
35. Frank A, Meissner J (1984) The influence of blending polystyrenes of narrow molecular weight distribution on melt creep flow and creep recovery in elongation. Rheol Acta 23:117–123
36. Minegishi A, Nishioka A, Takahashi T, Masubuchi Y, Takimoto J, Koyama K (2001) Uniaxial elongational viscosity of PS/a small amount of UHMW-PS blends. Rheol Acta 40:329–338
37. Sugimoto M, Masubuchi Y, Takimoto J, Koyama K (2001) Melt rheology of polypropylene containing small amounts of high-molecular weight chain. 2. Uniaxial and biaxial extensional flow. Macromol 34:6056–6063
38. Linster JJ, Meissner J (1996) Melt elongation and structure of linear polyethylene (HDPE). Polym Bull 16:187–194
39. Raju VR, Rachapudy H, Graessley WW (1979) Properties of amorphous and crystallizable hydrocarbon polymers. IV. Melt rheology of linear and star-branched hydrogenated polybutadiene. J Polym Sci Phys 17:1223–1235
40. Graessley WW, Raju VR (1984) Some rheological properties of solutions and blends of hydrogenated polybutadiene. J Poly Sci Polym Symp 71:77–93
41. Levine A, Milner S (1998) Star polymers and the failure of time-temperature superposition. Macromol 31:8623–8637
42. Gabriel C, Munstedt H (2003) Strain hardening of various polyolefins in uniaxial elongational flow. J Rheol 47:619–630
43. Kraus G, Gruver JT (1965) Rheological properties of multichain polybutadienes. J Polym Sci A 3:105–122
44. Gell CB, Graessley WW, Efstratiadis V, Pitsikalis M, Kadjichristidis N (1997) Viscoelasticity and self-diffusion in melts of entangled asymmetric star polymers. J Polym Sci B 35:1943–1954
45. Fetters LJ, Kiss AD, Pearson DS, Quack GF, Vitus FJ (1993) Rheological behavior of star-shaped polymers. Macromol 26:647–654
46. Ngai KL, Roland CM (1997) Terminal relaxation and diffusion of entangled three-arm star polymers: temperature and molecular weight dependencies. J Polym Sci B 35:2503–2510
47. Gell CB, Graessley WW, Efstratiadis V, Pitsikalis M, Kadjichristidis N (1997) Viscoelasticity and self-diffusion in melts of entangled asymmetric star polymers. J Polym Sci B 35:1943–1954
48. Archer LA, Varshney SK (1998) Synthesis and relaxation dynamics of multiarm polybutadiene melts. Macromol 31:6348–6355
49. Struglinski MJ, Graessley WW, Fetters LJ (1988) Effects of polydispersity on the linear viscoelastic properties of entangled polymers. 3. Experimental observations on binary mixtures of linear and star polybutadienes. Macromol 21:781–789
50. Watanabe H, Ykoshida H, Kotaka T (1988) Enganglement in blends of monodisperse star and linear polystyrenes. 1. Dilute blends. Macromol 21:2175–2184
51. Roovers J (1991) Melt rheology of highly branched polymers. J Non-Cryst Solids 131–133:793–798

52. Pakula T, Vlassopoulos D, Fytas G, Roovers J (1998) Structure and dynamics of melts with multiarm polymer stars. Macromol 31:8931–8940
53. Roovers J (1984) Melt rheology of H-shaped polystryrenes. Macromol 17:1196–1200
54. Archer LA, Varshney SK (1998) Synthesis and relaxation dynamics of multiarm polybutadiene melts. Macromol 31:6348–6355
55. Roovers J, Graessley WW (1981) Melt rheology of some model comb polystyrenes. Macromol 14:766–773
56. Walter P, Trinkle S, Mülhaupt R (2001) Influence of zirconocene structure and propene content on melt rheology of polyethene and ethene/propene copolymers. Polym Bull 46:205–213
57. Trinkle S, Walter P, Friedrich C (2002) Van Gurp-Palmen plot II—Classification of long chain branched polymers by their topology. Rheol Acta 41:103–113
58. Lohse DJ, Milner ST, Fetters LJ, Xenidou M, Hadjichristidis N, Mendelson RA, Garcia-Franco CA, Lyon MK (2002) Well-defined, model long chain branched polyethylene. 2 Melt rheological behavior. Macromol 35:3066–3075
59. Ye X, Sridhar T (2001) Shear and extensional properties of three-arm polystyrene solutions. Macromol 34:8270–8277
60. Stevens JC (1994) Insite(tm) catalysts structure/activity relationships for olefin polymerization. Stud Surf Sci Catal 89:277–284; (1996) Constrained geometry and other single site metallocene polyolefin catalysts: A revolution in olefin polymerization 101:11–20
61. Lai SY, Wilson JR, Knight JR, Stevens JC (1993) Elastic substantially linear olefin polymers. US Patent 5(272):236
62. Costeux S (2003) Statistical modeling of randomly branched polymers produced by combination of several single-site catalysts: toward optimization of melt properties. Macromol 36:4168–4187
63. Costeux S, Wood-Adams P, Beigzadeh D (2002) Molecular structure of metallocene-catalyzed polyethylene: rheologically relevant representation of branching architecture in single catalyst and blended systems. Macromol 35:2514–2528
64. Stadler FJ, Piel C, Klimke K, Kaschta J, Parkinson M, Wilhelm M, Kaminsky W, Münstedt H (2006) Influence of type and content of various comonomers on long-chain branching of ethane/α-olefin copolymers. Macromol 39:1474–1482
65. Yang Q, Jensen MD, McDaniel MP (2010) Alternative view of long chain branching formation by metallocene catalysts. Macromol 43:8836–8852
66. Wood-Adams P, Dealy JM, deGroot AW, Redwine OD (2000) Rheological properties of metallocene polyethylenes. Macromol 33:7489–7499
67. Torres E (2002) Extensional flow and rupture of molten polyethylenes, M Eng Thesis, McGill Univ
68. Wood-Adams P, Dealy JM (2000) Using rheological data to determine the branching level in metallocene polyethylenes. Macromol 33:7481–7488
69. He C, Costeux S, Wood-Adams P (2004) A technique to infer structural information for low level long chain branched polyethylenes. Polymer 45:3747–3754
70. Lai S, Plumley TA, Butler TI, Knight GW, Kao CI (1994) Dow rheology index (DRI) for Insite technology polyolefins. SPE ANTEC Tech Papers 40:1814–1815
71. García-Franco CA, Lohse DJ, Robertson CG, Georgon O (2008) Relative quantification of LCB in essentially linear polymers. Eur Polym J 44:376–391
72. Karjala TP, Sammler RS, Mangnus MA, Hazlitt LG, Johnson MS, Wang J, Hagen CM, Huang JWL, Reichek KN (2011) Detection of low levels of long-chain branching in polydisperse polyethylene materials. J Appl Polym Sci 119:636–646
73. Karjala TP, Sammler RS, Mangnus MA, Hazlitt LG, Johnson MS, Hagen CM, Huang JWL, Reichek KN (2008) Detection of low levels of long-chain branching in polyolefins. Soc Plast Eng ANTEC 887 AIP Conf Proc 1027:342–344
74. Malmberg A, Gabriel C, Steffl T, Münstedt H, Löfgren B (2002) Long-chain branching in metallocene-catalyzed polyethylenes investigated by low oscillatory shear and uniaxial extensional rheometry. Macromol 35:1038–1048

75. Benham E, McDaniel M (2002) Ethylene polymers HDPE encyclopedia of polymer science and technology. Wiley, New York
76. McDaniel MP, Rohlfing DC, Benham EA (2003) Long-chain branching in polyethylene from the Phillips chromium catalyst. Polym React Eng 11:101–132
77. Vega JF, Santamaria A (1998) Small-amplitude oscillatory shear flow measurements as a tool to detect very low amounts of long chain branching in polyethylenes. Macromol 31:3639–3647
78. Ghosh P, Dev D, Chakrabarti A (1997) Reactive melt processing of polyethylene: effect of peroxide action on polymer structure, melt rheology and relaxation behavior. Polymer 38:6175–6180
79. Lazar M, Kleinova A, Fiedlerova A, Janigova I, Borsig E (2003) Role of minority structures and mechanism of peroxide crosslinking of polyethylene. J Polym Sci A Polym Chem 42:675–688
80. Zhou W, Zhu S (1998) ESR study of peroxide-induced cross-linking of high density polyethylene. Macromol 31:4335–4341
81. Yoshii F, Makuuchi K, Kikukawa S, Tanaka T, Saitoh J, Koyama K (1996) High-melt-strength polypropylene with electron beam irradiation in the presence of pofunctional monomers. J Appl Polym Sci 60:617–623
82. Rätzch M (1999) Reaction mechanism in long chain branched PP. J Macromol Sci Pure Appl Chem 36:1759–1769
83. Kurzbeck C, Oster F, Münstedt H, Nguyen TQ, Gensler R (1999) Rheological properties of two polypropylenes with different molecular structure. J Rheol 43:359–374
84. Hingmann R, Marczinke BL (1994) Shear and elongational flow properties of polypropylene melts. J Rheol 38:573–587
85. Auhl D, Stange J, Munstedt H, Krause B, Voigt D, Lederer A, Lappan U, Lunkwitz K (2004) Long-chain branched polypropylenes by electron beam irradiation and their rheological properties. Macromol 37:9465–9472
86. Münstedt H, Laun HM (1979) Elongational behavior of a low density polyethylene melt II. Transient behavior in constant stretching rate and tensile creep experiments. Rheol Acta 18:492–504
87. Yamaguchi M, Takahashi M (2001) Rheological properties of low-density polyethylenes produced by tubular and vessel processes. Polymer 42:8663–8670
88. Wang J, Mangnus M, Yau W, deGroot W, Karjala T, Demirors M (2008) Structure-property relationships of LDPE. Soc Plast Engrs ANTEC 878–881
89. Lusignan CP, Mourey TH, Wilson RH, Colby RH (1998) Viscoelasticity of randomly branched polymers in vulcanization class. Phys Rev E 60:5657–5669
90. Janzen J, Colby RH (1999) Diagnosing long-chain branching in polyethylenes. J Mol Struct 485(486):569–583
91. Larson RG (2001) Combinatorial rheology of branched polymer melts. Macromol 34:4556–4571
92. Shroff RN, Mavridis H (1999) Long-chain branching index for essentially linear polyethylenes. Macromol 32:8454–8464
93. Shroff RN, Mavridis H (2001) Assessment of NMR and rheology for the characterization of LCB in essentially linear polyethylenes. Macromol 34:7362–7367
94. Robertson CG, García-Franco CA, Srinivas S (2004) Extent of branching from linear viscoelasticity of long-chain branched polymers. J Polym Sci B 42:1671–1684
95. Tsenoglou CJ, Gotsis AD (2001) Rheological characterization of long chain branching in a melt of evolving molecular architecture. Macromol 34:4685–4687
96. Vega JF, Santamaria A (1998) Small-amplitude oscillatory shear flow measurements as a tool to detect very low amounts of long chain branching in polyethylenes. Macromol 31:3639–3647

Chapter 8
Role of Rheology in Melt Processing

Abstract Rheological properties govern the behavior of polymers in melt forming operations. This chapter first explores melt behavior in channels, dies and extruders where the viscometric functions play a dominant role. Extrudate swell, wall slip, flow instabilities, and die build-up are described. The remaining operations dealt with are those in which elasticity is important. Processes for making sheet and film are described and processing problems related to rheology are explained. The roles of rheology in blow molding, injection molding, rotational molding and foam formation are taken up in the remainder of the chapter.

8.1 Introduction

This chapter treats primarily those aspects of melt forming processes in which rheology plays a significant role. There are a number of books on all aspects of polymer processing [1, 2], and it is not our intention to compete with them. The aspects of polymer melt rheology that are important in a given melt forming operation depend on the type of flow involved. For internal flows, i.e., flows through channels, dies, etc., knowledge of the viscosity as a function of shear rate is usually adequate. And in a process such as rotational molding the shear rate is very low, and it is the viscosity at very low shear rates that is important. In external flows, i.e., flow from a die into air, as in blow molding and film extrusion, viscoelasticity becomes important, because the melt is free to deform to relieve the stresses introduced by flow in the die. And when the melt is stretched, as in film blowing, extensional flow behavior is very important.

When developing a new resin for a certain application, the end product performance is the first consideration. Important factors include durability, strength, barrier properties, stress-crack resistance, weatherability, and optical properties. But it must also be possible to convert the resin into the end product using a reliable and economical forming process, and this depends on the properties of the

J. M. Dealy and J. Wang, *Melt Rheology and its Applications in the Plastics Industry*, Engineering Materials and Processes, DOI: 10.1007/978-94-007-6395-1_8,

melt. It is often impossible to optimize both of these requirements at the same time, and a compromise must be made. The objectives might be achieved by use of a blend, a filled resin, or coextrusion of two or more resins, although this may increase the cost the resin or of processing.

Assessing the processability of an experimental resin is a major challenge for an industrial rheologist. The ideal would be to predict rheological behavior based on the molecular structure of the polymer and then to mathematically simulate the forming operation to predict its processability. Molecular models for nonlinear behavior were mentioned in Chaps. 4 and 5; while significant progress has been made in predicting the behavior of linear polymers, long-chain branching is much more difficult to deal with. While progress has also been made in this field, we are still far from being able to deal with the structural polydispersity that is typical of commercial resins.

In the simulation of processes in which viscoelasticity must be taken into account, there are two difficulties. One arises from the lack of a reliable constitutive equation, and the other is dealing with mathematical singularities that occur at sharp corners in a flow field. The predicted stress becomes infinite at such points, which causes the simulation to crash. There are *ad hoc* methods for allowing the calculation to continue, although these programming tricks are not based on polymer physics. For example, Lee et al. [3] rounded the corners of their experimental channel slightly, and the model limited the stress that could be calculated to a reasonable value. They produced an impressive agreement with LDPE data of the entrance to and exit from a capillary inserted in a larger channel using the "eXtended" pom–pom model [4]. A basic question is whether the singularity arises from the nonlinearity of the model or a departure of the material from continuum behavior.

We must keep in mind that some melt flow phenomena cannot be described in terms of the continuum concept. An example is the behavior of a melt as it exits a die. Undesirable phenomena that can occur at this point are exit fracture (sharkskin) and die build-up (die drool.). Rupture cannot be described by a model that assumes the melt to be a continuum. While there are well-established models the fracture of elastic solids, these are not valid if there is significant dissipation, as in a melt. Die exit phenomena are discussed in this chapter.

Numerical simulations have proven to be of use for internal flows such as those in extruders, dies, channels and molds when there are no sudden changes of cross-section. For such simulations a temperature-dependent power law is often used. But as mentioned above when viscoelasticity plays a key role flow, simulations have proven much less useful in product and process design in spite of many years of strenuous efforts. As a result, the development of resins for use in such processes is more an art than a science. The only way to ensure that a new resin can be successfully converted is to observe its behavior in a full-scale test run. This is obviously a very expensive testing procedure, and it is not feasible to use it to evaluate large numbers of candidate materials.

In the first stage of resin development, only small quantities of material are available, and while it is not possible to do a conclusive processability evaluation, one wishes to use laboratory-scale measurements to at least rank candidate materials according their probable processability. This is where the art comes in. Those with long experience in the field may come to rely on one or two key lab tests, but there is often little scientific basis for this. Blow molding is a good example of this situation. Parison swell is a major factor in processability, but it is impossible to predict this reliably using numerical simulation or a lab test. It is easy to measure the diameter swell at the exit of a capillary, but blow molding dies are much more complicated than a capillary, and a great variety of die designs are used. A lab test may be of use to eliminate candidate materials that are unlikely to be successful, but once the search has been narrowed down, the remaining candidates must be evaluated using industrial blow-molding equipment. A similar challenge arises in the development of film blowing resins, as is explained in the section on that process.

8.2 Flow in Simple Channels and Dies

This section provides an introduction to flow in channels and dies. A detailed treatment of the design of extrusion dies can be found in the book by Michaeli [5].

8.2.1 Flow in a Circular Channel

We can use equations in Sect. 6.3.1 for capillary flow to derive the basic equations for flow of a power-law fluid in a circular channel. The shear stress for any fluid is a linear function of radius and for a power-law fluid is related to the shear rate, i.e., the velocity gradient, as follows:

$$\sigma(r) = \frac{r}{R}\sigma_\mathrm{w} = \frac{r}{R}\frac{\Delta PR}{2L} = \eta\left(\frac{\mathrm{d}v}{\mathrm{d}r}\right) = k\left|\frac{\mathrm{d}v}{\mathrm{d}r}\right|^{n-1}\left(\frac{\mathrm{d}v}{\mathrm{d}r}\right) \tag{8.1}$$

Now setting the velocity equal to zero at $r = R$ and integrating, we can obtain the velocity distribution.

$$v(r) = \left(\frac{\Delta PR}{2kL}\right)\frac{R}{(1/n)+1}\left[1-\left(\frac{r}{R}\right)^{(1/n)+1}\right] \tag{8.2}$$

The velocity can be integrated over the cross-section to obtain the volumetric flow rate.

$$Q = \frac{\pi R^3}{(1/n)+3}\left(\frac{\Delta PR}{2kL}\right)^{1/n} \tag{8.3}$$

8.2.2 Flow in a Slit Die

The equations for describing flow in a circular channel were presented in Sect. 6.3.1, and we present here a parallel treatment of flow in a slit. We define a slit as a rectangular channel having a width w that is at least ten times its thickness h. When W/h is large, the edges make a negligible contribution to the pressure drop, and the fully developed flow can be considered to be two-dimensional. The flow field in certain commercial forming operations approximates slit flow, and this geometry can also used for rheological measurements. It has certain advantages over a capillary, since flush-mounted pressure transducers can be used, and the two-dimensional flow field facilitates observation of the flow. We consider only fully developed flow, which means that there is no change of the velocity distribution in the flow direction.

A force balance on a rectangular element of fluid in the slit yields a relationship between the shear stress and the pressure gradient:

$$\sigma(y) = \frac{y \Delta P}{L} \tag{8.4}$$

where y is measured from the center plane of the slit as shown in Fig. 8.1, and L is the length of the die. Note that the shear stress and the pressure change over the slit length and are both negative quantities. The magnitude of the shear stress at the wall is:

$$\sigma_{\mathrm{w}} = -\sigma(y = h/2) = \left(\frac{-\Delta P}{L}\right)\frac{h}{2} \tag{8.5}$$

These equations are valid for any fluid, but in order to relate the shear rate at the wall to the flow rate, the viscosity must be known as a function of shear rate. For a Newtonian fluid, the magnitude of the shear rate at the wall is:

$$\dot{\gamma}_w = -\dot{\gamma}(y = h/2) = \frac{6Q}{h^2 W} \tag{8.6}$$

where Q is the volumetric flow rate.

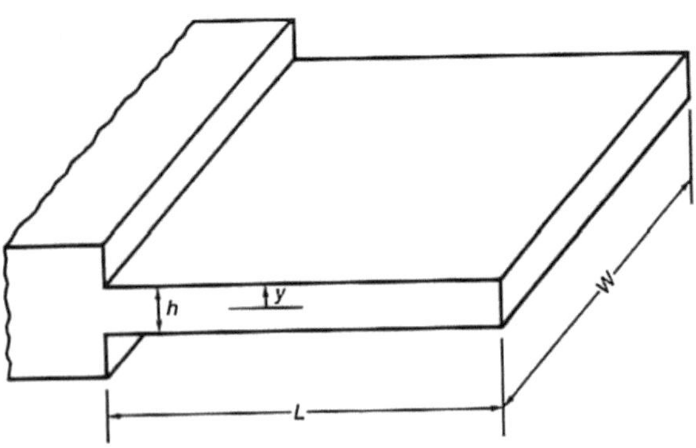

Fig. 8.1 Slit flow geometry showing the meaning of the symbols used in Eqs. (8.4–8.9)

For a power-law fluid (Eq. 2.6) the wall shear rate is:

$$\dot{\gamma}_w = \left(\frac{-h\Delta P}{KL}\right)^{1/n} \tag{8.7}$$

The velocity distribution is given by Eq. (8.8).

$$v(y) = \left(\frac{-y\Delta P}{KL}\right)^{1/n} \frac{hn}{n+1}\left[1 + \left(\frac{y}{n}\right)^{(1+n)/n}\right] \tag{8.8}$$

And the volumetric flow rate is:

$$Q = \frac{2h^2 W}{1/n + 2}\left(\frac{-\Delta Ph}{KL}\right)^{1/n} \tag{8.9}$$

8.2.3 Flow in Channels with Varying Cross-Section

8.2.3.1 The Lubrication Approximation

When analyzing the flow in a converging or diverging channel, there is a mathematical approximation for inelastic fluids in which the velocity distribution is assumed to be that of fully developed flow even though the cross section varies. This is the *lubrication approximation*, and is valid when both the Reynolds number and the rate of change of the size of the channel are small. The requirement of low Reynolds number is satisfied for a molten polymer because of its high viscosity. However, if the fluid is viscoelastic, there is an additional requirement that must be met to use the lubrication approximation, and that is that the Deborah number be small. If the Deborah number is not small, a viscoelastic model must be used to model the flow. For flow in a tapered channel, this requirement can be expressed as follows.

$$\mathrm{De} = \tau\left|\frac{\mathrm{d}V}{\mathrm{d}z}\right| \ll 1 \tag{8.10}$$

where τ is a relaxation time of the melt.

For flow in a tapered section of length L with inlet radius of R_o and outlet radius R_1 this can be expressed as:

$$\mathrm{De} = \frac{\tau Q}{R_o^3}\frac{|R_1 - R_0|}{L} \ll 1 \tag{8.12}$$

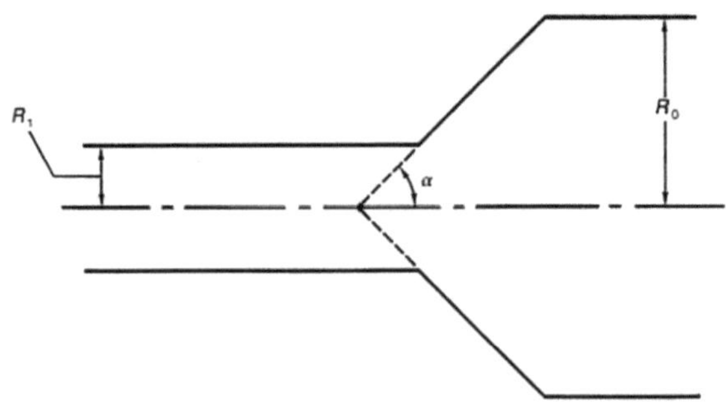

Fig. 8.2 Conical converging channel showing the symbols used in Eq. (8.13). The flow is from *right* to *left*

8.2.3.2 Flow in a Converging Channel

Using the lubrication approximation for flow through the conical channel shown in Fig. 8.2 the pressure drop for a power-law fluid can be shown to be:

$$-\Delta P = \frac{2K}{3n \tan \alpha}\left(\frac{1+3n}{4n}\right)^n \dot{\gamma}_A^n \left[1-\left(\frac{R_1}{R_0}\right)^{3n}\right] \tag{8.13}$$

where $\dot{\gamma}_A \equiv \frac{4Q}{\pi R_1^3}$.

As α becomes smaller, the magnitude of ΔP increases, because the tapered section is longer. However, it may still be desirable to use a small angle to reduce the severity of extrudate distortion that can occur at high flow rates (see Sect. 8.3.3). When the convergence angle is not small and the flow rate is high, the pressure drop is independent of convergence angle, and this reflects the fact that the lubrication approximation is no longer valid when α is not small.

When the Deborah number (Eq. 8.12) is not small, the flow can only be reliably modeled by use of a nonlinear viscoelastic constitutive equation. However, because of the complexities of this procedure, most die design formulas used in the plastics industry are based on the use of the lubrication approximation together with the power law model.

8.2.4 Flow in Noncircular Channels

While Newtonian fluids always have straight streamlines in straight channels of any shape, entangled polymers exhibit secondary flow, i.e., flow that is not co-linear with the axis of the die [6]. It has been shown that this occurs when the second normal stress difference is non-zero and varies with shear rate [7, 8]. Figure 8.3 shows cross-sections of extrudate at the entrance and exit of a square channel with a 13-ring feed-block [7]. A single polystyrene was used, but alternate layers were colored. A numerical prediction of this flow produced the streamlines shown in Fig. 8.4 [6]. These flows become significant at high flow rates in large channels.

Fig. 8.3 Cross-sections of extrudate at the entrance and exit of a *square die*. A 13-ring feed-block was used to create alternate layers of melt with and without colorant. The colored bands at the exit show that there is substantial secondary flow in the die and that the streamlines are not straight. From Debbaut et al. [6]

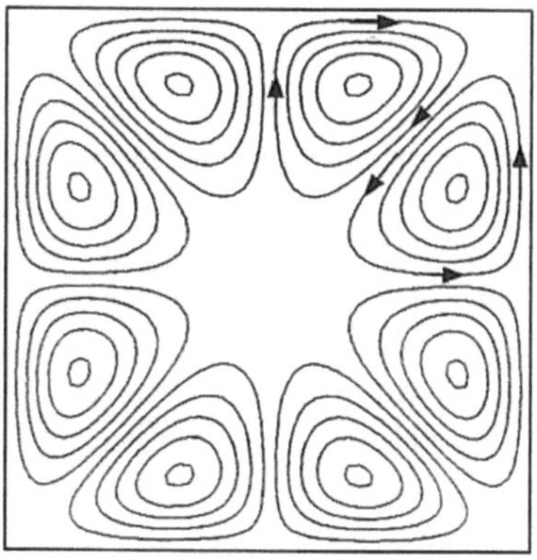

Fig. 8.4 A numerical prediction of the *streamlines* showing the secondary flow that results in the pattern seen in the melt exiting the *square die* of Fig. 8.3. From Dooley and Hughes [7]

8.2.5 Coextrusion Instabilities

When two or more different polymers are fed to a die to obtain a product with two or more layers, the interface between the two polymers is deformed by the flow in a manner governed by their rheological properties [9]. These are not prominent for slow flows in small dies, but they can be significant in a large die at high flow rates. For flow in a circular channel, the melt with the lower viscosity will preferentially flow toward the wall, eventually surrounding the more viscous one. This is called viscous encapsulation and is shown in Fig. 8.5 [9]. This process can be minimized by matching the viscosities of the two polymers, but this is complicated by the dependence of viscosity on shear rate and temperature. In noncircular dies, the second normal stress difference causes the more complex secondary flow described in the previous section, and this can result in distortions of the interface between the two polymers.

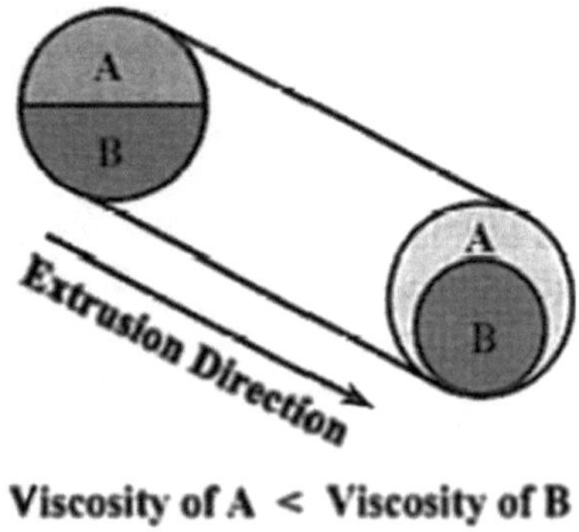

Fig. 8.5 Sketch showing viscous encapsulation. The melt with the lower viscosity preferentially flows toward the wall, eventually surrounding the more viscous one. From Dooley [9]

8.3 Flow in an Extruder

8.3.1 Basic Considerations

Extruders are essential elements of most melt processing operations, including pipe and profile extrusion, film and sheet extrusion, blow molding, wire coating, pelletizing, compounding and injection molding. We treat here only the role of rheology in extrusion; a number of books are available that describe all aspects of this important process, for example those by Chung [10] and Rauwendaal [11].

An extruder consists of three sections: feeding/melting/plasticating, compression, and metering. Melting/plasticating is a complex process in which solid is melted, and melt occupies an increasing fraction of the flowing solid–liquid mixture. In the compression zone solid is still present, and the diameter of the screw increases. Once melting is complete, what happens in the metering section is governed by melt rheology, principally the viscosity, with viscoelasticity playing a secondary role.

The earliest model of the flow in the metering section was developed before numerical simulation was an option and made use of equations that had analytical solutions. The melt was assumed to be a Newtonian fluid at constant temperature, and the channel depth was assumed to be small compared to the screw diameter. While more realistic models are now used, this one is still of use as an aid to understanding some basic flow phenomena. A simplified representation of the volumetric flow rate Q predicted by this model is shown below.

$$Q = K_1 N - \frac{K_2}{\eta}\frac{\Delta P}{\Delta z} \tag{8.14}$$

where K_1 and K_2 depend on the design of the extruder, N is the screw speed, and $\Delta P/\Delta z$. is the axial pressure gradient. Note that ΔP means P_{out}–P_{in} and is normally positive, and drag flow must overcome the pressure gradient in order for the extruder to act as a pump.

The first term in Eq. (8.14) is the drag flow Q_d due to the rotation of the screw, which must overcome the back flow resulting from the pressure term Q_p. The ratio of these two terms governs the shape of the velocity distribution.

$$\phi \equiv \frac{Q_p}{Q_d} \tag{8.15}$$

The flow is three dimensional, but for illustrative purposes, we show in Fig. 8.6 radial distributions of the axial component of dimensionless velocity for four values of this ratio. The velocity is a maximum at the surface of the screw (top) and zero at the barrel wall (bottom). In the distribution shown at top there is no back pressure, and we have pure drag flow. This would correspond to an extruder open at the end with no die. Moving down the figure, we see that for increasing pressure drop, there is more and more back flow. A net flow rate of zero implies a

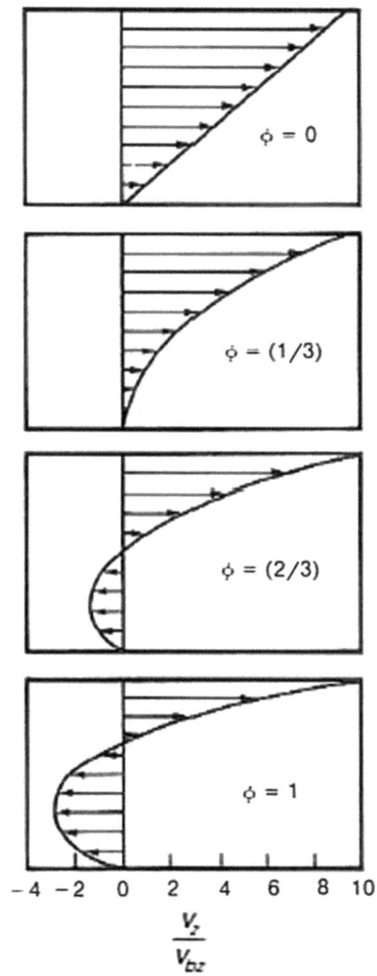

Fig. 8.6 Flow in an extruder for four values of φ, the ratio of pressure to drag flow. The ordinate is the relative distance from the root to the barrel, from 0 at the *bottom* to 1.0 at the *top*. The down-channel velocity v_z is divided by v_{bz}, which is the down-channel component of the relative boundary plane velocity

closed discharge, corresponding to a flow ratio ϕ of unity and to the bottom sketch in Fig. 8.6.

While the velocity distributions shown in Fig. 8.6 are of the correct general shape, the constant viscosity model is quite unrealistic, and the shear-thinning nature of melts must be taken into account if quantitative results are to be obtained. Rauwendaal [12] reported results of a simulation making use of the power-law model, generalized for two-dimensional flow. His results show that as the power-law index n becomes smaller, i.e., as the melt becomes more shear thinning, the pressure gradient contribution to the net flow becomes larger. Extrusion models now also account for leakage flow between the tip of the flight and the barrel and for thermal effects [11, 12]. The thermal effects are viscous dissipation and heat transfer with a heated or cooled barrel. Most of the power used to turn the screw is expended in viscous dissipation, and this energy input is often adequate for melting and heating, but to minimize temperature variations along the extruder, heating and cooling of the barrel may be necessary.

8.3.2 Effect of Wall Slip on Flow in an Extruder

On the basis of a numerical simulation [13], it was concluded that wall slip reduces mixing and pressurization efficiencies and increases leakage over flight tips, although slip on the screw may reduce power input for a given throughput. Chung, author of a book on extrusion [10], has expressed the following view. "The conveying capability for solid and melt arises from the shear stress at the barrel surface, and if slip occurs there, the conveying capacity will be reduced. It is known that output rate falls drastically if a lubricant such as silicone oil is added to the feed pellets. But if slip occurs on the screw, conveying capacity will be increased. The net effect of fluorocarbon processing aids is probably a slightly reduced output rate in a single-screw extruder."

8.3.3 Extrudate Phenomena

8.3.3.1 Sharkskin

Melt fracture is a term used for several phenomena that are quite different from each other in their origin and effects [14], and these were described in Sect. 6.3.2. One type mentioned there, gross melt fracture, which occurs at the entrance to a die, is not usually encountered in processing, but if it occurs in a pelletizer extruding into water, it can increase the moisture content of the pellets. One or more types of fracture might be expected to occur in high-speed wire coating, but the extremely high shear rate in the wall region results in intense viscous heating and a greatly reduced viscosity. Thus the stresses at the exit are relatively small, and fracture is avoided. There is negligible thermal degradation because of the very short residence time at high temperature.

The surface defect called sharkskin results from fracture at the exit of a die and can be a major problem in film extrusion and blow molding. This is now known to be caused by a fracture of the melt where the melt exits the die. Figure 6.9 is a sketch based on the observations of Moon et al. [15]. The melt at the wall just inside the channel is stationary, but the melt that was flowing near the wall suddenly finds itself free of the presence of the wall and accelerates, pulling on polymer that is still stuck to the wall. The resulting tensile stress can be large enough to cause a crack, and as the crack opens up the tensile stress is relieved, and the shear stress between the moving and stationary polymer starts the process over again.

The tendency of a polymer to exhibit this defect depends on its structure. For example LDPE is much less likely to undergo sharkskin than HDPE and LLDPE, possibly because of the high melt strength that is associated with its extensional flow behavior. Many schemes have been proposed to alleviate sharksin, including heating the die or altering its design. The most common approach is to modify the resin by either blending it with LDPE or adding a processing aid. Processing aids are generally fluoropolymers that coat the die wall and promote slip. In this way the melt at the wall is no longer restrained from moving out of the die with the bulk flow. Because these methods increase resin cost, it would be preferable to alter the polymerization process to produce a polymer resistant to exit fracture.

8.3.3.2 Extrudate Swell

At the exit of a die, the emerging stream of melt expands in cross-sectional area, a phenomenon called die swell or extrudate swell. If the flow channel is noncircular, as in profile extrusion, the extrudate will also undergo a change of shape. Such changes in extrudate shape and size are of practical importance in such processes as profile extrusion, blow molding and injection molding. In addition, swell has been used as a qualitative measure of melt elasticity for quality control purposes. Extrudate swell is an elastic response resulting from the stresses generated by shearing in the die. When the restraining influence of the die wall disappears, the melt is free to relieve these stresses. The presence of a slip coating at the die wall, as would result from the presence of certain processing aids, substantially reduces wall stress and shear rate and thus extrudate swell.

The simplest type of swell occurs at the exit of a circular die, such as an orifice plate or a capillary. There is no change of shape, and the swell can be described in terms of a single *swell ratio*, the ratio of the diameter of the extrudate to that of the die exit. Laun and Schuch [16] studied the effect of die geometry on this ratio and found that a diverging die decreases the swell, while a converging die increases it. We address in this section only capillary die swell; flow through annular dies will be dealt with in the section on blow molding.

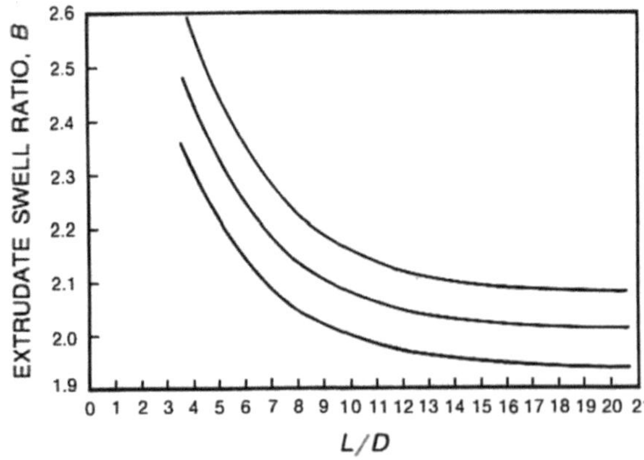

Fig. 8.7 Extrudate swell ratio versus *L/D* for capillary flow of a HDPE at 180 °C. The wall shear rates, from *top* to *bottom*, are 700, 400 and 200 s^{-1}. From Han et al. [17]

The variables commonly used to describe swell are the swell ratio *B*, the fully developed wall shear rate $\dot{\gamma}_w$, the length to diameter ratio *L/D*, the temperature *T*, and the time *t* that has elapsed since the melt left the die. Thus, for a given polymer, the relevant variables for a circular die are as follows:

$$B \equiv D/D_0 = f(\dot{\gamma}_w, L/D, T.t) \tag{8.16}$$

After a sufficient time, swell reaches an equilibrium value, and for a Newtonian fluid this ratio is 1.13, which is due to the redistribution of velocity. But for polymeric liquids it can be as high as two or three. This large swell ratio is a manifestation of molecular orientation generated by the flow in the die. At the inlet from a larger reservoir, streamlines converge rapidly, and this generates a high degree of stretching along streamlines. In a short die or orifice, this leads to a large swell at the exit, where the melt is free to deform in response to the anisotropic stresses associated with molecular orientation caused by stretching.

If the entrance leads to a capillary, there will be time for relaxation of the orientation generated at the entrance, and as the capillary is lengthened the degree of swell will decrease. At the same time, however, shearing in the die produces a moderate degree of axial orientation, and this will lead to a some extrudate swell. For a long capillary, therefore, swell approaches an ultimate value B_∞ that is independent of length and reflects only the shear-induced orientation generated in the capillary. Figure 8.7 shows the effect of *L/D* and wall shear rate on swell ratio [17].

For rapid evaluation for quality control purposes, the diameter of the extrudate at a fixed distance from the die exit can be measured by an optical method. Alternatively one can extrude into air and immediately measure the diameter of the frozen extrudate, or measure the extrudate diameter after annealing to permit the diameter to reach its fully swollen value.

Because swell is a viscoelastic phenomenon, it depends on time and temperature. If this dependence is of interest, the melt must be extruded into oil having the same density and temperature as the melt. This technique has been used to measure swell as a function of time for a capillary [18], a straight annular die [19] and for several converging and diverging dies [20]. For HDPE at 170 °C about 75 % of the swell occurred in the first few seconds, while the ultimate swell was only reached after several minutes. For a polypropylene at 190 °C, only 50 % of the swell occurred in the first few seconds, with equilibrium swell achieved only after more than 10 min. In a melt forming operation, if the extrudate is cooled immediately after exiting the die, it is only the swell that occurs immediately that is of interest.

For a given resin, if attention is restricted to the ultimate swell, and only capillaries with large *L/D* are used, Eq. (8.16) reduces to:

$$B_{\infty} = f(\dot{\gamma}_w, T) \tag{8.17}$$

Because commercial extrusion dies usually have their own temperature control system, it is of practical interest to examine the effect on the swell of a difference between the stock temperature and the die wall temperature. Cooling the die wall substantially increases the swell of some thermoplastics [21].

For noncircular dies as in profile extrusion, both the size and the shape of the extrudate vary with time after extrusion [22, 23]. Stevenson et al. [24] suggest a scheme for classifying dimensional changes for complex profile extrusions on the basis of size and shape factors. They also describe methods for controlling size and shape by manipulation of line speed, stock temperature and die temperature. The dependence of swell on molecular weight distribution is important, because it governs swell behavior in a linear resin. While it is generally accepted that increasing molecular weight increases swell at constant flow rate, no quantitative representation of this effect is possible, because it depends on many moments of the molecular weight distribution.

While capillary flow data can be used for comparing members of a family of resins, they cannot be related quantitatively to processing behavior, for example in blow molding, where parison swell is a very important. This issue is explored in the section on blow molding.

8.3.3.3 Die Build Up

A common problem in extrusion is die build-up (DBU), the accumulation of melt at the exit of a die. This material degrades with time and can affect the surface of the extrudate; pieces of this material can break off and attach to the extrudate. Other names for DBU are die-drool and die-bleed. When DBU becomes severe, the line must be shut down and the die cleaned. If this happens frequently it will result in increased production costs. Although much effort has been directed to identifying a cause of this phenomenon, no one "smoking gun" has been identified. Among the many contributing factors that have been proposed are:

Additives (e.g., talc, calcium stearate)
Molecular weight distribution
Moisture or oxygen in the resin

Thermo-oxidative degradation
Contamination
Quality of fabrication of die (surface finish, dimensional precision)
Die wall material of construction (surface energy)
Temperature of die.

It has been proposed that DBU is facilitated by extrudate swell. Patented methods for reducing DBU include increasing the land length or die gap and flaring the die exit, all of which reduce swell. Hogan et al. [25] observed die build-up of a polyolefin elastomer at the exit of a strand die and found no correlation between swell and DBU. However, they concluded that a key contributing factor was the time during which melt underwent high shear rates. By comparing the MWD of DBU material with that of feed resin they found that the DBU material contained a significant amount of oligomer that was not present in the feed. This would seem to be the result of thermo-mechanical degradation of the melt flowing slowly in the high-shear-rate region near the die wall. This material might volatilize at the exit and condense on the die face. Another possibility is fractionation of feed resin in the die with low molecular weight polymer migrating to the wall [26]. However, these conclusions may not apply to other resins under different conditions. Because this process involves the separation of a layer of polymer from the bulk, its mechanism might be related to that of surface fracture (sharkskin), and this idea has been explored by Giacomin and Schmalzer [27].

Many methods for suppressing DBU have been proposed, including:

Processing aid that promotes slip in the die
Special die design
Use of non-standard die material
Surface conditioning of steel die
Drying of resin.

The use of these methods has been described in detail by Gander and Giacomin [28]. They note, for example, that while the same processing aids that are useful for suppressing sharkskin have also been found to be helpful for DBU, the amount required is much less, from a tenth to a hundredth as much.

Capillaries and flat dies have been used to study DBU in the laboratory. A flat (rectangular) die has the advantage that it facilitates observation and the recovery of deposited polymer [29]. The capillary volumetric rate of build up q is sometimes normalized by dividing by the total volumetric flow rate Q [26], although this scaling may not be appropriate. The material accumulating on the die face comes from a thin layer of thickness h and should thus scale with the area of this annular region, which is πDh, whereas the total flow scales with the area, which is $\pi D^2/4$. If the build-up is associated with material flowing in the layer of thickness h with an average velocity of V_w, the rate of build up would be $\pi V_w hD$. Thus, changing the diameter has a larger effect on the total flow than on the build-up. A more meaningful dimensionless group would involve V_w and the average velocity in the die, which is $4Q/\pi D^2$. The wall velocity is $q/\pi hD$ so the ratio of interest is $qD/4Qh$, which involves the two dimensionless ratios, q/Q and D/h. The length h is unknown, but data can be used to calculate $qD/4Q$, which has units of length. This characteristic length might be used as an empirical tool for comparing data. If this length is relatively insensitive to the diameter D, it could be used to scale up data to larger diameters.

8.4 Sheet Extrusion/Film Casting

Sheet extrusion and film casting (chill role casting) are similar processes for producing a continuous sheet of polymer; both involve extrusion from a wide slot die with a narrow gap, usually downward. An extruded sheet may be cooled by quenching in water and produces relatively thick sheet that is further processed, for example by thermoforming. Cast film is cooled on a chill roll and squeezed by a rubber roll to produce a much thinner product, less than 0.25 mm, and is often multi-layered. The basic features of the process are shown in Fig. 8.8. The drawing length L is short, typically 25–65 mm. A greater variety of resins can be used for film casting than for film blowing, as high melt strength is not required, although Satoh et al. [30] found that strain-hardening decreased neck-in and edge bead size.

Sheet and film dies are of the coat-hanger or T-shaped type, with a typical width of one meter and a die opening of about one mm. It is desirable to design the die so that it produces a sheet of uniform thickness. The dependency of viscosity on shear rate is very important, and models are often based on the power-law. If a resin manufacturer wishes to supplant another supplier's product, it is essential to match the shear-rate dependence of viscosity. As was pointed out in Chap. 2, for linear polymers this dependency depends entirely on the molecular weight distribution, and the van Gurp-Palmen plot described in Sect. 3.16 has been found useful as a guide to MWD. Several strategies have been used to develop die design models,

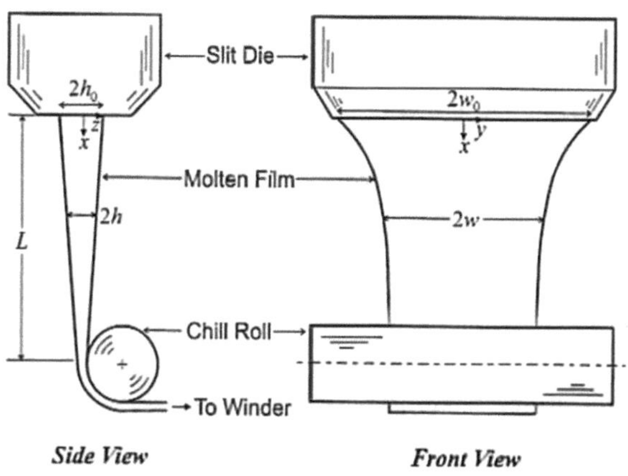

Fig. 8.8 Sketch showing essential features of the film casting process

and these are summarized by Tadmor and Gogos [2, p. 706]. However, these are not able to produce accurate, optimal designs. For example designing a die to provide uniform velocity distribution at the die lips does not yield a uniform web thickness, because swell depends not just on the exit flow but on previous strain history. For this reason sheeting dies have adjustable geometry for in-plant optimization. Viscoelasticity is of primary importance after melt leaves the die and experiences tensile stress, thickness swell and necking.

In film casting, the extrudate, called a web, is quenched on one or two chill rolls and wound up. The web is stretched, because the take-up velocity at the roll V_L is greater than the average melt velocity at the die lips V_0, and the ratio of these quantities V_L/V_0 is the principal stretch ratio, called the draw ratio or draw-down ratio DR. If the melt is assumed incompressible, the average thickness of the web at the chill roll h_L is related to the die gap h_0 as follows:

$$h_L = h_0(V_0/V_L) = h_0/\mathrm{DR} \tag{8.18}$$

The rheology-related phenomena that affect web behavior are edge bead, neck-in, draw resonance and web rupture. Extrudate swell occurs and plays some role in each of these phenomena.

Edge bead refers to the thickening of the web at its outer edges, which can be as much as five times as thick as the center of the web. Beads are trimmed off and go to waste or regrind. This phenomenon results from the combined effects of extrudate swell and drawdown. The melt at the ends of the die have been subjected to two-dimensional shearing, which enhances its swell compared to the more one-dimensional shearing away from the edges. If slip could be promoted in this region it would reduce the enhanced swell, but adding slip promoter would have an adverse effect on the flow distribution in the die.

The more important contribution to edge bead arises from the difference in the drawdown kinematics between the edges and the center of the web. The melt away from the edge is subjected to a transverse component of stress transmitted to it through the web from the melt flowing through the outer portions of the die, and this melt undergoes deformation that becomes more like planar extension as the center of the web is approached. Planer extension was described in Sect. 4.7; a rectangular sample of film is stretched in one direction, while shrinkage of the width is prevented. If the density is constant, the thickness is inversely proportional to the length in the stretch direction. For the present case, the ratio of the final thickness h_L to the initial thickness h_0 of a small element of film is inversely proportional the ratio of the take-up velocity V_L to the die exit velocity V_0, which is the draw ratio.

$$\frac{h_L}{h_0} = \frac{V_0}{V_L} \tag{8.19}$$

But the melt at the edges has a free surface on which the only stress that due to ambient pressure, and the stress field is closer to that of tensile extension. The transverse dimension of a segment of this material is inversely proportional to the square root of its length. Thus, the edge bead thickness at take-up d_L is related to the initial web thickness h_0 as follows:

$$d_L = h_0 \sqrt{V_0/V_L} \tag{8.20}$$

Combining the above two equations we find that the edge bead dimension is related to the thickness of the web away from the edges by Eq. (8.20).

$$d_L = h_L \sqrt{\frac{V_L}{V_0}} \tag{8.21}$$

Dobroth and Erwin [31] reported data for LDPE that agreed with this result. Various methods have been proposed to reduce the size of the bead [32], but it is usually removed by trimming.

Neck-in reduces the width of the sheet and is undesirable. Strain hardening reduces neck-in, and this usually implies the presence of long-chain branching. For example, polypropylene can be treated with peroxide to introduce branches, although this reduces the molecular weight somewhat.

Draw resonance is a flow instability that occurs in melt spinning and sheet extrusion at a critical draw ratio, even for Newtonian fluid [32]. It causes an oscillation in web thickness and to a lesser extent in width, with maxima in width corresponding to minima in thickness. A non-uniform film gauge is highly undesirable. Critical draw ratios for polyethylenes range from 15 to 35. Draw resonance is much more severe for linear than for branched polyethylene, and this was an initial barrier to the penetration of LLDPE into sheet extrusion markets. A proposed solution is to avoid the sudden cooling of the web at the chill roll by use of a linear jet of cooling air. Another is to blend LLDPE with LDPE. The modeling of draw resonance has not proven quantitatively useful because of deficiencies in viscoelasticity models and the lack of planar extensional flow data at high strain rates and temperatures.

As the drawdown ratio is increased, the deformation rate and tensile stress in the web also increase, leading eventually to rupture. By generating regions having less than average thickness, draw resonance exacerbates this problem.

8.4.1 Extrusion Coating

In extrusion coating a thin layer of polymer is coated onto a substrate at high speed. The substrate can be paper, aluminum foil or plastic film, and a wide range of products can be made, from thin film to thick board and including multi-layer coatings [33]. Extrusion coating is similar to chill-roll casting but runs under more intense conditions. The coating line runs much faster, up to 700 m/min, and the air gap between die lip and chill role is much larger than in film casting, typically between 150 and 250 mm as compared to 25–65 mm for cast film. A potential problem is rupture of the curtain before reaching the desired draw-down level. Neck-in is reduced by elasticity, but the ability to be drawn into a thin film without breaking is associated with more viscous behavior. The maximum speed a resin can run without rupturing the curtain is called the drawdown speed.

Coating runs at much higher temperatures than chill-roll casting, usually around 300 °C. This is to reduce viscosity so that the curtain can be stretched very thin (10–50 μm) before being coated on the substrate and to allow the line to run at high speed. For these reasons general purpose coating resins have relatively high melt index values, in the range of 4–25 g/10 min at 190 °C. The high temperature also promotes slight oxidation, which improves adhesion to the substrate.

The wider air gap results in larger neck-in, and the neck-in requirement is more stringent for extrusion coating resins than for cast film resins. Since the deformation of the web between die lip and chill roll is dominated by stretching, strain hardening at high strain rates is critical to minimize neck-in. Autoclave LDPE is the preferred resin for coating, since its highly branched structure provides strain hardening along with low zero-shear viscosity [34].

The behavior of the extruded melt curtain does not involve shear, and MI serves primarily as an indicator of molecular weight. Increasing M_w increases viscosity and melt strength but impairs draw-down ability. Clevenhag and Oveby [35] observed coating performance and measured MI and storage modulus for a large number of autoclave LDPEs, all having MI values in the normal range of coating resins. They found a substantial variation in both the rheology and processing among the resins but found little correlation between draw-down and MI. They observed that it is often necessary to evaluate LDPE coating resin processability by using a pilot or production scale coater.

When LDPE alone does not meet product specifications it is often blended with LLDPE to improve sealing performance, wear resistance or temperature resistance. Something that needs to be considered when blending LDPE with LLDPE is the flow activation energy. The viscosity of LDPE decreases more rapidly in going from the standard measurement temperature 190 °C to a processing temperature of 300 °C than a LDPE/LLDPE blend. So in order for the blend to have the same draw-down speed as LDPE, the blend often needs to have a lower viscosity at 190 °C, i.e., a higher melt index.

Draw resonance, which occurs in all drawing processes, can also be a problem in coating and manifests itself by a periodic fluctuation of neck-in and film gauge above a certain line speed, and low strain hardening or low melt strength has been associated with this phenomenon. Roberts et al. [36] describe a method for suppressing draw resonance.

Die designs are based on general principles similar to those for sheet extrusion but are more complicated. Automatic control of a flexible die lip may be used to maintain the desired product thickness.

8.5 Film Blowing

8.5.1 Introduction

The essential elements of film blowing are illustrated in Fig. 8.9. An extruder melts the resin and forces it through a screen pack and an annular die. The extruded melt, in the form of a tube, flows upward under the influence of a vertical, *machine direction* force, applied by means of nip rolls some distance above the die. There is stretching in this direction, and the ratio of the linear speed of the film through the nip rolls v_n, divided by the average melt velocity at the die lips v_{die}, is the *draw down ratio* (DDR).

When the process is started up air is injected through a hole in the die face, and when the extruded tube is threaded between the nip rolls to form an air seal, additional air is introduced to form the *bubble*. In this region the tube expands to a much larger diameter, and the ratio of this diameter to that of the die lips is the *blow up ratio* (BUR). This expansion of the molten tube as it moves into the inflated region generates stretching in the transverse (circumferential) direction. The bubble is cooled by means of an air ring that directs air along its outer surface.

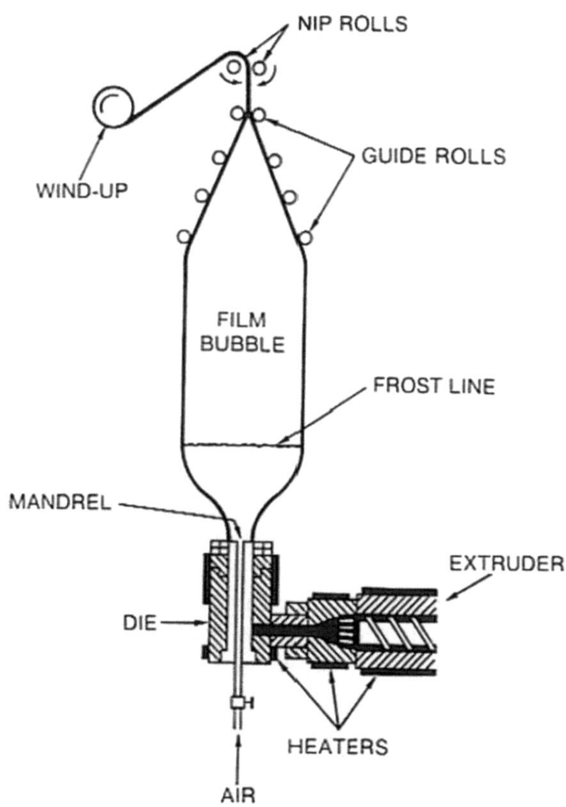

Fig. 8.9 Sketch showing the essential features of the film blowing process. The cooling is not shown

Additional cooling may be provided by the internal circulation of cold air. The *frost line* is actually a zone in which the final diameter is established and solidification is completed. This terminology arises from the loss of transparency due to crystallization in polyethylene film and not from the presence of condensed moisture on the film.

The film continues its upward movement until it is sufficiently cool to be flattened without sticking to itself or to the apparatus. The tube then passes through a frame or flattening ladder so that it can pass between nip rolls that generate machine direction tension. The flattened tube is then wound up and cut for shipment or for further processing. The width of the flattened tube is the *lay-flat width*, which is equal to π times the final bubble radius. Rheological properties play an important role in film blowing. They govern the shape and stability of the bubble and the onset of sharkskin (surface roughness). Because of the complexity of the flows involved, it is not generally possible to establish simple quantitative correlations between these phenomena and easily measured rheological properties. However, an understanding of how variations in the rheological behavior of melts can affect the processing and properties of blown film is essential if one is to achieve optimum results from this process.

The objective is to produce a film having uniform gauge and good optical and mechanical properties. Gauge nonuniformity is a serious defect that can lead to the rejection of finished film and usually results from bubble instability but can also be caused by extruder surging. Since the film is quite thin it is especially important to avoid the presence of unmelted material, gels or foreign matter, as these will be readily apparent in the finished product. In order to achieve good mechanical properties it is desirable that molecular orientation in the film be more or less balanced i.e., equal in the machine and transverse directions. This implies a small die gap and low DDR, but maximizing production rate may require relaxation of these conditions, resulting in preferred orientation in the machine direction and a *splitty film*, i.e., one that is weak in the transverse direction. For this reason, higher quality film has a more balanced orientation.

One problem that often occurs when linear polymers are being processed is sharkskin, which was described in Sect. 6.3.2. This results from surface fracture at the exit of the die and causes a marked deterioration in optical properties. More is said about this below.

The resins most used for blown film are polyethylene and polypropylene. Low-density polyethylene (LDPE) is easy to extrude, exhibits good bubble stability and yields a flexible film. It has been found that a narrow molecular weight distribution and a modest degree of long chain branching give good drawdown and optical properties, although this combination of characteristics does not give the best impact strength. A stronger and less flexible film is produced using high-molecular weight, high density polyethylene (HDPE), but it is more difficult to process and to make a film having balanced orientation. Because it is a copolymer, linear low density polyethylene (LLDPE) has the low crystallinity and flexibility of LDPE but better film strength. At the same time, it has some of the processing disadvantages of HDPE. Compared with LDPE, it has a higher viscosity at the high shear rates that

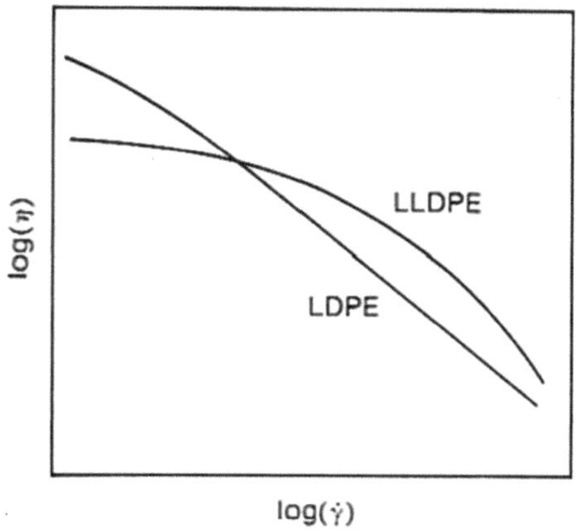

Fig. 8.10 Viscosity *curves* for *LLDPE* and *LDPE* film resins. The *LLDPE* is less shear-thinning and has a higher viscosity at shear rates that occur in an extruder and die

occur in the extruder and the die, thus increasing screw torque, barrel wear, and melt temperature. This is illustrated in Fig. 8.10, which shows typical viscosity versus shear rate curves for LDPE and LLDPE film resins. LLDPE also exhibits lower extensional stresses at the low strain rates typical of the bubble inflation region, and this renders the bubble less stable than LDPE but better in this regard than HDPE. Unlike LDPE, LLDPE is prone to sharkskin exit fracture.

Because of the complexity of the flow field and the strong effects of cooling and solidification on polymer properties, no simple cause-effect relationships are available to predict the process performance of a resin on the basis of a few easily performed laboratory tests. As in the case of blow molding, the final step in resin evaluation involves the use of a full-scale film line.

8.5.2 Flow in the Extruder and Die

The extrusion of LDPE is well understood, and screw design is a well established art. In the case of HDPE simple screw designs can produce inadequate solids conveying at moderate and high screw speeds. Thus, as screw speed increases, output rate and film properties deteriorate. Mixing sections can correct this, but their use leads to higher melt temperatures and degradation. This in turn reduces the maximum blow up ratio, which is needed to promote balanced orientation. Grooved barrels in the feed zone can improve solids conveying, as can barrier screws, which keep unmelted material separate from the melt and pushed up against the wall. This generates less shear and reduces melt temperature and pressure fluctuations.

LLDPE also poses extrusion problems compared to LDPE, which can be dealt with by the use of barrier screws [37]. Because of the higher viscosity at typical extrusion shear rates, as shown in Fig. 8.10, there is higher torque on the screw, accelerated barrel wear, and higher melt temperature.

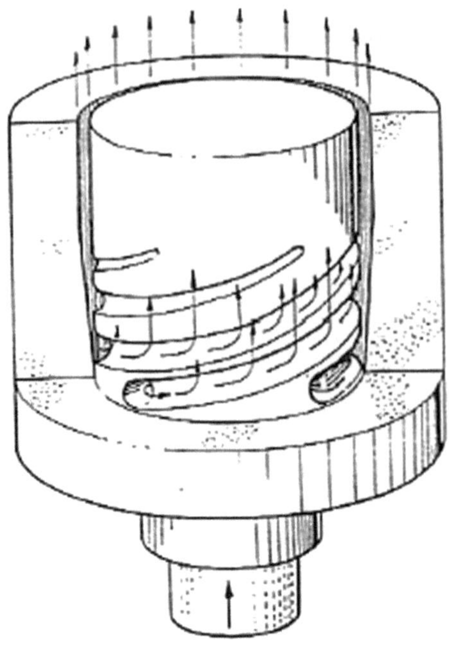

Fig. 8.11 *Spiral die* used to extrude melt for the film blowing process. The partial cross-section of the mandrel shows the main flow along the channel and the leakage flow over the land into the next channel

The die is designed to produce annular extrudate having no weld lines with uniform thickness and velocity around its circumference. The most popular type of film die, the spiral mandrel die, is shown in partial cross section in Fig. 8.11. The melt enters at the center and flows out through several feeding ports to a series of helical grooved channels cut into the wall of the mandrel. The cross section of these channels decreases in the flow direction, while there is an increasing gap between the land of the channels and the cylindrical outer wall of the die. Thus, an increasing quantity of the melt leaks over this land into the next channel, and there is thus substantial mixing. The depth of the helical channels ultimately reduces to zero, and the final section of the flow channel is an annulus whose gap varies in the axial direction, narrowing significantly near the exit.

The flow is complex, but a simplified method of calculation was used by Proctor [38] to calculate the details of the flow for a purely viscous material with a power-law viscosity. More elaborate methods for power law fluids that make fewer simplifying flow assumptions and require more computation were described later [39, 40], and several of these were compared by Perdikoulias et al. [41]. Another model incorporates viscoelastic material behavior [42].

Gates [43] discusses the problems involved in designing a spiral mandrel die for use with LLDPE. He says that the main problem is to avoid high shear rates, which result in high back pressure and melt temperature. Furthermore, he points out that there can be a substantial temperature variation across the diameter of the extruder adapter that feeds the die, and that the flow in the spiral mandrel die tends to produce a layering of polymer having a range of temperatures in the molten tube produced. This in turn can cause cyclic variations in gauge and frost line height. He suggests that this effect is minimized by use of a longer spiral wrap distance, leading to more port overlap. A longer spiral channel also yields performance that is less sensitive to changes in the viscosity of the resin.

In Sect. 6.3.2 we described extrudate distortions that can occur when melt exits a die. In the film blowing process, the most troublesome of these is sharkskin (surface melt fracture), because it results in film having a rough, unattractive surface. Sharkskin is especially troublesome with LLDPE, with the result that dies designed for LDPE can only be used at significantly reduced production rates. One solution is to use a wider die gap to reduce wall shear rate. However, this increases demands on the bubble cooling system and causes some deterioration of film properties. Other solutions proposed include a diverging die gap and heated die lips, but the usual approach, which does not require equipment modification, is the use of processing aid, which is usually a fluorocarbon elastomer that is added to the resin at time of processing or introduced as a masterbatch.

8.5.2.1 The Bubble Region

The melt leaving the die is suddenly free of restraints imposed by contact with the die wall and reacts quickly to its new environment. In particular, it swells in response to the molecular orientation induced by die flow, and it stretches in the machine direction in response to the tension imposed by the nip rolls. Its behavior in such a free surface flow is mainly dependent on its rheological properties, with thermal properties also playing an important role. The design of the die and film cooling system are also important.

For a given resin and film line, processing conditions that influence the flow in the bubble are:

1. Melt temperature
2. Output rate (melt flow rate)
3. Nip roll speed
4. Internal bubble pressure
5. Cooling air temperature and flow rate.

The draw down ratio and blow up ratio are usually used as process variable in place of nip roll speed and bubble pressure, respectively. Obviously many variables contribute to bubble behavior, and the systematic study of the effects of each of these would be an enormous task, and the result would be a huge mass of data that would be impossible to correlate without the guidance of a model. Our present understanding is based on inferences from small-scale studies in which only a few variables are examined. In addition, there is no reliable way to scale up results obtained with one apparatus so that they are relevant to an apparatus of a different size and design. We present here some empirical observations, but it is important to keep in mind that these do not constitute fundamental principles.

8.5.2.2 Bubble Cooling

The converter wants to maximize production, and this implies a high cooling rate. In early air-rings this rate was maximized by directing air at the bubble to generate stagnation flow, but this generated high pressure and severe buffeting. To avoid the destabilizing effects of this flow pattern, air flow was limited to low rates. This led to the use of a deflector to divert the air so that a *fan spray* flow pattern was generated. This permitted an increase in air flow, but the flow passages were not streamlined, and large-scale turbulence was generated. While this type of air ring was used for several decades, improved units were developed. The flow of cooling air next to the film has a significant effect on bubble shape [44], and this can be used to advantage. In the case of LLDPE the bubble is particularly sensitive to destabilizing forces, and the venturi effect has been used to control bubble shape while stabilizing the position of the bubble. Dual and multiple stacked air-rings are now used to generate streamlined flow without creating large suction at the base of the bubble. Zhang et al. [45] measured heat transfer rates at the bubble surface of a large-scale apparatus with a dual-lip air-ring.

8.5.2.3 Forces Acting on the Bubble

The bubble is supported from above by the nip rolls, and the position of the cooled film is stabilized by guides such as irises and cages, and by the collapsing ladder. However, at its base the bubble is held in place only by the tube of molten polymer coming from the die. This allows the bubble to move and deform in response to external forces, particularly those arising from the flow of cooling air. Disturbances of any kind can cause gauge variations and even bubble rupture. Therefore, it is of importance to understand all the forces that can act on the bubble.

Looking first at the entire bubble, there is an upward (machine direction) force exerted by the nip rolls. Bubble stabilizing devices and the collapsing frame will exert lateral forces on the bubble. At the bottom, viscous stresses act in the molten tube at the exit of the die to hold the bubble in place. Internal pressure supplies the driving force for inflation, while aerodynamic forces can be used to control bubble shape and stabilize its position. Finally, gravity acts directly on the bubble, so that the nip rolls not only produce the force that generates stretching in the bubble but must also support its weight.

8.5.2.4 Stresses in the Film

Internal stresses arise from the viscoelastic reaction of the melt to its deformation and act to stretch a fluid in the machine and transverse directions. Figure 8.12 shows the stresses in these directions acting on a small element of the bubble. The extensional strain rates: $\dot{\varepsilon}_M$ in the machine x_M direction and $\dot{\varepsilon}_T$ in the transverse x_T direction, can be written in terms of the velocity $v(t)$ and bubble radius $R(t)$ of a moving element of melt. The use of a frame of reference that tracks a fluid element is required when a constitutive equation is used to describe calculate the stresses in the melt. However, it is more convenient for describing experiments to use the variables $v_M(z)$ and $R(z)$ that refer to a point in the bubble at a fixed distance z above the die. Assuming axial symmetry, the strain rates in both frames of reference are given below.

$$\dot{\varepsilon}_M = \frac{1}{v_M(t)}\frac{dR(t)}{dt} = \cos\theta\frac{dv_M(z)}{dz} \tag{8.22}$$

$$\dot{\varepsilon}_T = \frac{1}{R(t)}\frac{dR(t)}{dt} = \frac{v_z(z)}{R(z)}\frac{dR(z)}{dz} \tag{8.23}$$

The stresses generated in each of these directions are related to strain rate histories of a fluid element by the viscoelastic properties of the polymer. The flow kinematics in the bubble varies from planar extension, if there is a region above the die where the diameter is constant, to biaxial extension in the inflation region. However, the only readily measured extensional flow in the laboratory is uniaxial extension, and this is the one used for correlating lab data with processability. Strain hardening is most apparent in uniaxial stretching but is thought to follow the same trend as in biaxial extension. The extensional flow properties of the resin govern its ability to avoid bubble instability and molecular orientation in the finished film.

8.5.2.5 Bubble Shape and Molecular Orientation

Bubble shape affects bubble stability and orientation of the final film; it depends on polymer, amount of air injected, DDR, BUR and air ring aerodynamics. The two bubble types used in film blowing are shown in Fig. 8.13. The *pocket* shape shown on the left is used for LDPE, LLDPE, polypropylene, and a number of other materials, while the *long stalk* shape shown on the right is used primarily for HDPE. The bubble shape plays an important role in generating molecular orientation, since transverse orientation is the direct result of the radial expansion that accompanies inflation. The principal factors affecting bubble shape are the amount of air injected and aerodynamic forces due to the flow of cooling air.

The long stalk process is used for HDPE in order to produce film having a more balanced orientation. First, it provides time for the relaxation of the molecular orientation generated by flow in the die. In addition, we note that in this type of operation most of the machine direction drawdown occurs before inflation, while transverse stretching occurs at a later time and at lower temperatures. In this way, the transverse extension makes a larger contribution to film orientation even though it may be smaller in total magnitude than the machine direction drawdown. Another reason for using a long stalk, and thus delaying inflation, is that a rapidly

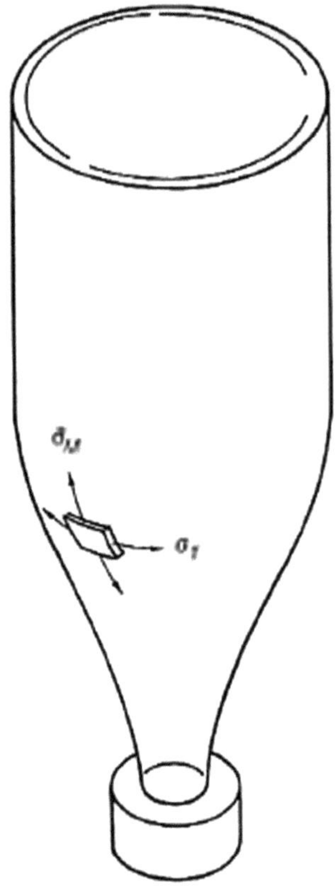

Fig. 8.12 Principal stresses in the blow-up region of the bubble. σ_M is the normal (tensile) stress in the machine direction, and σ_T is the normal (tensile) stress in the transverse direction

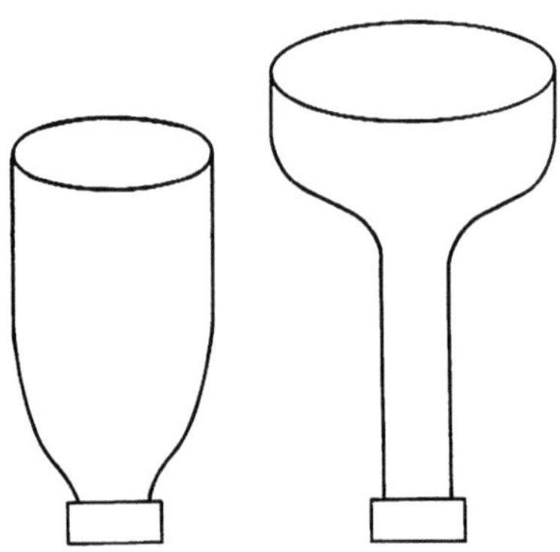

Fig. 8.13 *Bubble shapes* used in film blowing. The "pocket" shape on the *left* is used for LDPE, LLDPE and polypropylene, while the "long-stalk" shape on the *right* is used primarily with HDPE

thinning film of melt would be susceptible to instability and rupture at the high temperature of the extrudate.

Small variations in resin characteristics, for example those due to batch-to-batch differences, can cause significant variations in bubble shape [46], and these differences can be amplified by the bubble cooling process, which itself can have an important effect on bubble shape.

As mentioned above, molecular orientation in the finished film governs its tear and rupture strength, and orientation is the result of stresses generated during upward flow in the bubble. Detailed strain histories in the molten zone of low density polyethylene film bubbles have been determined experimentally [47, 48]. The machine direction strain rate is highest at some point above the die and then decreases monotonically up to the frost line. After a short zone in which extrudate swell is significant, the machine direction strain rate reaches its maximum value, while the transverse strain rate only becomes significant in the region of diameter increase where it reaches a maximum and then falls to zero in the neighborhood of the frost line. For large blow-up ratios, the transverse strain rate can rise far above the machine direction strain rate in the region of maximum rate of increase of the bubble radius.

However, the strain history must be considered in light of the time and temperature dependencies of viscoelastic behavior. Stresses present in the lower regions of the bubble will relax much more quickly than those introduced by deformation in the cooler regions, and stresses present as the melt begins to solidify are particularly effective in generating permanent orientation. However, high molecular weight linear resins have longer relaxation times, and this will enhance the effect of deformations occurring lower in the bubble. To summarize, the final orientation of the film depends in detail on the histories of both the stress field and the temperature as the polymer moves from the die to the frost line.

To determine the degree of molecular orientation in a sample of film one can use analytical methods such as infrared dichroism and x-ray diffraction, but a much simpler technique can provide useful information for LDPE. A circular film sample with the machine and transverse directions marked on its surface is placed in hot oil in a shallow dish. The sample shrinks, and the relative shrinkage in the two directions is indicative of the amount and direction of orientation. Liu and Harrison [49] compared their shrinkage data with those obtained using a polarized infrared technique and found a good correlation.

The relationship between melt deformation in the bubble and film orientation was studied by Farber and Dealy [47]. They measured deformation rates in a laboratory film line in the transverse and machine directions for an LDPE. Figure 8.14 shows strain rates in the two directions for a BUR of 2.2. The ordinate is time as seen by an element of melt as it moves through the bubble. Machine direction stretch rate is maximum just above the die and then decreases steadily to the frost line, while the transverse rate rises to a maximum in the region where the bubble diameter is increasing most rapidly. The ratio of the transverse shrinkage to that in the machine direction of the finished film was found to be 0.34. Thus, even though the deformation was mainly transverse over most of the melt path, the final orientation was in the machine direction. At the frost line, the curves are extrapolated to zero, but the melt was still under stress in this region, and while the strain occurring there was small it had a strong effect on final orientation.

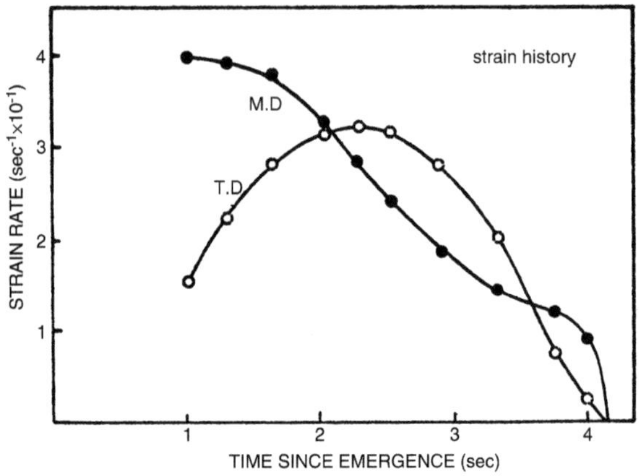

Fig. 8.14 Strain rates in the machine and transverse directions as functions of time following a melt element from the die to the *frost line* for an LDPE. The BUR was 3, and the ratio of shrink in the transverse direction to that in the machine direction was 0.34. While the rate in the transverse direction is much higher most of the time, the strain that occurs in the neighborhood of the *frost line*, although small, is much more effective in generating orientation, because the polymer is in its rubbery state there

8.5.2.6 Bubble Stability

Disturbances that can trigger bubble instability include air-ring flow, machinery vibrations, poor temperature control, and spikes in line voltage. Instabilities in bubble behavior are limiting factors in the production of blown film, and the limiting condition is usually associated with the maximum cooling rate that does not destabilize the bubble [50]. Instabilities can take many forms, including the following:

Oscillation of bubble diameter
Neck moves abruptly up and down at regular intervals
Frost line height oscillates
Circular motion or wobbling of bubble, which introduces a helical bulge in the film
Periodic thickness variation, similar in nature to the draw resonance that occurs in melt spinning [51].

Zhang et al. [45] carried out numerical simulations and experiments to determine the effect of air-ring aerodynamics on bubble behavior. Their results confirmed the strong interactions between cooling rate and bubble shape and how these affect bubble stability. Increasing cooling decreased the stability of a low BUR bubble but stabilized one with a high BUR.

The combination of instability, thin film and high tensile stress can result in rupture of the bubble and interruption of production. Melt strength is obviously important for good drawability. But the most common problem arising from instability is gauge variation in the finished film.

The ability of a resin to resist becoming unstable in response to a disturbance has been found to correlate with its extensional flow behavior. Strain hardening, particularly at low strain rates, appears to damp instabilities by suppressing further strain where a fluctuation has created a small reduction in film thickness and a locally higher strain rate. LDPE bubbles are much more stable than those of HDPE, although broadening the MWD of HDPE improves bubble stability somewhat. In start-up of steady simple extension, LDPE exhibits marked strain hardening, i.e., its tensile stress growth coefficient rises sharply above the linear viscoelastic curve at a Hencky strain that decreases with strain rate. In a comparison of two similar LDPE film resins [52], it was found that the resin with higher values of $\eta_E^+(t, \dot{\varepsilon})$ and lower maximum drawdown produced the more stable bubble. HDPE, on the other hand, is strain softening unless some very high MW material is present.

When LLDPE was introduced, it was intended as a replacement for LDPE, as it produced stronger film, and it was thus desirable to be able to process it using equipment already in place for the latter. However, this could only be achieved by reducing production rates, because LLDPE was more susceptible to gauge variation than LDPE although significantly better than HDPE. Improved bubble stability and higher production rates in LLDPE were achieved by use of air-rings that use aerodynamics to control bubble shape. Another way to stabilize LLDPE bubbles is to use a *booster*, often a peroxide cross-linking agent. Cross-linking can make LLDPE behave more like LDPE, although care must be taken not to produce sufficient cross-linking to interfere with the drawdown process or to produce gels. The most common way to improve bubble stability for LLDPE is to blend it with LDPE. Micic and Bhattacharya [53] studied blends of LLDPE with from 10 to 30 % of several LDPEs. The LDPEs varied in molecular structure, with MI values ranging from 0.3 to 1.7, MWDs from 7 to 30, and branching levels from 2.5 to 6 LCBs per 1,000 carbons. They measured extensional flow behavior of each of the blends using a sophisticated rotary-clamp instrument and observed their behavior in a film line. They found that the LDPE with the broadest MWD and the highest LCB level was the most effective in promoting strain hardening and bubble stability. They also reported that increasing the amount of LDPE in the blend effected these improvements up to the highest level used, which was 30 %.

Metallocene copolymers (mLLDPE) made more efficient use of comonomer than LLDPE made using Ziegler catalysts and offered much better control over MWD. However, their narrow MWD made them much less shear thinning so that they were more difficult to extrude. This drawback was eliminated by the development of the CGC catalyst system described in Sect. 7.2.3.1. The introduction of a small and well-controlled level of long chain branching made it possible to vary independently crystallinity and shear thinning.

8.5.3 Drawability

An important feature of the film blowing process is its ability to generate a high degree of drawdown, so that the gauge of the final film is much less than the die gap. However, a limit to this process is imposed by the strength of the melt in the bubble. For a given die, blow-up ratio, and throughput rate, if the nip roll speed is increased, a point is reached where the tensile stress in the bubble exceeds the cohesive strength of the melt, and rupture occurs. The ability of a resin to be drawn down can be characterized by the maximum drawdown ratio (DDR).

This limitation is particularly noticeable in the case of LDPE, and it is thought that this is related to the fact that this resin is strain hardening. This means that its tensile stress growth coefficient increases sharply above the linear viscoelastic curve as strain increases. If the principal stretching stresses in the process also increase sharply with strain, such a resin would be particularly subject to cohesive failure. However, it must be remembered that flow in the bubble is not a uniaxial deformation, and the correlation between strain hardening and the tendency to rupture is not direct.

Comparisons of similar LDPE film resins [46] and of HDPE with LDPE [16] showed that differences in maximum drawdown had no correlation with shear behavior but was related to some extent with extensional flow properties. However, the melt strength test described in Chap. 6 is the simplest method for detecting differences in drawability with similar resins [54].

8.5.4 Modeling the Film Blowing Process

Modeling this process to predict the performance of a given combination of resin, equipment, and operating conditions is extremely complex. The geometry of the bubble cannot be described simply and is dependent on resin properties, operating conditions and equipment design. The air-ring not only affects film temperature but also exerts aerodynamic forces on the bubble that can have a strong effect on bubble shape [44]. In addition, the air flow is turbulent and buffets the surface of the bubble. Flow in the die also affects film thickness uniformity, and this may be complicated by the presence of wall slip in HDPE [54]. A comprehensive model must account for air-ring aerodynamics, heat transfer, crystallization, and nonlinear viscoelastic behavior.

The deformation is close to planar extension just above the die and then becomes biaxial in the bubble but not equibiaxial. It starts out at the die as a hot viscous melt, but it cools rapidly becoming more and more elastic and finally crystallizes. For these reasons, the reliability of the model used for the melt viscoelasticity is of crucial importance. Continuum models in the form of differential equations are preferred, as these lend themselves to numerical analysis. Finally there is at present no way to measure the response of a melt to high-rate planar or biaxial stretching under controlled conditions. As a result, it is not possible to evaluate the ability of an empirical model to predict such flows. Another complication is that the melt becomes increasingly elastic as its temperature falls, and as it starts to solidify, a constitutive equation suitable for a homogeneous material will no longer be appropriate. Some models employ different rheological models in the two regions.

Pearson and Petrie [55] developed a coordinate system suitable for describing the complex shape of a bubble in 1970, and since then there have many efforts to do a useful numerical simulation. A final and especially challenging aspect of the problem is that the issue of most interest is the stability of the system, and this means that a steady-state axisymmetric solution will not be adequate. Jim et al. [56] have spent well over a decade on this problem. Meanwhile the group of Housiades [44] has focused their efforts on interactions between cooling air aerodynamics and bubble surface.

8.6 Blow Molding

8.6.1 Introduction

The principal types of blow molding are extrusion blow molding and injection blow molding. In the latter process a preform, often similar to a test tube with a threaded end, is injection molded and subsequently reheated and inflated inside a mold. This process affords excellent control of the thickness distribution in the preform and is used to make small containers with high quality finishes. It is also used in the stretch blow molding process used to make carbonated beverage bottles. Since there is no freely suspended parison, the issues addressed in this section, namely parison swell and sag do not arise.

In the extrusion blow molding process, a tube, called a parison, of melt is extruded from a die. The mold halves then close around the parison to pinch it off at one end, and, if a bottle is to be made, to form a threaded neck at the other. Then the parison is inflated to conform to the shape of the mold. The extrusion can be intermittent, halting while the parison is pinched, inflated and cooled, or it can be done continuously by the use of two or more moving molds. The extrusion process is faster and more economical than the injection process, and is preferred for manufacturing large items. Very large items can be made by extrusion blow molding, with parisons weighing as much as 100 kg or more. For such large objects the extruder is supplemented by a ram accumulator that permits rapid formation of the parison.

There are two types of extrusion blow molding, reciprocating, involving periodic extrusion, and shuttling, which involves continuous extrusion. In the reciprocating process, melt accumulates at the end of the extruder as the screw backs up, and the screw then moves forward to generate the parison. For very large containers melt collects in a separate accumulator. Then melt is pushed through the die by a ram, generating a relatively high flow rate in the die and more parison swell than in the continuous process. While the parison is being clamped and inflated, the extruder is generating a new batch of melt. In the continuous extrusion process, once a parison has been formed it is shuttled aside for molding, and a new parison is formed immediately. The flow rate in the die is lower than in the case of reciprocating extrusion, resulting in smaller swell and allowing more time for parison sag.

The viscoelastic properties of the melt play a key role in blow molding, particularly in the extrusion blow molding process. The pressure profiles and flow patterns in the extruder and die are governed by the viscous properties and the shape of the flow passages. However, between the time the melt exits the die and the time it touches the cooled mold walls, it moves and changes its shape entirely as a viscoelastic response to three factors:

1. Molecular orientation generated by flow in the die, which results in parison swell,
2. Gravity, which causes "drawdown" or "sag" of the parison, and
3. Blow pressure, which causes parison inflation.

8.6.2 Flow in the Die

Approximate methods are used to calculate the important aspects of flow in a blow molding die. The calculation of forces and velocity distributions requires only the viscosity as a function of shear rate and temperature; the flow is usually assumed to be isothermal [57–59]. Winter and Fritz [59] used the power-law to calculate the velocity distribution of velocity in the die gap. The resulting model was then used to design dies having a uniform velocity distribution, and dies fabricated based on these designs were found to have good flow distribution. Another use of models is to calculate the total force exerted by the melt on the mandrel [58]. For a converging die, the normal pressure acts in a direction opposite to the shear force. This makes it possible to design a die so that the net force on the mandrel is quite small. This simplifies die design and makes it possible to use a stepper motor rather than a servo-hydraulic actuator to drive a movable mandrel for purposes of parison programming.

Swell is one manifestation of the flow that occurs in the die, and other effects include weld lines, shear modification, and extrudate distortion. If there are spider legs to hold the mandrel in place, weld lines will be formed as the melt flows around the legs. The reentanglement of molecules across weld lines occurs slowly, leaving lateral weakness in the parison and in the molded product in the neighborhood of a weld line. Shear modification can reduce elasticity and viscosity, especially in the case of branched polymers.

Finally, the melt leaving the die may exhibit sharkskin, described in Sect. 6.3.2, which affects the surface finish of a molded container. It becomes noticeable above a critical wall shear stress and can limit the extrusion rate. Extrudate distortion is most severe for narrow MWD, high viscosity resins at high shear rates. A very rough estimate of the critical stress for surface melt fracture is 0.1 MPa. In the development of a new blow molding resin, if the stress in the die is estimated to approach this value, measurement of the critical stress by capillary rheometry is advised. Sharkskin can sometimes be eliminated by increasing the temperature or reducing the extrusion rate, although these actions increase the cycle time. The use of an internal heater in the mandrel is thought to reduce melt fracture on the inside of the parison.

In the case of large or irregularly shaped objects, it is desirable to be able to change the parison thickness during extrusion, and this is done by use of parison programming to alter the geometry of the die to produce a parison having the desired profile. However, parison programming cannot compensate for resin properties that are basically unsuited for use in a given machine, and for this reason it will be useful to examine in more detail the processes of parison swell and sag.

8.6.3 Parison Swell

The melt deformation that occurs in the die generates molecular orientation, which manifests itself at the die exit as extrudate swell, a phenomenon first described in Sect. 8.3.3.2. Flow at the inlet of a die, where the streamlines are converging rapidly, involves a high rate of stretching in the flow direction. This results in molecular orientation, and if the melt were permitted to exit immediately, as in the case of flow through an orifice plate, there would be a very high degree of swell. However, if the entrance is followed by a long straight section, for example a capillary or straight annular die, molecular relaxation will lead to the decay of orientation, and as the die becomes longer, the degree of swell falls. At the same time, however, shearing in the die produces some axial orientation, and this can result in significant swell.

Furthermore, if the die has an expanding or contracting section, this will induce orientation that depends on the details of the die design. For these reasons, parison swell is very sensitive to the shape of the die channel. To describe the swell of a tubular extrudate it is necessary to use two swell ratios, one for diameter swell and one for thickness swell. These are defined as follows; the symbols used are defined in Fig. 8.15.

$$B_D \equiv D_P/D_0 \tag{8.24}$$

$$B_H \equiv h_p/h_0 \tag{8.25}$$

There is also a weight swell B_w that is the weight of a given length of parison, divided by the weight of the same length of a parison having the same inner and

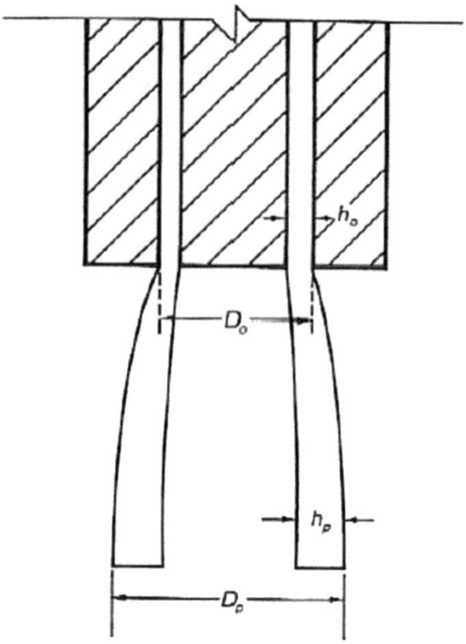

Fig. 8.15 Definitions of the two swell ratios for extrudate from a die with annular lips. The thickness swell $B_H = h_P/h_O$, and the diameter swell $B_D = D_P/D_O$

outer radii as the die. If the density of the melt does not change significantly while the parison is being formed, the weight swell is equal to the area swell, B_A.

$$B_W = B_A \equiv A_P/A_0 \tag{8.26}$$

If the parison wall thickness is small compared to its diameter, the area swell is approximately equal to the product of the diameter and thickness swells.

$$B_A \cong B_D B_H \tag{8.27}$$

Both the quality and cost of a blow molded container depend on the parison swell ratios. If the diameter swell is too small, incomplete handles, tabs or other unsymmetrical features may result. On the other hand, if the diameter swell is too large, polymer may be trapped in the mold relief or pleating may occur. Pleating, in turn, can produce webbing in a handle. Weight swell governs the weight and thus the material cost of the molded product. The optimum is the minimum weight that provides the required strength and rigidity.

Because swell is a manifestation of viscoelasticity, it is time dependent. For example, for high density polyethylene at 170 °C, 70–80 % of the swell occurs during the first few seconds after the melt leaves the die, while the remainder occurs over a period of 2–3 min [19]. Garcia-Rejon and Dealy studied the swell behavior of a set of commercial polyolefins extruded from an annular die [19] by extruding into hot oil having the same temperature and density as the melt, so that the swelling process could be monitored as a function of time in the absence of sag. Figure 8.16 shows diameter swell versus time for a polyethylene at three flow rates, and we see that it takes several minutes to reach steady state. The same study

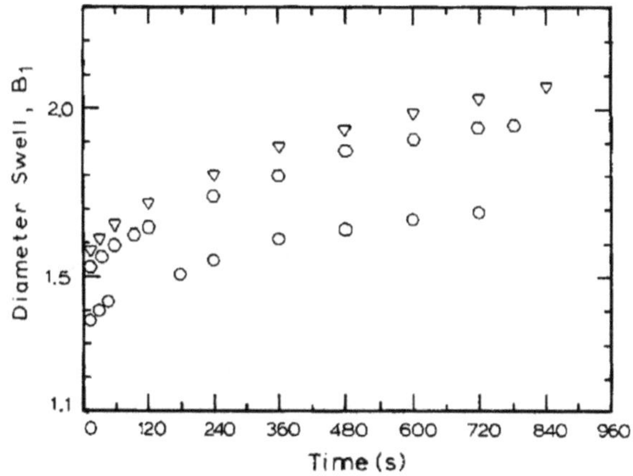

Fig. 8.16 Isothermal diameter swell versus time for an annular die in the absence of sag for HDPE 28 at 190 °C at wall shear rates of 19 (*circles*) 109 (*hexagons*) and 231 s^{-1} (*deltas*). There is a nearly instantaneous swell followed by a slow approach to an equilibrium swell. From Garcia-Rejon and Dealy [19]

found that for polypropylene at 190 °C, only 50 % of the swell occurred in the first few seconds, while more than 10 min were required to reach the ultimate value [19]. While short-time swell is more directly relevant to the case of short parisons [60], ultimate swell is easier to determine quantitatively.

Another source of time dependency in the case of intermittent extrusion is the *cuff effect*. The first melt to be extruded after a period of no flow has had an opportunity to relax in the die and experiences relatively little shear just prior to being extruded. Thus, there is less molecular orientation, and less swell, than for melt that has just flowed through the entire length of die. This means that the swell is least at the bottom of the parison (the "cuff") increasing to a maximum and then decreasing. Of course sag will cause swell to decrease toward the top of the parison.

Swell increases as the temperature decreases, although it takes place somewhat more slowly. Since the time during which the parison is exposed to the air before the mold closes is rather short, and the air surrounding the parison is warm and relatively still, the decrease in temperature that occurs during parison formation is generally rather small, usually less than 5 °C [61]. For this reason, parison formation is often assumed to be an isothermal process. Swell increases as the flow rate increases due to enhanced molecular orientation in the die. Swell varies greatly from one polymer to another and is strongly affected by the extent of branching, molecular weight, and molecular weight distribution. For linear polymers, a broader MWD generally results in higher swell. However, resins with very similar measured molecular weight distributions can have significantly different swell behavior [20], and this probably reflects the fact that swell is highly sensitive to small amounts of high molecular weight or branched material.

Koopmans [62] studied the effect of small amounts of high molecular weight material on the swell of HDPE blends at the exit of a capillary at a wall shear rate of 300 s^{-1}. He found that the molecular weight of the high molecular weight component was very important and concluded that it is misleading to use a single parameter such as M_w/M_n as a measure of molecular weight distribution. The apparent decrease in swell that Koopmans observed as the molecular weight of the heavier component was further increased may be due to the time dependency of the swell. Raising the molecular weight increases the ultimate swell but decreases the rate of approach to a steady state value. Highly branched polymers tend to swell more, but it is not possible to generalize when both branching and MWD are altered simultaneously.

Because swell is an elastic recoil process that results from molecular orientation in the die, the shape of the die channel has a strong effect on both diameter and thickness swells. The simplest case is a long straight annular die. Here we have only shear flow, and we would expect to see some orientation in the axial direction. This suggests that there should not be preferential orientation in either the radial or hoop direction and that the diameter swell should thus be equal to the thickness swell.

$$B_{\mathrm{D}} = B_{\mathrm{H}} \tag{8.28}$$

This is referred to as isotropic swell. If Eqs. (8.26) and (8.27) are valid, (8.28) implies that:

$$B_{\mathrm{W}} = B_{\mathrm{D}}^{2} \tag{8.29}$$

For anisotropic swell, it has been suggested [63] that the following empirical relationship might be useful.

$$B_{\mathrm{W}} = B_{\mathrm{D}}^{\mathrm{m}} \tag{8.30}$$

We expect to find that $m > 2$ for a diverging die for which diameter swell is suppressed

Figure 8.17 shows equilibrium thickness swell versus diameter swell from a straight, annular die for three polyethylenes, 22, 26 and 27 and a polypropylene, 28 [19]. The three HDPEs were made by different companies but were marketed for making the same type of bottle. The HDPEs have similar behavior, which is quite different from the PP. The lines are plots of Eqs. (8.29) and (8.30). A question of interest is the degree to which capillary swell data can be related to parison behavior. Figures 8.18 and 8.19 show diameter and thickness swells versus capillary swell for the three HDPEs and the PP. The ranking of the HDPEs is the same, while the PP is clearly from a different family.

For more complex dies like those used for blow molding, the swell ratios are strongly influenced by two die geometry features, the angle of divergence or convergence of the outer die wall, and the variation of gap spacing along the flow path. A diverging die stretches the melt in the hoop direction, and this should reduce diameter swell by counteracting the axial orientation generated by the shear flow. Orbey and Dealy [20] used the apparatus described above to study the effect of die geometry on swell. Cross-sections of their dies are shown in Fig. 8.20. Note that in the 10° converging die the gap narrows, while in the 20° converging die the die walls are parallel. The three HDPEs studied were made by different manufacturers but were all intended to make bottles for non-stress-cracking fluids.

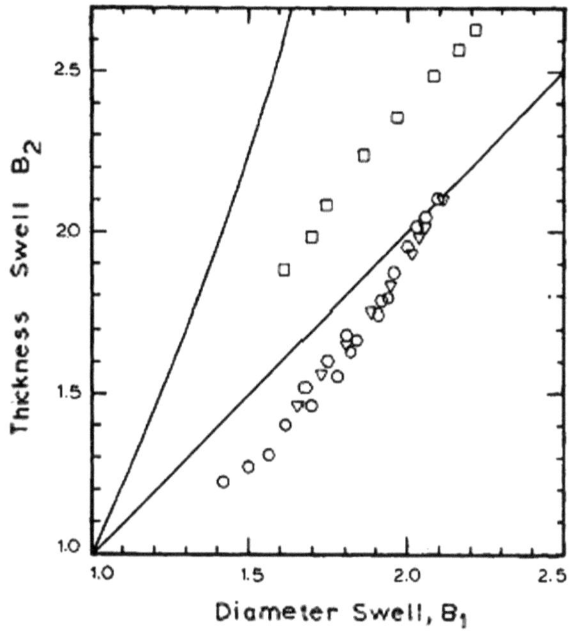

Fig. 8.17 Equilibrium thickness swell versus equilibrium diameter swell for an annular die in the absence of sag for three similar HDPE resins and a polypropylene resin (*squares*). The *straight line* is Eq. (8.28), and the *upper curve* is Eq. (8.29). From Garcia-Rejon and Dealy [19]

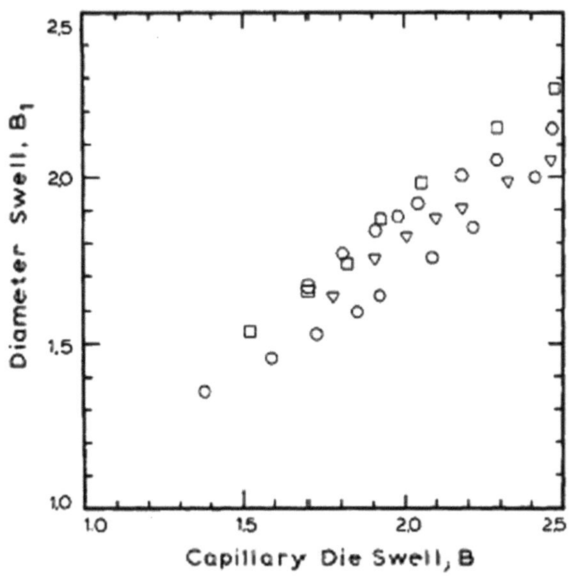

Fig. 8.18 Equilibrium diameter swell from an annular die versus capillary swell for three similar HDPE resins and a polypropylene resin (*squares*). There is clearly no correlation between the two swell ratios. From Garcia-Rejon and Dealy [19]

Resins 22A and 29 were made by solution polymerization using a modified Z-N catalyst, while 27 was polymerized in a fluidized bed. Figures 8.21 and 8.22 show the effect of die design on diameter and thickness swells. As expected the type of molecular orientation introduced in the die has a powerful effect on swell behavior.

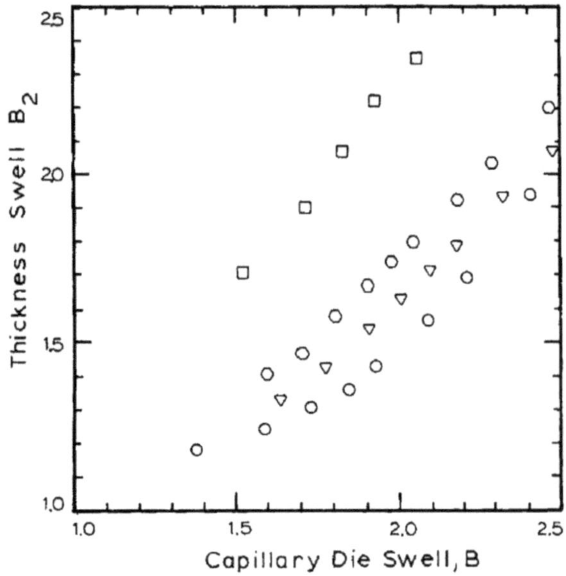

Fig. 8.19 Annular thickness swell versus capillary swell for three similar HDPE resins and a polypropylene resin (*squares*). There is clearly no correlation between the two swell ratios. From Garcia-Rejon and Dealy [19]

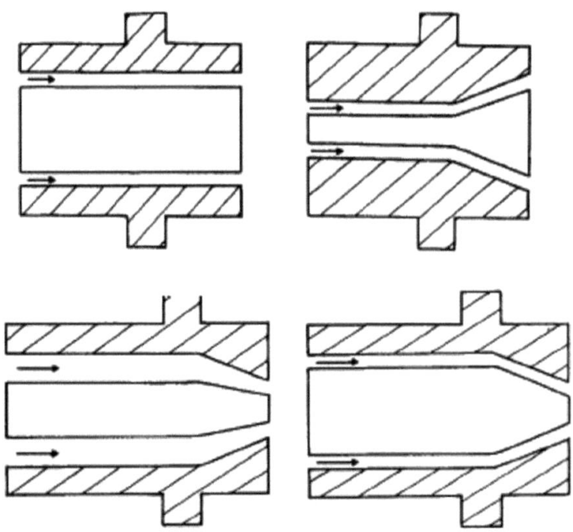

Fig. 8.20 Dies used to study the effect of die geometry on extrudate swell. Clockwise from *upper left*: annular die, diverging die, 10° converging die, 20° converging die. From Orbey and Dealy [20]

Flow in the straight die generates some orientation in the axial direction, which is expected to result in swelling that is isotropic in the plane normal to the flow axis. And this implies that the diameter swell should be about the same as the thickness swell so that Eq. (8.29) is valid. Figure 8.23 shows that this is what was observed for two resins, although resin 29 deviated somewhat from isotropy. The diverging die is expected to generate stretch in the hoop direction and deceleration along streamlines. The resulting orientation is expected to suppress diameter swell, and this is what we see in Fig. 8.24. The dashed line corresponds to $m = 4$ in

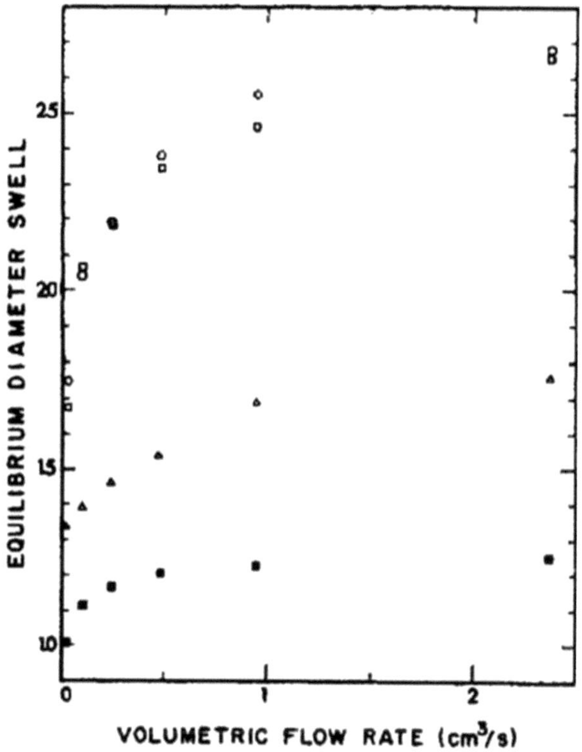

Fig. 8.21 Equilibrium diameter swell versus flow rate (cm^3/s) for a HDPE for four dies: *straight annulus, open triangles*; 10° converging, *open squares*; 20° converging, *open circles*; diverging, *solid squares*. From Orbey and Dealy [20]

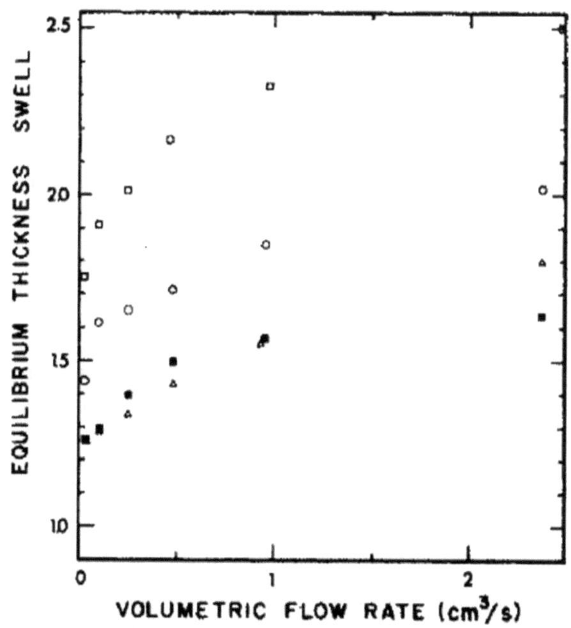

Fig. 8.22 Equilibrium thickness swell versus flow rate (cm^3/s) for a HDPE for four dies: *straight annulus, open triangles*; 10° converging, *open squares*; 20° converging, *open circles*; diverging, *solid squares*. From Orbey and Dealy [20]

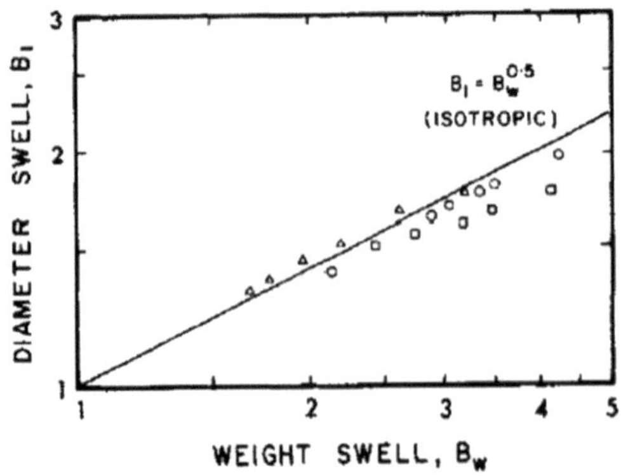

Fig. 8.23 Equilibrium diameter swell versus weight swell for the annular die. The line is for isotropic swell, Eq. (8.29). The three HDBEs exhibit similar behavior, while the polypropylene swell (*open squares*) deviates somewhat from isotropy. From Orbey and Dealy [20]

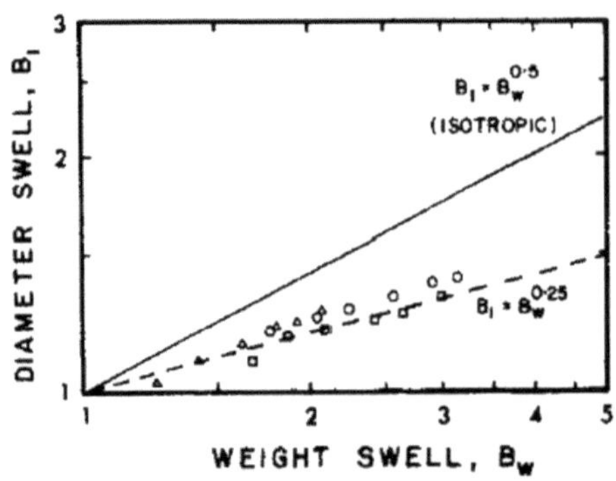

Fig. 8.24 Equilibrium diameter swell versus weight swell for three similar HDPE resins. The *solid line* is a plot of Eq. (8.29) (*isotropic*), and the *dashed line* is Eq. (8.30) with $m = 4$. From Orbey and Dealy [20]

Eq. (8.30), which is greater than one as expected. A very strongly diverging die should be able to produce diameter swell ratios less than one.

The thickness swells for the straight, diverging and 20° converging dies are all similar, reflecting the fact that these three dies all have parallel walls and the same annular gap. The 10° converging die, on the other hand, converges in the thickness direction, promoting orientation along streamlines and generating larger thickness and diameter swells. Blow molding dies often have gaps that are narrowest at the die lips, as this enhances the stability of the exit flow and reduces the tendency of the melt will pull away from one of the die walls before it reaches the exit. Orbey and Dealy [20] found that their converging die with a narrowing gap produced extrudate exhibiting nearly isotropic swell corresponding to $m = 2$ in Eq. (8.30).

Figures 8.25, 8.26 compare the behavior of three resins in the 20° converging die. While the three resins were marketed for the same usage, it is clear that switching from one to another would require adjustment of the processing machinery.

The method used in the two studies described above was designed to measure the time-dependency of swell in the absence of drawdown due to gravity, but this is a complex technique that is rarely used. The most common method used in industry makes use of a *pinch-off* or *pillow mold*. Figure 8.27 is a photo of such a

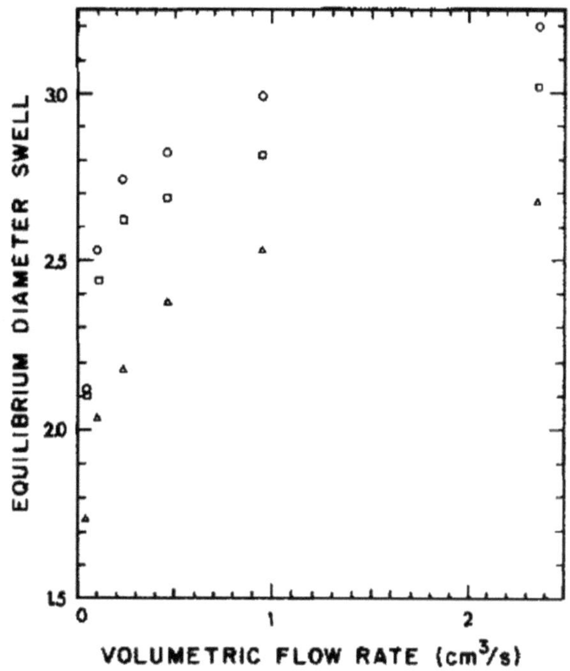

Fig. 8.25 Equilibrium diameter swell versus flow rate (cm^3/s) for the 20° converging for three similar HDPEs. While the three resins are similar in density, melt index and polydispersity, their swell behaviors are markedly different. From Orbey and Dealy [20]

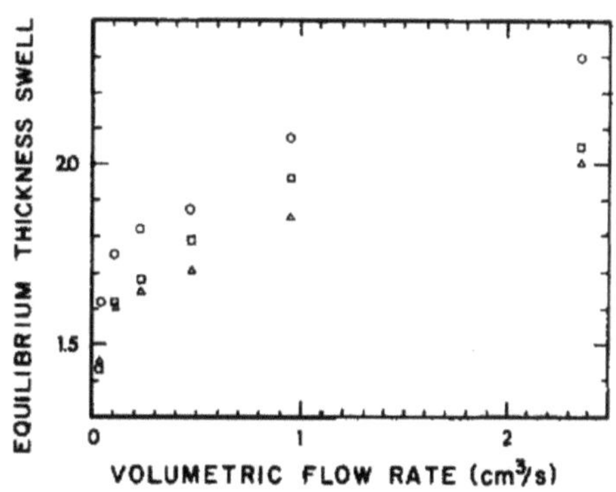

Fig. 8.26 Equilibrium thickneess swell versus flow rate (cm^3/s) for the 20° converging for three similar HDPEs. While the ranking of the three resins for this die is the same as that for diameter swell (Fig. 8.25) there is no quantitative correlation between the two swell ratios. From Orbey and Dealy [20]

device [64]. While this technique can produce unreliable results [65], it continues to find use for the rapid evaluation of commercial resins [64].

It would obviously be very useful to be able to predict swell by use of a numerical simulation. A long-term effort at the National Research Council of Canada has led to the development of a useful model [66]. An integral constitutive equation is used to describe nonlinear viscoelastic behavior, and like other continuum mechanics models it is empirical and of limited accuracy. However, the resulting software package, BlowParison®, has found application in the plastics industry.

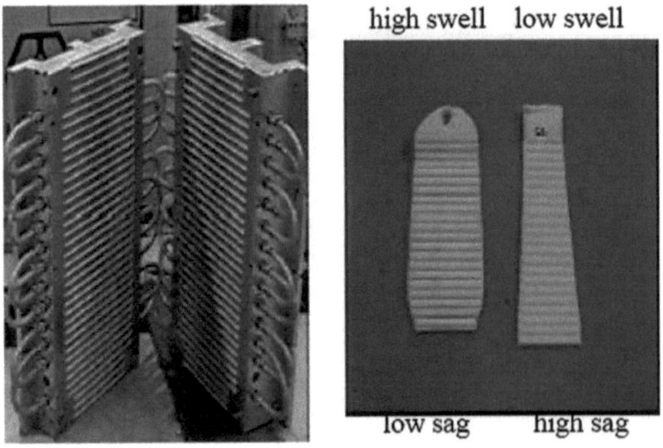

Fig. 8.27 Photo showing a pinch-off mold used to evaluate extrudate swell of blow molding resins. From Jivraj et al. [64]

8.6.4 Parison Sag

As soon as melt leaves the die it is subjected to gravity, which stretches it downward, but this occurs simultaneously with swell that causes the length to decrease with time. When extrusion stops, at the *drop time*, swell continues and can sometimes cause the parison to recoil or bounce back. After this, there is continuous sag or draw-down until inflation takes place; draw-down involves extensional flow. For long parisons, sag can lead to significant variations in thickness and diameter along the parison and in an extreme case can cause the parison to break off. Since the low stress viscosity provides a rough guide to parison melt strength, PE blow molding resins are usually *fractional* MI grades having melt index values less than one. Sag becomes more severe as temperature increases and is a major concern in the case of large parisons.

In addition to the low-stress viscosity, the amount of sag depends on drop time and parison length. The drawdown stress at the top of the parison is its weight per unit area, which is equal to (pgL), where ρ is the density of the melt, g is the acceleration due to gravity, and L is parison length. If sag is small, the case of most interest, the maximum amount of sag occurs near the middle of the parison and is equal to ($\rho g L t_d/2\eta$), where t_d is the drop time, and η is a relevant viscosity.

While a long drop time usually means that viscosity contributes more than elasticity to total sag, if the drop time is short compared to the relaxation time of the melt, initial resistance to sag will be governed by viscoelasticity [67]. Thus, resistance to sag cannot always be quantitatively correlated with viscosity. Ajroldi [67] used linear creep compliance, as calculated from a tensile relaxation modulus, to predict sag, while Sebastion and Dearborn [68] suggested that the extensional stress growth function could be correlated with sag behavior. However, since sag is neither a constant stress nor a constant strain rate process, neither of these material functions can be quantitatively related to sag. In order to compare resins according to their sag behavior, Swerdlow et al. [69] simply clamped a weight onto the end of a specified length of extrudate from a melt indexer and measured the resulting sag during a prescribed time.

8.6.5 Pleating

Pleating (also called *draping* or *curtaining*) is a buckling of the parison that occurs when the melt in the upper portion of the parison is unable to withstand the compressive hoop stresses due to the weight of the parison suspended below it. This phenomenon is illustrated in Fig. 8.28. Pleating is usually undesirable, for example it can cause cause webbing in handles, but it is virtually unavoidable for very large parisons.

Obviously, large diameter swell and a small die gap are factors that will increase the severity of pleating. It is not clear exactly which rheological properties govern the ability of a melt to resist pleating, but the low-shear-rate viscosity is a rough guide to melt stiffness. Thus, high viscosity resins are less likely to pleat, and increasing the temperature increases the likelihood of pleating.

8.6.6 Parison Inflation

The behavior of the parison during the inflation process is a manifestation of its extensional flow rheological properties. It has been observed [70] that the parison does not inflate uniformly and tends to bulge out in the center, especially in the case of PET resins [71]. Blow-outs can occur if the ratio of the mold diameter to the parison diameter (the blow-up ratio) is too high.

The deformation involves stretching, but it is more like planar or biaxial extension, depending on the mold shape, than uniaxial extension. However, uniaxial extension data (Chap. 6) are thought to be relevant to inflation [68]. In particular, it is thought that strain hardening implies that a resin will be easy to inflate and unlikely to exhibit blow-out, even when the blow-up ratio and inflation pressure are high. Strain softening on the other hand is thought to imply unstable

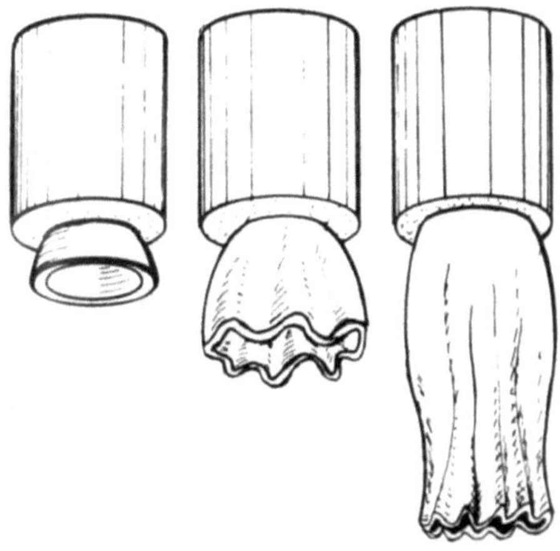

Fig. 8.28 Sketch showing pleating (or curtaining) of a large blow molding parison

inflation and an increased likelihood of blow-outs. This hypothesis is consistent with the observation that LDPE is easier to inflate than HDPE. Numerical simulations of the inflation stage often make use of rubber elasticity models rather than viscoelastic liquid constitutive equations.

8.6.7 Blow Molding of Engineering Resins

In order to take advantage of the economies of extrusion blow molding, there is a growing interest in its use for the processing of engineering resins. However, many of these materials were originally developed for injection molding, and they tend to have properties that make them unsuitable for blow molding. Specifically they have low viscosity, and if semicrystalline they have high melting points and a narrow melting range. Moreover, they are often hygroscopic and sensitive to thermal degradation. Some of these disadvantages can be overcome by modifications of the process. Hygroscopic resins must be thoroughly dried immediately prior to processing, and temperatures must be closely controlled. Extruder screws having a low compression ratio and a long transition zone are often advantageous, and it is especially important to streamline all flow channels to avoid stagnation regions.

The low viscosity poses a problem at the parison formation stage because of excessive sag and the inability of the melt to be blown into deep pockets without tearing. Accumulators are helpful, as they decouple the extrusion speed from the parison formation speed and thus reduce the hang time prior to inflation. Methods for reducing sag include "preblow," which supports the parison by an upward jet of air prior to inflation, and "prepinch," which involves a partial inflation of the parison while it is still forming.

To minimize equipment modifications, however, and to make possible the blow molding of large products, it is highly desirable to modify the resin to increase its viscosity at low strain rates. This can be accomplished by altering the molecular weight distribution, blending, cross-linking or using a comonomer or an additive.

8.6.8 Stretch Blow Molding

In the stretch blow molding process, the parison is preformed in an injection molding or extrusion operation and is reheated before inflation. However, the preform is not heated to a temperature above its melting point but only to a temperature sufficient for rubbery behavior. The proper selection of the reheat time, which governs the inflation temperature, is of central importance. The essential feature of the process is the generation of a high degree of molecular orientation in the inflation step. To accomplish this one must ensure that nearly all of the stretching work is taken up by the polymer as elastic strain rather than viscous dissipation. This only occurs in a fairly narrow range of temperatures, which is between 90 and 115 °C for PET.

The rubbery state temperature range can be altered by changing the formulation of the resin. In the case of polypropylene a comonomer can be used, and in the case of PET, diethylene glycol can be added. Absorbed water can also act as a plasticizer, reducing the optimum temperature for orientation [72]. However, such changes will also affect other characteristics of the resin and can sometimes result in a deterioration of mechanical properties.

Molecular weight also influences the temperature range for processing [73]. In the case of PET, the inherent viscosity (I.V.) is used as an indicator of molecular weight, and PET resins for bottles normally have an I.V. between 0.72 and 0.85. Cakmak et al. [74] studied the kinematics of the stretch blow molding of PET. PET is nearly Newtonian above its melting point and is thus unsuitable for extrusion blow molding. However, the use of a multifunctional comonomer that produces long chain branching and broadens the MWD can render PET suitable for extrusion blow molding.

8.6.9 Resin Selection

The selection of a polymer for a given blow molding application is governed primarily by product requirements such as gas permeability, strength, and resistance to solvent and high or low temperatures. Once a polymer is selected, its processability must be considered, and this depends primarily on its rheological properties. If the behavior of the neat resin is a problem, it may be possible to modify it by means of additives or by altering the molecular weight distribution.

Because of the diversity of blow molding processes and of shapes and sizes of blow molded articles, it is not possible to specify a general rheological profile for a blow molding resin. Several process requirements must be met simultaneously, and the rheological behavior necessary to meet one requirement may conflict with that needed to meet another. Ideally one would design the machinery and tooling to match the resin characteristics and optimize the entire process. However, it is often impossible to do this especially when the resin must be processable using existing equipment, and then one must identify rheological behavior that can accommodate all of these requirements.

Schaul et al. [75] described one approach to this problem when the only machine settings that could be varied were the shot pressure and die gap. The resin selected had to meet specifications for parison diameter, bottle weight, melt fracture, and pleating, but some of these requirements were in conflict. Decreasing the die gap, for example, increased parison diameter, but this could cause curtaining. An understanding of each stage of the process is necessary for defining the effects of rheological characteristics if an empirical trial-and-error technique is to be avoided.

Parison formation is the most critical stage of the process and also the most difficult to model. A given rheological property affects more than one aspect of parison formation. For example, the viscosity at the shear rate in the die governs the pressure required to extrude the parison. And if extrusion rate is limited by the pressure capability of the extruder, this limitation will govern the time required to form the parison, i.e., the drop time. If this time is too long, excessive sag will occur, which will affect the diameter, length, and appearance of the parison. Another possible result of a high viscosity is that the critical stress for extrudate distortion will be exceeded.

8.7 Injection Molding

The equations provided in this section are intended to facilitate understanding. All those required for the detailed numerical simulation of all stages of the injection molding process are given by Kennedy [76].

8.7.1 Flow in Runners

A simplified analysis of flow in a circular runner can be made using the equations in Sect. 8.2 for flow in a circular channel. At the beginning of a shot the runner is empty, and the length of runner filled with melt $Z(t)$ increases with time. Equation (8.3) can then be written as follows:

$$Q(t) = \frac{\pi R^3}{(1/n) + 3} \left(\frac{\Delta PR}{2kZ(t)} \right)^{1/n} \tag{8.31}$$

For the simplest case of isothermal flow k and n are constant. As the runner fills with melt, if the pressure is held constant the flow rate will decrease rapidly, while if the flow rate is to be kept constant the pressure must increase. If the runner is cool the viscosity will increase, and solidification will take place at the wall, reducing the effective radius and slowing the flow. This can be avoided if the runners are heated. Shear rates in a runner can be well above 1,000 s^{-1}, which will result in significant viscous heating and a reduction in viscosity.

8.7.2 Mold Filling

Viscosity is the key rheological property governing mold filling, while viscoelasticity governs the residual stresses in the cooled part [77]. Melt index is universally used to specify ease of flow, even when it is not directly relevant to a

Fig. 8.29 A *spiral mold* using to evaluate the mold-filling properties of injection molding resins. (Photo courtesy of Master Unit Die Products Inc.)

given process. For example, where a typical film resin might have an MI of about 1.0, a high-flow molding resin for thin parts might have an MI of 100 or more. But MI is not a reliable guide to molding performance, because the stresses and flow rates that arise in a melt indexer, even at high load, are much lower than those that occur in a mold. The behavior at high flow rate is sensitive to the molecular weight distribution, and while the melt flow ratio MFR, for example I21/I2, provides some information about MWD, it is too coarse a measure. While the ability of a melt to fill a mold before solidifying depends also on its thermal properties, these are much less sensitive to molecular weight than viscosity.

A widely used tool for evaluating injection molding resins is the spiral flow mold, an example of which is shown in Fig. 8.29. (Made in accord with ASTM standard test method D3123 for thermosetting molding compounds) Melt is injected at the center of the spiral, and the distance traveled before solidification prevents further flow is reported in terms of numbers etched in the mold cavity. This is called the *flow length* or *spiral flow number* SFN. Within a single family of resins, i.e., polyethylenes made in the same reactor using the same catalyst, SFN increases with MI, although SFN goes up much more slowly than MI. But between families, there is no correlation at all. This is illustrated in Table 8.1 based on

Table 8.1 Melt indexes and spiral flow numbers for four families of injection molding resins

PE family	Melt index	Spiral flow number
HDPE-A	2.2	9.2
	6.0	12
	20	14
	34	16
HDPE-B	3.8	8.1
	4.5	8.4
	6.6	10
	18	13
LLDPE	21	13
	32	16
	135	27
LDPE	9.0	14
	25	17
	38	21

information from an actual polyethylene manufacturer. Numbers are shown for four families of resins. A single number is rarely, if ever, adequate to describe the flow behavior of a melt a forming process and the SFN is no exception.

In going from a test mold to a production mold, for example, the relative roles played by rheological and thermal properties may be altered. Or the relative importance of the roles played by various regions of the viscosity-shear rate curve may change. As a result, one may obtain a ranking of several resins that is not a correct indication of the relative ease with which they will fill a given production mold. The empirical moldability tests are therefore most useful when used to compare several resins of the same family, for example several linear polyethylenes having similar molecular weight distributions but different average molecular weights.

As might have been anticipated, the single most important physical property contributing to flow length is the high shear rate viscosity, particularly in the range of 1,500–2,500 s^{-1}, although this range will depend on the resin, injection pressure, and mold dimensions.

Of course the flow length depends also on the thermal properties of the melt. In his analytical model of the flow in a cooled spiral mold at fixed wall temperature, Richardson [78] obtained the following result:

$$L = Ch\left[\frac{h^2}{\alpha}\left(\frac{\Delta P}{k}\right)^{1/n}\right]^{n/(1+n)} \tag{8.32}$$

where:

L Flow length

h Smaller dimension of a rectangular spiral mold

α Themal diffusivity

k Power-law constant
C Constant of order 0.1.

Experimental results yield values of L corresponding to values of C between 0.2 and 0.4 [79, 80]. Since thermal diffusivity is much less sensitive to molecular weight distribution than viscosity, the high shear rate viscosity is the governing property. Hieber et al. [79] carried out a numerical simulation of the flow in a spiral mold and used it together with experimental data to estimate the four parameters of a temperature and pressure-dependent power-law model:

$$\eta\,(T,P,\dot{\gamma}) = A\exp(T_a/T)\exp(bP)\,\dot{\gamma}^n \tag{8.33}$$

Since there is no universally accepted spiral mold design or set of molding conditions, a moldability index has meaning only within a given laboratory. In fact, it is nearly impossible to specify a truly universal flow length test, because there are so many variables, and some of these cannot be satisfactorily controlled [81]. The design of machinery is obviously a key factor, and it is not practical to specify in detail, the injection molding machine that must be used for the test. For a given machine and mold design, there are two sets of operating conditions: machine settings and mold conditions. While mold conditions are probably more important, the conditions actually seen by the melt cannot be fixed by merely specifying these. For example, the temperature of the cavity wall depends not just on the temperature and flow rate of the mold coolant, but also on melt temperature and the injection rate. It is also not practical to fix the mold pressure. Fritch [81] gives the following advice regarding flow length testing:

1. Stick to one machine and mold
2. Keep the hydraulic pressure constant
3. Melt temperature is more important than mold temperature, but a single temperature is not adequate
4. Injection rate influences the shear rate in the mold, and a single value is not adequate.

Since there is no way to scale up a single spiral mold test to various specific molding situations, it is not possible to select in advance "relevant" standard values for the test variables. Therefore, Fritch recommends that tests be carried out at three melt temperatures and three injection rates for a total of nine tests. Obviously, the values selected should cover the ranges of melt temperature and mold flow rate that are likely to occur in commercial processing.

Another approach to evaluating the processability of a molding resin is to measure the pressure required to fill a standard mold. Furches and Kachin [82] compared the results of several rheological tests often used to evaluate injection molding resins.

A phenomenon that can lead to a complex pattern of weld lines is *jetting* in which the melt jets to the end of the cavity without wetting the walls near the gate. In one manifestation of this phenomenon, the cavity fills by a piling up of the jet at

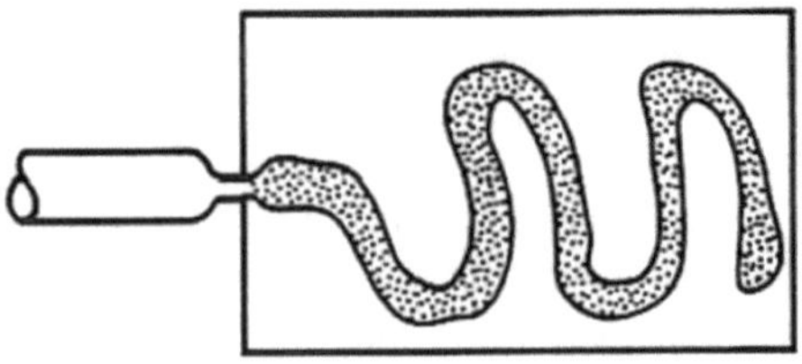

Fig. 8.30 Sketch showing the jetting phenomenon in injection molding

the end of the cavity to form many folds, as sketched in Fig. 8.30. Or jetting may cease when the stream hits the opposite wall, followed by normal filling. Jetting can occur when the gate faces a mold wall that is far away and when the flow rate is high. Jain et al. [83] found that the LDPE, polypropylene, polystyrene, polyamide and polybutylene terephthalate that they used did not exhibit jetting but that ABS, polyacetal and polycarbonate did. Oda et al. [84] suggest that jetting occurs when the swell at the exit from the gate is insufficient to cause immediate contact between the melt and the cavity walls. Jain et al. [83] reported that Oda's criterion was useful only when the thickness at the gate matched the cavity thickness but that, in general, there is no reliable way to predict when jetting will occur. Jetting can sometimes be eliminated by redesigning the mold so that the gate is opposite a nearby wall or by using a fan-shaped gate or by using a larger gate.

Surface defects called tiger stripes or flow marks are alternating smooth and dull bands in injection molded parts. Bogaerds et al. [85] reviewed the extensive work on this problem and concluded there is at present no clear understanding of its origin.

8.7.3 Formulation and Selection of Injection Molding Resins

For rapid filling at modest melt temperatures and pressures, it is clearly advantageous that the melt viscosity be as low as possible. The cycle time is thus short, and thin runners can be used. A low viscosity is especially advantageous for the molding of parts with thin walls. For these reasons, there has been a trend toward the marketing of molding resins with ever decreasing viscosity. However, the increase in melt index is usually accomplished by reducing the molecular weight, and this has the undesired effect of decreasing the toughness of the finished part, as indicated by its impact strength. In order to minimize the sacrifice in mechanical properties for a given application, the molder looks for the lowest melt index material that will give satisfactory flow in the mold. This requires the availability of many grades of resin, and this is disadvantageous to the economics of resin manufacturing. One approach to this problem is to eliminate the highest molecular weight fractions by narrowing the molecular weight distribution, as is done to make controlled-rheology resins.

8.8 Rotational Molding

Rotational molding is used principally to make large, hollow objects [86]. Everything happens in the mold, from melting of resin to final part cooling, and it is thus a batch rather than a continuous process. The mold is rotated inside an oven and is thus heated only from the outside. The most used resins are polyethylene and polypropylene. PEs for rotational molding have MI (190 °C) values in the range of 2–10, while the MFIs (230 °C) of PPs are in the range of 5–10. In order of decreasing processability these are: LLDPE > HDPE > PP, while for the PEs, product properties usually go in the opposite direction. For thermoplastics the feed is a fine powder, but liquid PVC plastisol is also used. Particle density is roughly half the melt density. Polyethylene may be cross-linked to make large containers.

The only forces affecting resin behavior are gravity and interfacial tension. The rheological properties of importance are low-shear-rate viscosity and low frequency viscoelasticity. For thermoplastics the steps in the process are:

1. Loading of powder
2. Particle softening to tack temperature
3. Melting
4. Coalescence
5. Densification
6. Cooling
7. Removal of product.

The *tack temperature*, at which powder particles begin to stick to the wall and to each other, is very close to the melting point. The temperature of the air in the mold rises more slowly than the wall, and as the melt layer at the wall builds up, the rate of temperature increase decreases. Coalescence, driven by interfacial tension, now acts to form bridges or necks between particles, and the result is a three-dimensional network. This leads to densification of the bed with air trapped in void spaces [86]. Elasticity hinders the early stages of neck growth, and a high low-shear-rate viscosity inhibits the later stages. Continued capillary action leads to the collapse of the network, trapping air bubbles. Diffusion of air into the melt involves viscoelastic expansion. Thus, capillary action, network collapse, and air dissolution act to densify the melt and form a monolithic layer. Models that have been proposed for these processes are described by Crawford and Throne [87]. If cross-linking is used, it takes place at this time, followed by curing. ASTM-2765 is a standard test for the extent of cross-linking that takes place during rotational molding.

The low-shear-rate viscosity is of central importance for rotational molding resins. For practical purposes, we will speak of the zero-shear viscosity, since it is a well-defined property. While MI is the property usually used to describe melt flow, its relationship with η_0 depends strongly on molecular weight distribution and long-chain branching, as was shown in Fig. 6.13. Taking into account the entire molding cycle, a high η_0 (or less precisely a low MI) can result in poor flow, longer cycle time, poor replication of the mold surface, and trapped air bubbles. Meanwhile, if η_0 is too low (or less precisely MI is too high) it can result in flash at the parting line and low mechanical strength of the molded article.

Elasticity hinders the early stages of neck growth, which implies that it is desirable to have a low value of G' or a high value of tan δ at a low frequencies, probably below 0.1 s^{-1}. Van Gurp-Palmen plots (see Sect. 3.16) have been found useful in this regard.

8.9 Foam Extrusion

Foaming is a complex process, and foam quality depends strongly on all the phenomena listed below. We discuss here only the use of a *physical blowing agent* [(PBA), also called a *physical foaming agent* (PFA)].

Large reduction of viscosity due to the plasticizing effect of the PBA.
Diffusion of PBA out of the melt phase.
Large, rapid increase in viscosity as the foaming agent diffuses out of the melt phase.
Effect of viscosity and extensional viscosity on development of foam structure.
Effect of crystallization on foam structure.

Furthermore, molecular weight, MWD and long-chain branching affect most of these phenomena. Noting that well-controlled experiments to study all the possible interactions between them are difficult, if not impossible, to carry out it is not surprising that there have been few reports of useful data. We provide here a brief review of some of these reports. A thorough discussion of the role of rheology in foam extrusion can be found in the book chapter by Gendron [88].

The action begins in the extruder, where the dissolution of the blowing agent leads to a strong reduction in viscosity. Evidence of this can be seen in Fig. 2.8, which shows the effect of dissolved carbon dioxide on the viscosity of a HDPE. For polystyrene the reduction in viscosity can approach a factor of 100. The pressure falls sharply at the exit of the die, and the blowing agent (PBA) starts to diffuse out of the melt and nucleate to form bubbles, which develop into cells imbedded in a matrix having a complex structure. The stability of this structure is crucial to successful foam formation. With the loss of PBA the viscosity increases markedly, and this helps to stabilize the structure. The deformation in the cell walls is mainly biaxial extension, and strain hardening has been found to be beneficial to successful foaming [89, 90]. This subject was addressed by Münstedt et al. [89] who studied the effect of extensional flow behavior on foaming. They studied blends of linear PP with long-chain branched PP and also blends of LLDPE with LDPE. They found that strain hardening resulting from LCB was very beneficial to the quality of the foams produced. For the PP blends they found that the expansion ratio, the density of bulk polymer over the density of foam, increased steadily with the addition of LCB-PP up to a level of about 50 %.

Cell rupture must be avoided before crystallization freezes the structure, because it causes cell coalesce and allows gas to escape. Melt strength, as measured using the Rheotens device described in Chap. 6, or an extensional rheometer, has been found useful by some in the foam industry as a guide to the ability of the cell structure to resist rupture [90].

Results have been recently reported on the role of rheology in poly (lactic acid) extrusion foaming [91], compression foam molding [92], and the injection molding of foam [93].

References

1. Agassant JF, Avenas P, Sergent JP, Caarreau PJ (1991) Polymer processing: principles and modeling. Hanser Publishers, Munich
2. Tadmor Z, Gogos CG (2006) Principles of polymer processing, 2nd edn. Wiley Interscience, Hoboken
3. Lee K, Mackley MR, McLeish TCB, Nicholson TM, Harlen OG (2001) Experimental observation of transient "stress fangs" within flowing molten polyethylene. J Rheol 45:1261–1277
4. Verbeeten WMH, Peters GWM, Baaijens FPT (2002) Viscoelastic analysis of complex polymer melt flows using the extended pom–pom model. J Nonnewton Fluid Mech 108:301–326
5. Michaeli W (2003) Extrusion dies for plastics and rubber, 3rd edn. Hanser, Munich
6. Debbaut B, Avalosse T, Dooley J, Hughes K (1997) On the development of secondary motions in straight channels induced by the second normal stress difference: experiments and simulations. J Nonnewton Fluid Mech 69:255–271
7. Dooley J, Hughes K (2000) Measurement of layer deformation in coextrusion using unique feedblock technology. SPE-ANTEC technical papers 46, 36
8. Yue P, Dooley J, Feng JJ (2008) A general criterion for viscoelastic secondary flow in pipes of noncircular cross section. J Rheol 52:315–332
9. Dooley J (2005) Coextrusion instabilities. In: Hatzikiriakos SG, Migler KB (eds) Polymer processing instabilities. Marcel Dekker, New York
10. Chung CI (2011) Extrusion of polymers: theory and practice, 2nd edn. Hanser Publishers, Munich
11. Rauwendaal C (1986) Polymer extrusion. Hanser Publishers, Munich
12. Rauwendaal C (1986) Throughput-pressure relationships for power-law fluids in single-screw extruders. Polym Eng Sci 26:1240–1244
13. Zi Z, Gotsis AD (1992) Numerical study of slip at the walls in the extruder. Int Polym Process 7:132–139
14. Koopmans R, Doelder J, Molenaar J (2011) Polymer melt fracture. CRC Press, New York
15. Moon D, Rur AJ, Migler KB (2008) Multi-sample micro-slit rheometry. J Rheol 52:1131–1142
16. Laun HM, Schuch H (1989) Transient elongational viscosities and drawability of polymer melts. J Rheol 33:119–175
17. Han CD, Charles M, Philipoff W (1970) Rheological implications of the exit pressure and die swell in steady capillary flow of polymer melts. I. The primary normal stress difference and the effect of *L/D* ratio on elastic properties. Trans Soc Rheol 14:393–419
18. Utracki LA, Bakerdjian Z, Kamal MR (1975) A method for the measurement of the true die swell of polymer melts. J Appl Polym Sci 19:481–501

19. Garcia-Rejon A, Dealy JM (1982) Swell of extrudate from an annular die. Polym Eng Sci 22:158–165
20. Orbey N, Dealy JM (1984) Isothermal swell of extrudate from annular dies; Effects of die geometry, flow rate and resin characteristics. Polym Eng Sci 24:511–518
21. Henderson AM, Rudin A (1986) Effects of die temperature on extrudate swell in screw extrusion. J Appl Polym Sci 31:353–365
22. White JL, Huang D (1981) Extrudate swell and extrusion pressure loss of polymer melts flowing through rectangular and trapezoidal dies. Polym Eng Sci 21:1101–1107
23. Huang DC, White JL (1979) Extrudate swell from slit and capillary dies: an experimental and theoretical study. Polym Eng Sci 19:609–616
24. Stevenson JF, Lee LJ, Griffith RM (1986) Multidimensional control of profile extrusion. Polym Eng Sci 26:233–238
25. Hogan TA, Wall P, Dems BC (2008) Investigation of the relationships between die build up and die swell. Polym Eng Sci 48:334–343
26. Musin J, Zatloukal M (2012) Experimental investigation of flow-induced molecular weight fractionation for two linear HDPE polymer melts. Chem Eng Sci 81:146–156
27. Giacomin AJ, Schmalzer AM (2012) Die drool theory. J Polym Eng 33(1):1–18
28. Gander JD, Giacomin AJ (1997) Review of die lip buildup in plastics extrusion. Polym Eng Sci 37:1113–1126
29. Van den Bossche L, Georgon O, Donders T, Focquet K, Dewitte G, Briers J (2000) Dyneon Tech Rep 16, Belgium
30. Satoh N, Tomiyama H, Kjiwara T (2001) Viscoelastic simulation of film casing process for a polymer melt. Polym Eng Sci 41:1564–1579
31. Dobroth T, Erwin L (1986) Causes of edge beads in cast films. Polym Eng Sci 26:462–467
32. Bezigian T (ed) (1999) Extrusion coating manual, 4th edn. Tappi Press, Atlanta
33. Co A (2005) Draw resonance in film casting. In: Hatzikiriakos SG, Migler KB (eds) Polymer processing instabilities. Marcel Dekker, New York
34. Wang J, Mangnus M, Yau W, deGroot W, Karjala T, Demirors M (2008) Structure-property relationships of LDPE. Soc Plast Eng ANTEC:878-881
35. Clevenhag P-A, Oveby C (2005) Rheological indicators to predict the extrusion coating performance of LDPE. TAPPI J 4:21–23
36. Roberts EH, Lucchesi PJ, Kurtz SJ (1986) Draw resonance reduction in melt embossing and extrusion coating resins. Adv Polym Technol 6:65–71
37. Steward EL, Cline AW (1987) Barrier screw hikes quality of HMW-HDPE blown film. Plast Eng, September '87:45
38. Proctor B (1972) Flow analysis in extrusion dies. SPE J 28 (Feb) 34
39. Rauwendaal C (1987) Flow distribution in spiral mandrel dies. Polym Eng Sci 27:186–191
40. Vlcek J, Vlachopoulos J, Perdikoulias J (1988) Determination of output uniformity from spiral mandrel dies. Int Polym Process 2 ¾:174–181
41. Perdikoulias J, Vlcek J, Vlachopoulos J (1987) Polymer flow through spiral mandrel dies: a comparison of models. Adv Polym Technol 7:333–341
42. Kalyon DM, Yu JS, Du C-C (1987) A distributed model of flow in a spiral mandrel die. Polym Proc Eng 5(2):179–207
43. Gates PC (1987) Film quality improvement through effective die design. SPE ANTEC Tech. Papers XXXIII: 181–183
44. Housiades KD (2011) A mathematical model to study the effect of the air jet in the film blowing process. Polym Eng Sci 41:301–1315
45. Zhang Z, Lafleur PG, Bertrand F (2006) Effect of aerodynamics on film blowing process. Int Polym Process 21:527–535
46. Meissner J (1975) Basic parameters melt rheology, processing and end-use properties of three similar low-density polyethylene samples. Pure Appl Chem 42:553–612
47. Farber R, Dealy J (1974) Strain history of the melt in film blowing. Polym Eng Sci 14:435–440

48. Huang TA, Campbell GA (1985) Deformational history of LLDPE/LDPE blends on blown film equipment. Adv Polym Technol 5(3):181–192
49. Liu T, Harrison IR (1988) Shrinkage of low-density polyethylene film. Polym Eng Sci 28:517–521
50. Butler RI (2000) Blown film bubble instability induced by fabrication conditions. SPE ANTEC 46:156–164
51. Jung HWJ, Hyun JC (2005) Fiber spinning and film blowing instabilities. In: Hatzikiriakos SG, Migler KB (eds) Polymer processing instabilities. Marcel Dekker, New York
52. Winter HH (1983) A collaborative study on the relationship between film blowing performance and rheological properties of two low-density and two high-density plolyethylene samples. Pure Appl Chem 55:943–976
53. Micic P, Bhattacharya SN (2000) Rheology of LLDPE, LDPE and LLDPE/LDPE blends and its relevance to the film blowing process. Polym Int 49:1580–1589
54. Higuchi H, Fujikawa S, Sato M, Koyama K (2004) Thickness uniformity of HDPE blown film: relation to rheological properties and density. Polym Eng Sci 44:965–972
55. Pearson JRA, Petrie CJS (1970) The flow of a tubular film, part 1. Formal mathematical representation. J Fluid Mech 40:1–19; Part 2, Interpretation of the model and discussion of solutions. J Fluid Mech 42:609–625
56. Jim MK, Lee JS, Jung HW, Hyun JC (2011) Frequency response analysis of nonisothermal film blowing process using transient simulations. J Appl Polym Sci 123:3028–3035
57. Kim HT, Darby JP, Wilson GF (1973) Study of the variables affecting pressure drop and temperature rise in blow molding dies. Poly Eng Sci 13:372–381
58. Pritchatt RJ, Parnaby J, Worth RA (1975) Design considerations in development of extrudate wall-thickness control in blow molding. Plast Polym 43:55–64
59. Winter HH, Fritz HG (1986) Design of dies for the extrusion of sheets and annular parisons: the distribution problem. Polym Eng Sci 26:543–553
60. Kalyon DM, Kamal MR (1986) An experimental investigation of capillary extrudate swell in relation to parison swell behavior in blow molding. Polym Eng Sci 26:508–516
61. Cogswell FN, Webb PC, Weeks JC, Maskell SG, Rice PDR (1971) Scientific design of fabrication processes-blow molding. Plast Polym 39:340
62. Koopmans RJ (1988) Die swell–molecular structure model for linear polyethylene. J Polym Sci A 26:1157–1164
63. Henze ED, Wu WCL (1973) Variables affecting parison diameter swell and their correlation with rheological parameters. Polym Eng Sci 13:153–159
64. Jivraj N, Sehanobish K, Ramanathan R, Garcia-Rejon A, Carmel M (2001) Large part blow molding of HDPE resins: parison extrusion behavior and its relationship with resin rheological parameters. Soc Plast Eng ANTEC 876–880
65. Kalyon D, Tan Kamal MR (1980) The dynamics of parison development in blow molding. Polym Eng Sci 20:773–777
66. Yousefi A-M, den Doelder J, Rainville M-A, Koppi KA (2009) A modeling approach to the effect of resin characteristics on parison formation in extrusion blow molding. Polym Eng Sci 49:251–263
67. Ajroldi G (1978) Determination of rheological parameters from parison extrusion experiments. Polym Eng Sci 18:742–749
68. Sebastion DH, Dearborn JR (1983) Elongation rheology of polyolefin's and its relation to processability. Polym Eng Sci 23:572–575
69. Swerdlow M, Cogswell FN, Krul N (1980) Plast Rubber Process
70. Ryan ME, Dutta A (1982) The dynamics of parison free inflation in extrusion blow molding. Polym Eng Sci 22:569–577
71. Erwin L, Pollack MA, Gonsalez H (1983) Blowing of oriented PET bottles: predictions of free blown size and shape. Polym Eng Sci 23:826–829
72. Jabarin SA, Lofgren EA (1986) Effects of water absorption on physical properties and degree of molecular orientation of poly (ethylene terephthalate). Polym Eng Sci 26:620–625

73. Fifer RL (1981) PET stretch blow molding: experimental data showing container physical characteristics related to a variable container size. SPE ANTEC Tech Pap 27:696
74. Cakmak M, White JL, Spruiell JE (1985) an investigation of the kinematics of stretch blow molding poly(ethylene-terephthalate) bottles. J Appl Polym Sci 30:3679–3695
75. Schaul JS, Hannon MJ, Wisbrun KF (1975) Analysis of factors determining parison properties in high shear rate blow molding. Trans Soc Rheol 19:351–377
76. Kennedy P (1995) Flow analysis of injection molding. Hanser Publishers, Munich
77. Kamal MR, Goyal SK, Chu E (1988) Simulation of injection mold filling of viscoelastic polymer with fountain flow. AICHE J 34:94–106
78. Richardson SM (1985) Injection moulding of thermoplastics, II. Freezing-off in cavities. Rheol Acta 24:509–518
79. Hieber CA, Chiang HH, Ricketoson RC, Jong WR, Wang KK (1987) Melt viscosity characteristics via spiral mold. SPE ANTEC Tech Pap 33:938–941
80. Hull AM, Richardson SM, Selopranoto JH (1986) Plast Rubber Proc Appl 6:189
81. Fritch LW (1986) Honing molding parameters by measuring flow length. Plast Eng 42:41–44
82. Furches BJ, Kachin GA (1989) SPE Tech Pap 35:1663
83. Jain NS, Barry CMF, Barry MF (2001) Criteria for flow instabilities in end-gated injection molds. Soc Plast Eng ANTEC 47:471–475
84. Oda K, White JL, Clark ES (1976) Jetting in injection mold filling. Polym Eng Sci 16:585–592
85. Bogaerds ACB, Peters GWM, Baaijens FTP (2005) Instabilities in injection molding. In: Hatzikiriakos SG, Migler KB (eds) Polymer processing instabilities. Marcel Dekker, New York
86. Kontopoulou M, Vlachopoulos J (2001) Melting and densification of thermoplastic powders. Polym Eng Sci 41:155–169
87. Crawford RJ, Throne JL (2002) Rotational molding technology. PDI-William Andrew Publishing, Norwich
88. Gendron R (2005) Rheological behavior relevant to extrusion foaming. In: Gendron R (ed) Thermoplastic foam processing. CRC Press, Boca Raton
89. Munstedt H, Kurzbeck S, Stange J (2006) Advances in film blowing, thermoforming, and foaming by using long-chain branched polymers. Macromol Symp 245–246:181–190
90. Stange J, Münstedt H (2006) Rheological properties and foaming behavior of polypropylenes with different molecular structures. J Rheol 50:907–923
91. Mihai M, Huneault MA, Favis BD (2010) Rheology and extrusion foaming of long-chain-branched poly (lactic acid). Polym Eng Sci 50:630–642
92. Zhang Y, Kontopoulou M, Ansari M, Hatzidiriakos S, Park CB (2011) Effect of molecular structure and rheology on compression foam molding of ethylene-α-olefin copolymers. Polym Eng Sci 51:1145–1154
93. Qin X, Thompson MR, Hrymak AN (2007) Rheology studies of foam flow during injection mold filling. Polym Eng Sci 47:522–527

Appendix A
Structural and Rheological Parameters of Several Polymers

These data are extracted from a much more extensive tabulation published as: Fetters LJ, Lohse DJ, Colby RH (2005) Chain dimensions and entanglement spacings. In: Mark, JE (ed) Physical Properties of Polymers Handbook. Springer, Berlin.

Sources of all the original data are given in the Handbook article.

Polymer	T (K)	ρ (g/cm^{-3})	G_N^0 (MPa)	$\langle R^2 \rangle_0/M$ (Å^2)	$M_e' \times 10^{-3}$ Eq. (3.72)	$M_C' \times 10^{-3}$
PE	413	0.785	2.6	1.25	1,040	
PE	443	0.768		1.21	980	3,480
a-PP	298	0.852	0.48	0.678	4,390	
a-PP	463	0.765	0.42	0.678	7,010	
i-PP	463	0.765	0.43	0.694	6,850	
s-PP	463	0.766	1.35	1.03	2,180	
PIB	298	0.918	0.34	0.570	6,690	13,100
cis-PI	298	0.910	0.58	0.679	3,890	
cis-PBd	298	0.900	0.76	0.758	2,930	8,200
PBd-30	298	0.894	0.98	0.813	2,260	5,600
a-PMMA	413	1.13	0.31	0.390	12,500	29,500(1)
a-PS	413	0.969	0.20	0.437	16,600	33,000(2)
i-PS	413	0.969	0.19	0.420	17,500	
a-PVA	333	1.08	0.35	0.490	8,540	24,500(3)
PDMS	298	0.970	0.20	0.422	12,000	24,500

J. M. Dealy and J. Wang, *Melt Rheology and its Applications in the Plastics Industry*, Engineering Materials and Processes, DOI: 10.1007/978-94-007-6395-1,

Polymers Listed

PE	Polyethylene
a-PP	Attactic polypropylene
i-PP	Isotactic polypropylene
s-PP	Syndiotactic polypropylene
PIB	Polyisobutylene
cis-PI	cis polyisoprene
cis-PBd	1,4 polybutadiene, 96 % cis content
PBd-30	Polybutadiene with 30 % vinyl content
a-PMMA	Atactic poly (methyl) methacrylate
a-PS	Atactic polystyrene
i-PS	Isotactic polystyrene
a-PVA	Atactic poly (vinyl acetate)
PDMS	Poly (dimethylsiloxane)

Notes

490 K
183 K
428 K

Appendix B
The Displacement Gradient Tensor

As the first step in establishing a quantitative measure of the strain that occurs in a fluid as a result of a large deformation, we need to describe the position of a particle of fluid as it moves during a deformation process. This is easily done by writing down the coordinates of the particle as functions of time. However, we need to keep track of which particle we are tracking. This can be done by "labeling" each particle with its position vector, **x**, at some reference time, t_1. Thus, giving the components of $\mathbf{x}(t_1)$ identifies a particular fluid particle and distinguishes it from all other particles. At some other time t_2 the coordinates of this particle will be given by the following *displacement functions*:

$$x_1 = x_1[t_2, x(t_1)] \tag{B.1a}$$

$$x_1 = x_1[t_2, x(t_1)] \tag{B.1b}$$

$$x_1 = x_1[t_2, x(t_1)] \tag{B.1c}$$

These functions tell us the location at any time t_2 of the fluid particle that was located at $\mathbf{x}(t_1)$ at time t_1.

As an example of the use of the displacement functions to describe a deformation, consider simple shear flow in which the shear strain γ is some specified function of time $\gamma(t)$, with $\gamma(0) = 0$. The displacement of a typical fluid particle is shown in Fig. B.1. We locate the origin for the position vectors of all fluid particles at zero on the lower, stationary plate. Since we are considering only one or two specific particles in the remainder of our discussion, it will not be necessary to include the particle label $\mathbf{x}(t_1)$ as an argument of the displacement functions. Thus, the shear strain can be expressed in terms of the displacement functions as follows for the particular particle shown:

$$\gamma(t_2) - \gamma(t_1) = \frac{x_1(t_2) - x_1(t_1)}{x_2(t_1)} \tag{B.2}$$

where $[x_1(t_2) - x_1(t_1)]$ is the distance moved in the x_1 direction by a fluid particle that is located at a distance x_2 from the stationary plate, during the time interval

J. M. Dealy and J. Wang, *Melt Rheology and its Applications in the Plastics Industry*,
Engineering Materials and Processes, DOI: 10.1007/978-94-007-6395-1,

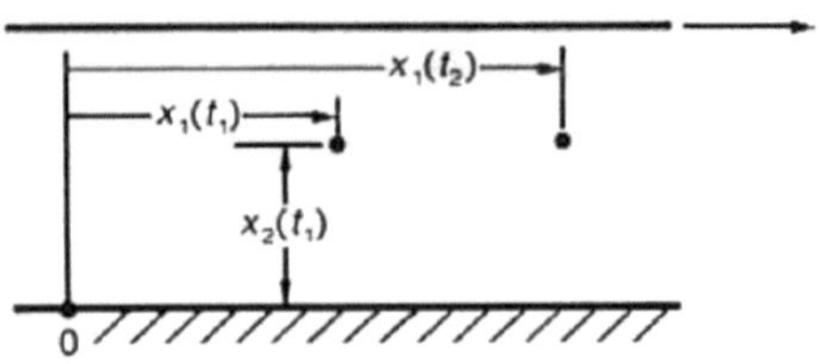

Fig. B.1 Fluid particle displacement in simple shear

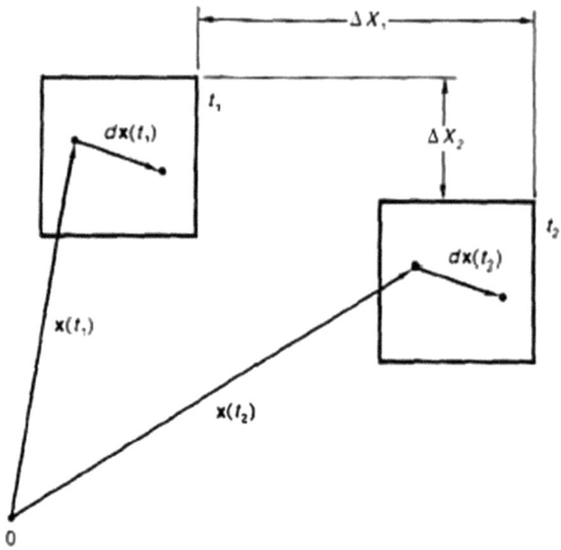

Fig. B.2 Position and displacement vectors at times t_1 and t_2 of two particles in an element of fluid undergoing solid body translation

between t_1 and t_2. The displacement functions for this deformation can then be expressed in terms of the shear strain as follows:

$$x_1(t_2) = x_1(t_1) + x_2(t_1)[\gamma(t_2) - \gamma(t_1)] \tag{B.3a}$$

$$x_2(t_2) = x_2(t_1) \tag{B.3b}$$

$$x_3(t_2) = x_3(t_1) \tag{B.3c}$$

To describe the deformation of a fluid element, we need to examine the relative displacement of two fluid particles. Consider two neighboring particles that are separated by the vector $d\mathbf{x}(t_1)$ at time t_1 and by $d\mathbf{x}(t_2)$ at time t_2, as shown in Fig. B.2. If we could define a quantity relating these two vectors, it might be useful to describe the strain that has occurred between times t_2 and t_2.

One such a quantity is the *displacement gradient tensor*, whose components $F_{ij}(t_1, t_2)$, are defined below.

$$F_{ij}(t_1, t_2) \equiv \frac{dx_i(t_2)}{dx_j(t_1)} \tag{B.4}$$

Another tensor that relates the two vectors $\mathbf{dx}(t_1)$ and $\mathbf{dx}(t_2)$ is the inverse of the displacement gradient tensor, which is defined as follows:

$$F_{ij}^{-1}(t_1, t_2) \equiv \frac{dx_i(t_1)}{dx_j(t_2)} \tag{B.5}$$

The use of the displacement gradient tensor to determine the components of $\mathbf{dx}(t_2)$ given $\mathbf{dx}(t_1)$ is demonstrated below:

$$dx_i(t_1) = F_{i1}(t_1, t_2)dx_1(t_1) + F_{i2}(t_1, t_2)dx_2(t_1) + F_{i3}(t_1, t_2)dx_3(t_1) \tag{B.6}$$

The inverse tensor can be used in a similar manner to determine $\mathbf{dx}(t_1)$ given $\mathbf{dx}(t_2)$.

To see if these tensors might be useful measures of strain, we consider first the *solid body translation* of the fluid element shown in Fig. B.2. In this type of motion, there is no deformation of the fluid element, i.e., no strain. To see if the components of the displacement gradient tensor or its inverse reflect this, we first write the displacement functions for this motion.

$$x_1(t_2) = x_1(t_1) + \Delta X_1 \tag{B.7a}$$

$$x_2(t_2) = x_2(t_1) - \Delta X_2 \tag{B.7b}$$

$$x_3(t_2) = x_3(t_2) \tag{B.7c}$$

Now we want to evaluate F_{ij}, and its inverse for this flow. We note that the two vectors $\mathbf{dx}(t_1)$ and $\mathbf{dx}(t_2)$ are equal; this implies (from Eqs. B.4, B.5 and B.7) that the components of $F_{ij}(t_1, t_2)$ and its inverse have the special, simple form shown below:

$$F_{ij}(t_1, t_2) = F_{ij}^{-1}(t_1, t_2) = \begin{bmatrix} 1 & 0 & 0 \\ 0 & 1 & 0 \\ 0 & 0 & 1 \end{bmatrix} \tag{B.8}$$

where the quantity on the right is the *unit tensor*. When a vector is multiplied by the unit tensor it is unchanged, and since there is no strain in this sample, that is exactly the way we want a strain tensor to behave. Thus, both F_{ij} and its inverse show promise as possible measures of finite strain.

As a further test of the usefulness of these tensors as measures of strain, consider the *solid body rotation* shown in Fig. B.3. Again there is no deformation, i.e., no strain. However, in this case the vectors $\mathbf{dx}(t_1)$ and $\mathbf{dx}(t_2)$ are *not* equal and neither F_{ij} nor its inverse reduces to the unit tensor. Thus, neither of these tensors is a useful measure of finite strain. However, they can be used to define tensors that do meet our requirements.

Two measures of strain that satisfy our criteria and that are particularly useful in polymer rheology are the Cauchy tensor: $C_{ij}(t_1, t_2)$ and the Finger tensor: $B_{ij}(t_1, t_2)$. The time arguments have the following significance: t_1 is the time at which a material element is in its reference configuration, and t_2 is the time at which the

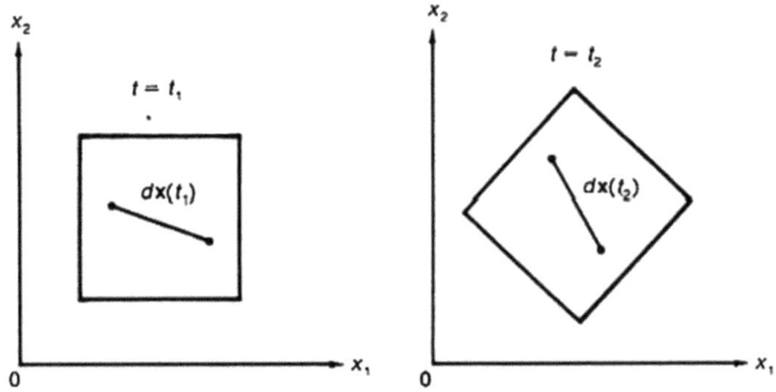

Fig. B.3 Displacement vectors at times t_1 and t_2 of two particles in a fluid element undergoing solid body rotation

Fig. B.4 Principal component of a position vector for simple extension of a sample with one end fixed at $x_1 = 0$

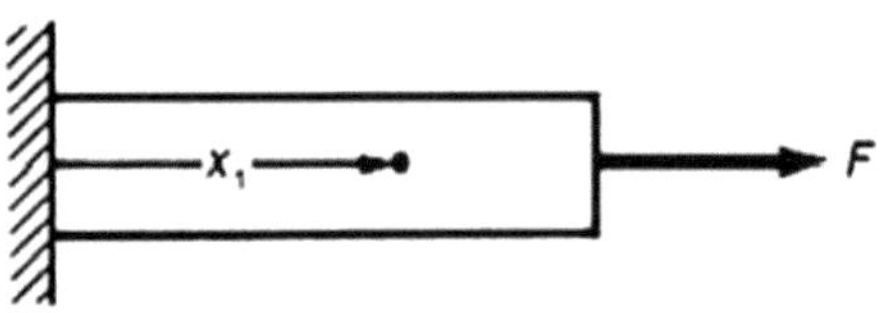

strain is evaluated, *relative* to the configuration at time t_1. The components of these two tensors can be represented in terms of the deformation gradient and its inverse.

$$C_{ij}(t_1, t_2) = F_{1i}F_{1j} + F_{2i}F_{2j} + F_{3i}F_{3j} \tag{B.9}$$

$$B_{ij}(t_1, t_2) = F^{-1}_{i1}F^{-1}_{j1} + F^{-1}_{i2}F^{-1}_{j2} + F^{-1}_{i3}F^{-1}_{j3} \tag{B.10}$$

To find the components of these tensors for simple shear and simple extension, we start by writing the components of $F_{ij}(t_1, t_2)$ and $F^{-1}_{ij}(t_1, t_2)$ for these flows. For simple shear from Eqs. (B.3–B.5) we have:

$$F_{ij}(t_i, t_j) = \begin{bmatrix} 1 & [\gamma(t_2) - \gamma(t_1)] & 0 \\ 0 & 1 & 0 \\ 0 & 0 & 1 \end{bmatrix} \tag{B.11}$$

$$F^{-1}_{ij}(t_i, t_j) = \begin{bmatrix} 1 & [\gamma(t_1) - \gamma(t_2)] & 0 \\ 0 & 1 & 0 \\ 0 & 0 & 1 \end{bmatrix} \tag{B.12}$$

The components of the Cauchy tensor for simple shear (Eq. 5.34) are obtained by combining Eq. (B.9) with Eq. (B.11), while the components of the Finger tensor (Eq. 5.35) are obtained by combining Eq. (B.10) with Eq. (B.12).

For simple extension, with x_1 taken to be the principal stretch direction (see Fig. B.4), the distances of a fluid element from the $x_1 = 0$ plane at times t_1 and t_2 are related to the Hencky strain ε, evaluated at these two times, by:

$$\ln\left[\frac{x_1(t_2)}{x_1(t_1)}\right] = \varepsilon(t_2) - \varepsilon(t_1) \tag{B.13}$$

The displacement functions are:

$$x_1(t_2) = x_1(t_1)\exp[\varepsilon(t_2) - \varepsilon(t_1)] \tag{B.14a}$$

$$x_2(t_2) = x_2(t_1)\exp\{(-1/2)[\varepsilon(t_2) - \varepsilon(t_1)]\} \tag{B.14b}$$

$$x_3(t_2) = x_3(t_1)\exp\{(-1/2)[\varepsilon(t_2) - \varepsilon(t_1)]\} \tag{B.14c}$$

And the components of the displacement gradient tensor and its inverse are:

$$F_{ij}(t_1,t_2) = \begin{bmatrix} e^{[\varepsilon(t_2)-\varepsilon(t_1)]} & 0 & 0 \\ 0 & e^{\{(-1/2)[\varepsilon(t_2)-\varepsilon(t_1)]\}} & 0 \\ 0 & 0 & e^{\{(-1/2)[\varepsilon(t_2)-\varepsilon(t_1)]\}} \end{bmatrix} \tag{B.15}$$

$$F_{ij}^{-1}(t_1,t_2) = \begin{bmatrix} e^{[\varepsilon(t_1)-\varepsilon(t_2)]} & 0 & 0 \\ 0 & e^{\{(-1/2)[\varepsilon(t_1)-\varepsilon(t_2)]\}} & 0 \\ 0 & 0 & e^{\{(-1/2)[\varepsilon(t_1)-\varepsilon(t_2)]\}} \end{bmatrix} \tag{B.16}$$

The components of the Cauchy and Finger tensors are obtained from B.9 and B.10 and are shown in Chap. 5 as Eqs. (5.36) and (5.37).

Author Index

Engineering Materials and Processes, DOI: 10.1007/978-94-007-6395-1,

Subject Index

J. M. Dealy and J. Wang, *Melt Rheology and its Applications in the Plastics Industry*, Engineering Materials and Processes, DOI: 10.1007/978-94-007-6395-1,

FSC
www.fsc.org
MIX
Papier aus verantwortungsvollen Quellen
Paper from responsible sources
FSC® C105338

If you have any concerns about our products,
you can contact us on
ProductSafety@springernature.com

In case Publisher is established outside the EU,
the EU authorized representative is:
Springer Nature Customer Service Center GmbH
Europaplatz 3, 69115 Heidelberg, Germany

Printed by Libri Plureos GmbH
in Hamburg, Germany